# Competition over Content

—◆—

## *Negotiating Standards for the Civil Service Examinations in Imperial China (1127–1279)*

Harvard East Asian Monographs 289

Publication of this work was subsidized in part by funding from the Achilles Fang Prize, awarded occasionally to a doctoral dissertation on the traditional Chinese humanities or related cultural developments throughout East Asia that continues the tradition, which Achilles Fang exemplified, of rigorous textual research. Gift of the students and friends of Achilles Fang.

# Competition over Content

———◆———

## Negotiating Standards
## for the Civil Service Examinations
## in Imperial China (1127–1279)

Hilde De Weerdt

Published by the Harvard University Asia Center
Distributed by Harvard University Press
Cambridge (Massachusetts) and London 2007

Printed in the United States of America

The Harvard University Asia Center publishes a monograph series and, in coordination with the Fairbank Center for East Asian Research, the Korea Institute, the Reischauer Institute of Japanese Studies, and other faculties and institutes, administers research projects designed to further scholarly understanding of China, Japan, Vietnam, Korea, and other Asian countries. The Center also sponsors projects addressing multidisciplinary and regional issues in Asia.

Library of Congress Cataloging-in-Publication Data

De Weerdt, Hilde Godelieve Dominique.

Competition over content : negotiating standards for the civil service examinations in imperial China (1127–1279) / Hilde De Weerdt.

p. cm. -- (Harvard East Asian monographs ; 289)

Includes bibliographical references and index.

ISBN 978-0-674-02588-2 (cl : alk. paper)

1. Civil service--China--Examinations--History. 2. China--History--Song dynasty, 960–1279. I. Title.

JQ1512.Z13E872767 2007

351.51076--dc22

2007010676

Index by David Prout

♾ Printed on acid-free paper

Last figure below indicates year of this printing

17 16 15 14 13 12 11 10 09 08 07

*To Mimi Borremans–De Weerdt*

*and Mary Lucal*

I thank the editors and publishers of *The Journal of Song-Yuan Studies*, *Extrême-orient, Extrême-occident*, and *Chūgoku shigaku / Studies in Chinese History* for allowing me to reproduce or summarize parts of the articles I published in these journals.

I gratefully acknowledge support for the dissertation that lies at the foundation of this book from the Japanese Ministry of Education / Monbushō, the American Council of Learned Societies in cooperation with the Chiang Ching-Kuo Foundation, the Pacific Cultural Foundation, and The Woodrow Wilson National Fellowship Foundation. Substantial rewriting and rethinking was facilitated by a Center for Chinese Studies Postdoctoral Fellowship at the University of California at Berkeley and a Harvard-Yenching Postdoctoral Fellowship in Print Culture and Library Studies.

H.D.W.

# Contents

## Appendixes

## Reference Matter

# Tables and Figures

## Tables

## Figures

# *Chronology*

Note: This table lists all emperors; era names are listed only when they are mentioned in the main text.

### Song Dynasty

Northern Song (960–1126)

    Taizu 太祖 (r. 960–76)

    Taizong 太宗 (r. 976–97)

    Zhenzong 眞宗 (r. 998–1022)

    Renzong 仁宗 (r. 1023–63)
        Qingli 慶曆 (1041–48)

    Yingzong 英宗 (r. 1064–67)

    Shenzong 神宗 (r. 1068–85)

    Zhezong 哲宗 (r. 1086–1100)
        Yuanyou 元祐 (1086–94)

    Huizong 徽宗 (r. 1101–25)

    Qinzong 欽宗 (r. 1126–27)

Southern Song (1129–1276/1279)

    Gaozong 高宗 (r. 1127–62)
        Jianyan 建炎 (1127–30)
        Shaoxing 紹興 (1131–62)

Xiaozong 孝宗 (r. 1163–89)
   Qiandao 乾道 (1165–73)
   Chunxi 淳熙 (1174–89)

Guangzong 光宗 (r. 1190–94)

Ningzong 寧宗 (r. 1195–1224)
   Qingyuan 慶元 (1195–1200)
   Kaixi 開禧 (1205–7)

Lizong 理宗 (r. 1225–64)

Duzong 度宗 (r. 1265–74)

Gongdi 恭帝 (r. 1275–76)

Duanzong 端宗 (r. 1276–78)

Wei wang 衛王 (r. 1278–79)

# *Abbreviations*

The following abbreviations are used in the notes and Works Consulted. For complete bibliographic citations for these titles, see the Works Consulted, pp. 431–78.

| | |
|---|---|
| *Bishui* | Liu Dake, *Bishui qunying daiwen huiyuan xuanyao* |
| *BMF* | *Yongjia xiansheng bamian feng* |
| *BSYS* | *Bulletin of Sung and Yuan Studies* |
| *ECCP* | Library of Congress, Asian Division, and Arthur W. Hummel, eds., *Eminent Chinese of the Ch'ing Period* |
| *HJAS* | *Harvard Journal of Asiatic Studies* |
| *JAS* | *Journal of Asian Studies* |
| *Jie jiang wang* | *Qunshu huiyuan jie jiang wang* |
| *JSYS* | *Journal of Sung-Yuan Studies* |
| *LXSC* | Wei Tianying, *Lunxue shengchi* |
| *MB* | Association for Asian Studies, Ming Biographical History Project Committee; L. Carrington Goodrich; and Fang Chaoying, eds., *Dictionary of Ming Biography* |
| *PEW* | *Philosophy East and West* |
| *SB* | Herbert Franke, ed., *Sung Biographies* |
| *SHY* | Xu Song, *Song huiyao jigao*; *SHY*, XJ refers to the section on selection, *xuanju*; *SHY*, XF to the section on law, *xingfa* |

| | |
|---|---|
| *SHY* (AS) | Xu Song, *Song huiyao jigao,* electronic edition by Zhongyang yanjiuyuan (Academia Sinica) and Harvard University |
| SKQS | Siku quanshu |
| *SKTY* | Ji Yun et al., eds., *Siku quanshu zongmu tiyao* |
| *SRZJZL* | Chang Bide et al., eds., *Songren zhuanji ziliao suoyin* |
| *SRZJZL-BB* | Li Guoling, ed. *Songren zhuanji ziliao suoyin bubian* |
| *SS* | Tuo-tuo, ed., *Song shi* |
| *SSJSBM* | Feng Qi and Chen Bangzhan, *Song shi jishi benmo* |
| *SYXA* | Huang Zongxi, Quan Zuwang, et al., *Song Yuan xue'an* |
| *WXTK* | Ma Duanlin, *Wenxian tongkao* |
| *XZZTJ* | Bi Yuan, *Xu Zizhi tongjian* |
| *Xiangshuo* | Lü Zuqian, *Lidai zhidu xiangshuo* |
| *YSJ* | Ye Shi, *Ye Shi ji* |
| *Yuanliu zhilun* | Lin Jiong and Huang Lüweng, *Gujin yuanliu zhilun* |
| *ZZYL* | Zhu Xi, *Zhuzi yulei* |

# Competition over Content

———————

*Negotiating Standards*
*for the Civil Service Examinations*
*in Imperial China (1127–1279)*

# Introduction

Last winter, in the tenth month, I managed to be sent from the provinces to participate in the examinations at the capital. In the spring of this year, in the third month, I obtained the [advanced scholar, or *jinshi*] degree. . . . Now, wealth and status such as yours are hard to attain. I hope that you, Censor, would look for a wife for me this year, and that you would look for a position for me next year. . . . Maybe you forget a lot, and maybe extraordinary talent is uncommon, but if for once the unexpected were to occur and I were to join you at the top of the bureaucracy, then when you looked at me from the corner of your eyes, you would have regrets and you would apologize to me. How could I be affected by you!

> —From a letter by Wang Lengran,
> early eighth century

Renzong paid attention to serious scholarship and was devoted to the way of government. He severely censured frivolous and superficial writing. Originally Liu Sanbian, the advanced scholar (*jinshi* 進士), liked to compose licentious songs. His compositions were transmitted everywhere. He once composed a song titled "The Crane Rises Up in the Sky" that concluded as follows: "I may exchange my empty title for some drinking and some soft singing." When the emperor announced the results of the examinations in the front hall, he had dropped him from the list on purpose. [The emperor] said: "Just go and do some drinking and enjoy some soft singing! Why would you want an empty title!" Then in 1034 Liu succeeded in the examinations. Later he changed his name to [Liu] Yong. Only then was he able to be promoted and serve in the bureaucracy.

> — Wu Zeng, *Nenggai zhai*
> *man lu*, 16.418

EPIGRAPHS: Wang Dingbao, *Tang zhiyan*, 2.12b–13b; cited in Fu Xuanzong, *Tangdai keju yu wenxue*, 172–73. Wu Zeng cites the lyric in full; cited in part in Yü Ying-shih, *Zhu Xi de lishi shijie*, 1: 290–91.

## *The Civil Service Examinations:*
## *Continuity and Change*

Regularly held, written civil service examinations—from their first use for the recruitment of government officials during the Sui 隋 dynasty (581–617) until their cancellation in 1905, shortly before the collapse of the imperial order—were instrumental in the creation and maintenance of political unity across the Chinese territories. Emperors intent on reducing the power of alternative sources of political authority such as aristocratic birth or military power promoted the use of written examinations in order to recruit men likely to put loyalty to the dynasty ahead of family interests or military ambition. The establishment of a tiered system of local, capital, and court examinations during the Song dynasty (960–1279) mirrored the hierarchy of bureaucratic control over the Chinese territories and tied literate elites across the empire to the various levels of government and ultimately the court. Participation in the examinations turned literate elites into state subjects at the local and national levels.

The civil service examinations further solidified political unity through the social and cultural effects of their continued use. Access to the financial and social resources required for participation in the examinations, in the form of subsidies for years of study or social status qualifications, excluded the majority of the population; by the same token, literati (or scholar-official) status became associated with participation in the examinations. Although the odds of passing were consistently low and the prospects for a career in government slim even for successful candidates, the numbers of those sitting for the examinations increased as participation per se became a status marker.

The centrality of examination participation to the status of scholar-officials was the result of a transformation in elite Chinese society. The adoption of the civil service examinations under the Sui and Tang 唐 (618–907) dynasties[1] and their promotion at the cost of other channels

---

1. Historians typically trace the origins of regular civil service examinations held across the empire and for the recruitment (not simply the promotion) of officials to the Sui dynasty. For a slightly different reading of the examinations held under the Han and Six Dynasties, see Dien, "Civil Service Examinations."

for recruitment under the Song dynasty restructured the value system of elite society. In the older system, family pedigree was primary and intellectual capital secondary; between the ninth and the twelfth centuries, family pedigree and social status became dependent on intellectual capital acquired through participation and success in the examinations. It is this transformation that explains why elite families continued to invest in the civil service examinations throughout the nineteenth century in spite of the long history of discontent and open criticism of the failure of the examinations to fulfill their promise to recruit the best for officialdom.

This restructured value system transformed the everyday life of literate elites. Each stage in the typical life of the literatus was marked by the centrality of the examinations. Literati education, from childhood through adolescence, focused on the knowledge and skills tested in the examinations. In adulthood, the literatus remained engaged in the examinations as participant, teacher, examiner, or as a father overseeing the education of male offspring. The centrality of the examinations in elite family life produced a disposition among elite men and women to valorize education. This carried over into other areas of social life. Elite marriage strategies were adjusted to reflect the recognition of degreeholders. In the religious field, elite aspirations were expressed in prayers to gods wielding control over the fate of examination candidates. In urban areas, the nomenclature of the examinations was applied to the naming of streets and commercial items.

The centrality of the examinations also shaped elite behavior and expectations. As the examinations became central to elite status, the state granted more privileges to participants, ranging from protection against physical punishment to labor exemptions to the codification of a distinctive dress code.[2] Successive dynastic regimes ranked examination candidates and graduates in a proliferating hierarchy according to the degree of their success and measured their privileges accordingly. As illustrated by the two stories translated above, the privileges generated corresponding sets of expectations.

The story of Wang Lengran 王冷然 (ca. 698–742) dates to the early eighth century, to a time when the aristocratic families of the north

---

2. Takahashi, "Sōdai no shijin mibun ni tsuite"; Min, "The Sheng-yuan-Chien-sheng Stratum in Ch'ing Society."

were contesting the use of examinations. It was one of many in the collection of examination anecdotes compiled by Wang Dingbao 王定保 (870–940) in the mid-tenth century. In this and other stories circulating about Tang dynasty advanced scholars (*jinshi*), holders of the most prestigious examination degree were portrayed as uninhibited and lacking in the decorum that was the hallmark of aristocratic families. After his success in the examinations in 717, Wang Lengran sent a blunt letter to his old acquaintance Gao Changyu 高昌宇 (fl. 710s), then active at court as a censor. Wang's lack of decorum was evident not only in his straightforward appeal for Gao's help in securing a position but also in his request to arrange a wedding for him first. The letter was very likely Wang's way of poking fun of an "old friend" who had failed to recommend him. After Wang had graduated without the help of this friend, he set out to shame him into helping him translate his degree into official emoluments.[3]

Within the context of Tang elite society, Wang Lengran's letter was indicative of the rising status of advanced scholars. Since the reign of Empress Wu 武 (r. 690–705), advanced scholars had gained access to leading positions at court. However, the letter also expressed the frustration of the advanced scholars. As a new subgroup in the political elite, they were still marginal to the political scene, which continued to be dominated by the aristocratic families from the north. The aristocratic families based their claims to power on pedigree and the protection of aristocratic traditions. One century after Wang Lengran's letter to Gao Changyu, Prime Minister Li Deyu 李德裕 (787–850) argued that the examinations produced men who excelled in literary skill but who were unable to operate at court because they lacked the knowledge of court etiquette that was second nature to those raised in aristocratic households.[4]

The story of Liu Sanbian 柳三變 / Liu Yong 柳永 (990–1050), included in Wu Zeng's 吳曾 (?–after 1170) mid-twelfth century collection of anecdotes and observations, illustrates both the higher sociopolitical status the advanced scholar enjoyed by the eleventh century and the corresponding expectations. Liu Sanbian's initial behavior is reminis-

---

3. For this story, see Moore, *Rituals of Recruitment*, 77–79.

4. Des Rotours, *Le traité des examens*, 204–5. Cf. Twitchett, ed., *The Cambridge History of China*, vol. 3, *Sui and T'ang China*, 652–53.

cent of the activities for which Tang advanced scholars were notorious. In representations of Tang dynasty advanced scholars, insolence was the rule. Following the first Song emperor's revaluation of the civil service examinations in the creation of a unified empire, however, the title of advanced scholar conveyed respect, and those who vied for it or held it were expected to behave as prospective civil servants of the dynastic state. In the eyes of Song rulers and political elites, the behavior of Tang advanced scholars was a token of their lack of dynastic loyalty and evidenced in their service to the regional warlords who eventually overthrew the Tang.[5]

Zhao Kuangyin 趙匡胤 (927–76), the founder of the Song dynasty, reinstituted the civil service examinations upon his ascension to the throne in 960. He and his successors expanded the role of the examinations to the extent that they became the primary channel for entry into officialdom in the eleventh and twelfth centuries.[6] The primacy of the examinations was especially evident in the assignment of top posts in the central government. According to one estimate, 72 percent of twelfth-century chief and assistant councilors held the *jinshi* degree.[7] The importance the Song dynasty attached to the examinations throughout its reign was unprecedented and transformed Chinese culture.

The number of literati participating in the examinations continued to grow at exponential rates, even as the territories of the Song state diminished by about half in the twelfth century. John Chaffee estimates that the number of candidates taking the qualifying prefectural examinations increased from 20,000 to 30,000 in the early eleventh century to 79,000 one century later and reached 400,000 or more by the mid-

---

5. Yü Ying-shih, *Zhu Xi de lishi shijie*, 1: 272–312.

6. Protection privileges increased during the late twelfth and early thirteenth centuries (Chaffee, *The Thorny Gates*, 29). Despite the increased ability of high officials, many of whom we may surmise held degrees, to obtain low-ranking official positions for male family members, Chaffee's work demonstrates that the declining employment opportunities for *jinshi* degree-holders without powerful family connections did not impact the popularity of the examinations among literati. The number of candidates preparing for and sitting the examinations continued to grow throughout the twelfth and thirteenth centuries, reaching 400,000 by the mid-thirteenth century (ibid., 35).

7. Ibid., 29, citing Sudō, *Sōdai kanryōsei to daitochi shoyū*, 20–25. Note that the percentage Chaffee derived from Sudō's numbers covers the period from 1127 to 1194 and excludes those councilors for whom no information was available.

thirteenth century.[8] Preparation for and participation in the civil service examinations had by then become hallmarks of scholar-official status.

Intellectually, the civil service examinations defined educational standards for literate elites across the empire. By determining the format of the examinations, emperors and court officials prescribed training in particular literary and administrative genres. They also endorsed changing configurations of the classical canon and, at times, other texts such as commentaries on the Classics. The imposition of curricular standards contributed to the dissemination of a shared language for the practice of poetry, classical exegesis, and the discussion of history and government among examination candidates and graduates. Fluency in this language determined scholar-official status. And as the civil service examinations became central to scholar-official status during the Song dynasty, literati competition over the definition and redefinition of examination standards restructured the examination field. The principal goal of this book is to explicate the restructuring of the examination field in twelfth- and thirteenth-century Song China.

### *From Northern Song to Southern Song*

Like human relationships in the social, political, and intellectual fields more generally, those in the field of the civil service examinations changed considerably during the Southern Song dynasty (1127–1279). During the first half of the Song dynasty, the court, the capital, and local communities were dominated by an elite whose marriage alliances and political networks spanned the extent of the Song empire and whose ambitions focused on the court and the capital. Intellectual eminence was associated with top court officials like Ouyang Xiu 歐陽修 (1007–72), Su Shi 蘇軾 (1037–1101), Sima Guang 司馬光 (1019–86), or Wang Anshi 王安石 (1021–86), or with those with powerful relations at court such as Cheng Hao 程顥 (1032–85) and Cheng Yi 程頤 (1033–1107). The aristocratic traditions of the past and the Song founders' policies of centralization created tight bonds between the court and the ruling elite.

After the capture of the capital Kaifeng 開封 by the armies of the expanding Jurchen Jin empire (1115–1234) in 1127, the imprisonment of

---

8. Chaffee, *The Thorny Gates*, 35.

the retired and reigning emperors Huizong 徽宗 (r. 1101–25) and Qin-zong 欽宗 (r. 1126–27), and the turmoil and dislocation caused by the flight of hundreds of thousands of court servants, officials, soldiers, and commoners, the court in its new capital of Lin'an 臨安 (Hangzhou 杭州) was no longer the sole or even the main focus of literati ambition. Literati elites settled outside of the capital, intermarried with other local families, invested in local welfare projects, and prided themselves, along with other local elite families, on their reputations as managers of local society.[9] At the same time, in the late twelfth and thirteenth centuries, Neo-Confucian beliefs and practices, emphasizing moral self-cultivation and its translation into local educational and social welfare institutions, gradually spread among elite families.

Corresponding changes occurred in the civil service examinations, and those changes in turn reinforced the broader social and political changes just outlined. The civil service examinations resemble other Northern Song institutions such as the government archives and the Memorials Office (Jinzouyuan 進奏院), which was charged with the compilation and distribution of the court gazettes. These institutions existed during the Tang and Five Dynasties (907–60) but were re-engineered as part of the centralization campaign launched by the first Song emperors. They were key institutions in the reassertion of impe-rial control over the regional warlords who successfully competed with

---

9. For the paradigmatic articulation of this interpretation of Song history, see Hymes, *Statesmen and Gentlemen*. Beverly Bossler (*Powerful Relations*) modified this inter-pretation of the Northern to Southern Song transition and concluded that the distinc-tion between Northern and Southern Song elites as presented in the works of Hartwell and Hymes was more apparent than real. The families of top-ranking officials contin-ued to marry outside local networks in Southern Song times, and the lower bureaucracy during the Northern Song was staffed by elites that acted like Southern Song local elites. Bossler argued that the shift identified in earlier scholarship was largely the result of historiographical developments. Bossler's caveat about the differences in the Northern and Southern Song historiographical record does, however, not amount to a rejection of the localist paradigm. She pointed out that the transition from the Northern to the Southern Song was marked by significant social and political changes that gradually transformed life in the provinces. Whereas the top members of the political elite settled in Kaifeng during the Northern Song period, they dispersed to larger cities after the court moved to Lin'an and never returned to the capital as a resident class of profes-sional bureaucrats. For further discussion, see De Weerdt, "Amerika no Sōdaishi ken-kyū ni okeru kinnen no dōkō."

the court between the eighth and the tenth centuries and the elites who served them. Gradually these institutions took on a second role in communications between the court and the provinces. They developed into sites through which local elites gathered information about the court, discussed it, and fed it back to the center.[10]

The institutional history of the examinations illustrates the changing relationship between court and literati from the Northern to the Southern Song period. Throughout the Northern Song, the court frequently revised the organizational framework of the examinations. The revisions reflected both the Song emperors' policies of centralization and the conflicts between opposing court factions. These revisions broke with Tang and Five Dynasties precedents and established parameters for the civil service examinations that held until the nineteenth century.

First, in 1071, the court, then under the control of the famous reformist councilor Wang Anshi, cancelled all degrees in "various fields" (*zhuke* 諸科; a generic name for separate examinations on the Classics, dynastic histories, the ritual canon, and law). Thereafter candidates in the regular examinations competed for the "advanced scholar," or *jinshi*, degree only. Second, in contrast to Tang practice, the regular *jinshi* examinations had been held at three levels since the first decades of the Song dynasty. The prefectural examinations (*jieshi* 解試 "forwarding examination") were held every third year in the fall. In the early years of the Song dynasty, there was some fluctuation in the frequency with which the examinations were held, but in 1066 the three-year interval became standard. This practice prevailed throughout the rest of imperial Chinese history.[11] Those who passed, the *juren* 舉人 "presented men," were as yet unqualified for office. They proceeded to the capital to take the departmental examination (*shengshi* 省試) in the early spring. Those who passed this examination, organized by the Ministry of Rites under the Department of State Affairs (Shangshu sheng 尚書省), advanced to the palace examination (*dianshi* 殿試) in the late spring.[12]

---

10. De Weerdt, "'Court Gazettes' and 'Short Reports.'"

11. Chaffee, *The Thorny Gates*, 51.

12. For a brief introduction to the examination routine in Song times, see Hirata, *Kakyo to kanryōsei*.

The palace examination was a third institutional innovation of the Northern Song dynasty. Theoretically the emperor himself presided. It originated in the first emperor's self-proclaimed desire to create a personal bond between himself and his officials and to instill loyalty in the bureaucracy.[13] Fourth, the format of the *jinshi* examinations underwent major changes. Northern Song court officials incessantly debated the pros and cons of different examination genres and their place in the sequence of tests to be completed over the course of each examination. Their proposals led to frequent and abrupt alterations in examination procedures during the eleventh and early twelfth centuries. In the last century of the Northern Song period, for example, rules concerning the position of poetry in the examinations changed six times (see Chapter 5).

The activity of the Northern Song court contrasts sharply with the retreat from centralization and institution-building on the part of the Southern Song court. Between 1150 and 1279, the focus of this study, there were no significant changes in the organizational framework of the civil service examinations. The layout of the *jinshi* examinations at both the local and the metropolitan levels remained the same throughout the second half of Song rule. According to the regulations issued by the court in 1145,[14] which shaped procedures through the end of the dynasty, both the prefectural and the departmental examinations consisted of three sessions.

Candidates who opted for the poetry track were required to write one regulated verse (*shi* 詩) and one regulated poetic exposition (*fu* 賦) in the first session, one exposition (*lun* 論) in the second session, and

---

13. Even though the palace examination itself became an enduring legacy to all subsequent imperial regimes, Zhao Kuangyin's assertion of imperial control over recruitment was not honored by his successors. Institutionally, the emperor's prerogative to fail candidates was rescinded under Emperor Renzong 仁宗 (r. 1023–63) in 1057. The first emperor reigning in the southern capital, Gaozong 高宗 (r. 1127–62), further renounced the right to change the ranking of the candidates as they appeared on the list prepared by the examination officials (Chaffee, *The Thorny Gates*, 23; Araki Toshikazu, *Sōdai kakyo seido kenkyū*, 303, 332; Ning, "Songdai gongju dianshi ce yu zhengju," 154).

14. Already in 1128 the court decided to accept candidates in both the poetry and the Classics tracks. Gaozong reconfirmed this decision in 1145 (Araki Toshikazu, *Sōdai kakyo seido kenkyū*, 393–94; *SHY*, XJ, 4.21b). I thank John Chaffee for drawing my attention to the earlier decree.

three policy response essays (*ce* 策) in the last session. Those who chose
the "meaning of the Classics" (*jingyi* 經義) track wrote three essays on
the classic in which they specialized, one essay on *The Analects* (*Lunyu*
論語), and one on the *Mencius* (*Mengzi* 孟子) for the first part; then, like
their colleagues in the poetry track, one exposition in the second session; and, finally, three policy response essays in the final session.[15]
Candidates in either track who sat for the palace examination had only
to write a response to one policy question.

The acquisition of the knowledge and skills tested in the different
sessions of the examinations shaped childhood education (*xiaoxue* 小學).
The ability to compose poetry was required for the poetry session;
knowledge of the Classics was tested in the meaning of the Classics and
exposition sessions; the memorization of historical events and anecdotes recorded in the histories and philosophical texts was a precondi-
tion for success in the policy response session. Children were taught
such skills in official elementary schools or in private schools run by
lineages, local communities, or individual teachers. Private tutoring,
frequently by family members, was common either as an alternative to
schooling or in preparation for attendance at a government school.
Elementary school curricula presupposed basic literacy in the form of
the ability to read and write a limited set of characters.[16]

Preparation for the examinations intensified after students entered
secondary education (*daxue* 大學). Age limits for the beginning and
ending of both elementary and secondary education were inconsistent.
Some students started the second stage of education at the age of
twelve, whereas others began at the age of fifteen.[17] From early adolescence on, students continued to study the Classics, histories, and
philosophical texts and practiced responding to questions on these
sources in the formats tested in the examinations. This stage in the life
of the student had no fixed end. Barring early graduation or the renun-

---

15. Chaffee, *The Thorny Gates*, 5; Araki Toshikazu, *Sōdai kakyo seido kenkyū*, 394; *SHY*,
XJ, 4.21b–22a, 28b.

16. On childhood education during the Song period, see Zhou Yuwen, *Songdai ertong
de shenghuo yu jiaoyu*; and Yuan Zheng, *Songdai jiaoyu*, chap. 5.

17. Zhou Yuwen, *Songdai ertong de shenghuo yu jiaoyu*, 119. Cf. Chen Wenyi, *You guanxue
dao shuyuan*, 307–8, 327–29; and Chaffee, *The Thorny Gates*, 5–6.

ciation of the examination path to success, students could continue to prepare for and participate in the examinations until old age.

The curriculum of this second stage in the education of the literati was a central concern of the activist emperors and court officials of the Northern Song period. Before the invention of modern media, standardized testing would have been impracticable. The license given individual examiners to design questions and oversee grading was a constant source of conflict in the highly factionalized atmosphere of Song court politics. The continual shifting and anonymity of the examiners prevented the formation of curricular monopolies and thus reinforced the impersonal power of the state. Some Northern Song politicians, nevertheless, conceived of the civil service examinations as an instrument for the indoctrination of a standardized state curriculum. Most famous were the efforts of Wang Anshi. In the 1070s, Wang Anshi commissioned a set of new commentaries on the Classics and a new dictionary, distributed them to government schools, and made them a core requisite for the civil service examinations. Commercial printers were quick to follow the court's lead and sold printed editions of the state-imposed curriculum (Chapter 5).

As the examinations became central to scholar-official status, students preparing for the examinations kept close watch on curricular standards, which were more likely than the institutional framework to change. As the Southern Song court withdrew from curricular leadership, the examinations became the site of intense competition among scholars, especially among those of the literate elite who turned to teaching. Teachers translated competing intellectual and political agendas in their examination preparation classes and textbooks. Given the new relationship between the examinations and the scholar-officials, they were not merely defining curricula for examination preparation; curricular standards also reflected elite conceptions of statesmanship and local leadership.

## The Examination Field

This book analyzes examination curricula by investigating a wide range of primary sources produced in the twelfth and thirteenth centuries that fall broadly within the genre of the examination manual (Table 1 in Appendix B). By "examination manual," I refer to a wide array of textbook genres, including encyclopedias, anthologies, rhyme dictionaries, examination guides, classical texts and commentaries, and histories. Encyclopedias (*leishu* 類書) are classified collections of primary source texts and interpretive essays. Anthologies (*zongji* 總集) are collections of full-length essays, typically by reputable authors, arranged by author or genre. I read these materials in conjunction with essay questions and responses included in the collections of individual authors and official reports and private commentary on examination preparation and examination essay writing.

The central questions this book addresses converge on the subject of how standards were set for the examinations. How did changes in curriculum and examination criteria take place? How did occupational, political, and intellectual groups shape curricular standards for examination preparation and evaluation criteria for examination writing? How did examination standards shape the political and intellectual agendas of the groups involved?

I ask these questions as a way to reframe debates about the civil service examinations and their place in the imperial Chinese order. Since the early twentieth century, three major theses have captured scholarly attention. The first posits that the examinations ought to be perceived in terms of the establishment and consolidation of hegemonic state power. In the 1920s, Naitō Konan advanced the first version of this thesis. He argued that the establishment of a three-tiered examination system (local, departmental, and palace examinations) as the main channel of official recruitment and the elaboration of procedures intended to maximize fairness provided Song emperors with a neutralized bureaucracy. Naitō saw the subordination of this impersonal bureaucracy to imperial power as a major factor in the transition from a medieval aristocratic Chinese society to a modern centralized state during the Song reign.[18] The paradigm of imperial absolutism (and political modernity) has since been questioned in research on power relations at

---

18. For a study in English on the hypothesis, see Miyakawa, "An Outline of the Naitō Hypothesis." Araki Toshikazu (*Sōdai kakyo seido kenkyū*, preface, introduction, and *passim*) cast his work on the Song examinations in the framework of this thesis—even though he did not identify it with Naitō. Araki was a student of Miyazaki Ichisada, the second head of the Kyoto school founded by Naitō.

court, in the bureaucracy, and in local society.[19] In recent years, this thesis has resurfaced in a different guise. Benjamin Elman has proposed that the civil service examinations primarily consolidated dynastic state power not as a political instrument of imperial absolutism but as an educational strategy ensuring political legitimacy.[20] Iona Man-Cheong adds that, in the eighteenth century, political legitimacy was achieved through the subjectification of the individual to the state, a process engendered by years of preparation for and participation in a proliferating number of examinations.[21]

I generally accept the Althusserian argument that participation in the examinations identified examination candidates as subjects of the state, and I will argue that it applies to Song as well as Qing times. Participation in the examinations implied recognition of the legitimacy of the dynastic state. This recognition did not, however, automatically extend to the emperors and court officials in charge of the dynastic state. More work is needed on the question of how political subjectivity was articulated, interpreted, and modified among an intellectually and politically divided scholar-official class. The field of the civil service examinations overlapped with that of court and bureaucratic politics. Factionalism spilled over into examination preparation, and partisan agendas shaped examination writing. The field of civil service examinations further accommodated agents (such as private teachers and commercial printers) and agendas external to court and bureaucratic politics. The historical question of how examination standards were set and altered therefore requires that the examinations be perceived as a site for competition among differing political and intellectual agendas within the parameters of the imperial state rather than as an arena for the celebration of an all-encompassing imperial ideology.[22]

A second thesis that has animated debate in Chinese social history since the 1940s posits that the civil service examinations resulted in

---

19. See, e.g., Chaffee, *Branches of Heaven*; Davis, *Court and Family in Sung China*; and Hymes, *Statesmen and Gentlemen*.

20. Elman, "Political, Social, and Cultural Reproduction"; idem, *A Cultural History of Civil Examinations*.

21. Man-Cheong, *The Class of 1761*.

22. De Weerdt, Review: *The Class of 1761*.

high degrees of social mobility from the Southern Song through the Qing dynasty. Based on statistical analysis of lists of successful candidates, Edward Kracke, Jr. (in 1947), and Ho Ping-ti (in 1962) concluded that staggering numbers of successful *jinshi* candidates, over 50 percent in the Southern Song, 49.5 percent in the Ming, and 37.6 percent in the Qing, came from families with no immediate patrilineal forefathers in the bureaucracy.[23] In the 1970s and 1980s, critics of the mobility thesis pointed out the underestimation of the scope of significant family ties in the earlier studies and subordinated the significance of examination degrees in obtaining and maintaining power to other social factors, such as wealth, landholding, and marriage.[24]

In a landmark study of the Song civil service examinations, John Chaffee integrated the problematic evidence on social mobility into a study of the multiple social functions of the examinations over the course of the Song dynasty. He argued that, as the dynasty progressed, although the number of candidates soared to an estimated 400,000 and fairness, chances of success, and opportunities for entering the bureaucracy declined, the examinations retained their appeal to an increasingly localized elite because of their social function as a status marker: "The institutions, ceremonies, symbols, and stories surrounding the examinations set off officials from commoners, literati from non-literati."[25] Due to the examinations, learning, which was not reducible to the other social factors mentioned above, gained an unprecedented priority in the determination of elite status. Even though I recognize the contribution local history has made to our understanding of the nature of Chinese elites and agree that elite status accrued to groups other than the literati, I propose that the centrality of examination learning (which is not necessarily examination success) to the status of Chinese literati, which accompanied both the expansion of examination participation and the elite turn toward localist strategies, was essential to the restructuring of

---

23. Kracke, "Family Versus Merit"; Ho, *The Ladder of Success in Imperial China.*

24. Hymes, *Statesmen and Gentlemen*; Beattie, *Land and Lineage in China.* For review articles summarizing this debate, see Ebrey, "The Dynamics of Elite Domination in Sung China"; and Waltner, "Building on the Ladder of Success."

25. Chaffee, *The Thorny Gates,* 188.

the field of the civil service examinations in the twelfth century examined here.

The cultural characteristics and effects of the examinations have come to the forefront of academic debate since the 1980s. Most influential, and most controversial, in the new wave of research on the cultural dimensions of the examinations is Benjamin Elman's work. His articles and hefty monograph on "the cultural history of the civil examinations" between 1400 and 1900 stand out as the only systematic discussion of the various manifestations of late imperial examination culture, ranging from language acquisition to religious practices to the art of interpreting dreams to predict examination success.[26] The controversy surrounding Elman's work centers on the thesis that, as a mechanism of selection, the examination system was, inherently, a "process of social, political, and cultural reproduction of the status quo."[27] In brief, the political legitimation of the dynasty through education (political reproduction) and the perpetuation of the elite through their control over the cultural resources required for examination success (social reproduction) necessitated cultural reproduction, which Elman defines as the perpetual internalization and externalization of "orthodox schemes of classical language, thought, perception, appreciation and action."[28] Cultural reproduction was more than the result of attempts by representatives of the imperial state to control literati culture; literati interests, Elman emphasizes, were also represented in the cultural arena of the examinations.[29]

The history preceding the cultural reproduction of classical and Neo-Confucian discourse in the Ming and Qing dynasties covered in Elman's work suggests that reproduction works better as an explanation of the social history rather than the cultural history of the examinations, or, more specifically, their intellectual history, which is the focus of this book. The reproduction thesis is essentially a response to, and denial of,

---

26. Elman, esp. "Political, Social, and Cultural Reproduction"; and idem, *A Cultural History of Civil Examinations*.

27. Elman, *A Cultural History of Civil Examinations*, xxix. For challenges to Elman's interpretation, see, e.g., the reviews by Chow, Langlois, and Magone.

28. Elman, "Political, Social, and Cultural Reproduction," 20.

29. Elman, *A Cultural History of Civil Examinations*, xxiv.

the social mobility thesis; it argues that the cultural requirements of the examinations effectively excluded the majority of the population throughout imperial Chinese history. Few would challenge this view. The question of how curricular change was achieved, as it was in the emergence and ascendancy of Neo-Confucian ideology in examination curricula in the twelfth and thirteenth centuries, however, cannot be answered adequately with the model of cultural reproduction. The distinction between literati and commoners, as well as distinctions among literati groups, resulted from cultural reproduction as well as cultural innovation.

My approach to the questions raised at the beginning of this section is informed by a particular conceptualization of the civil service examinations. I conceive of the civil service examinations as a bounded cultural space in which students, teachers, emperors, examiners, court and local officials, literati intellectuals, editors, and printer/publishers in effect negotiated standards for examination preparation and examination essay writing. This cultural space, or field, was separate from the field of court and bureaucratic politics and the field of literati intellectual culture. It operated according to its own norms, even though those norms overlapped with those of both the political and the intellectual fields. "Field" in this sense is a heuristic device, one that allows for the systematic investigation of not only the rules and relationships that shaped examination preparation and examination essay writing and changes in them but also the various kinds of relationships between the examination field and other social, political, and cultural fields.

This understanding of the civil service examinations has been enriched by the work of sociologist Pierre Bourdieu.[30] Bourdieu defined fields as sites in which individuals compete over different forms of capital, that is, economic, cultural (language or knowledge), or symbolic (prestige) resources. The positions individuals assume in a particular

---

30. For a comparable application of Bourdieu's concept of "field" to the imperial civil service examinations, see Chow, *Publishing, Culture, and Power in Early Modern China*, esp. 11–12, 154. Whereas Chow emphasizes the subversive potential of the expansion of commercial printing in the sixteenth- and seventeenth-century examination field, this study aims to explain the processes of negotiation between scholar-officials and the court and among intellectual and political associations that shaped examination preparation.

field are structured—they are determined by the distribution of the kind of capital at stake in that field. Regardless of the positions of individuals in this structured space and regardless of their goals (preservation of the status quo or change), all of them share the basic rules governing the field. The field thus resembles a game. At the very least, the individuals or agents must think the game and the stakes at play in it are worthwhile.[31]

The application of the notion of field as a heuristic device to examination preparation and examination essay writing helps us reconstruct the history of the types of agents involved, their specific interests, their positions and possible changes to them, their relationships with other agents, and the conventions that governed participation in the setting of examination standards. Examination standards were shaped by different types of agents, such as emperors and their entourages, examiners, teachers, examination graduates and candidates, literati intellectuals, editors, and publishers. The actions of these agents affected both examination preparation (study activities that led to participation in prefectural, departmental, and palace examination sessions) and examination writing (the writing of actual examination essays). I conceive of examination preparation and examination writing as two generic categories of exchanges in the examination field. As explained in Chapter 2, the examinations tested competence in two areas that defined the scholar-official: textual exegesis and the discussion of government. We can call these two areas of competence forms of examination capital. Specific interests in examination capital diverged as different agents espoused different modes of interpretation in textual exegesis and the discussion of government. In the twelfth century, for example, the court officials serving as examiners supervising the departmental examinations called for a classical mode of interpretation, endorsing in broad terms pre-Song texts and modes of explanation, the "Yongjia" 永嘉 teachers adopted a historical mode of interpretation in both exegesis and the discussion of government, and Learning of the Way (Daoxue 道學) teachers advocated moral philosophy and moral judgment as

---

31. Bourdieu, *Distinction, The Field of Cultural Production*, and *Language and Symbolic Power*. I found John Thompson's introduction to the last title especially useful in configuring Bourdieu's disparate explanations of the "field" concept.

the guiding principles for both. The following chapters attempt a history of the interests of these groups, the positions they occupied, and the relationships among them.

The preceding clarification of terms is meant to inform readers of the concepts that have helped me organize and explain the relevance of the wide range of source materials for this project. It is by no means intended as an unmediated application of Bourdieuan sociology to twelfth- and thirteenth-century Song society. There are, for example, clear differences between Bourdieu's understanding of "field" and my use of the term. In Bourdieuan sociology, the actions of individuals in particular fields are directly related to their *habitus* (or predispositions), and those predispositions are in turn shaped by class background. Class background, or at least difference in class background, is an inapplicable variable in the context of the imperial Chinese civil service examinations. The lack of social background data on examination candidates and graduates, teachers, editors, and printer/publishers precludes such an analysis. Consequently, agents in the examination field represent occupational roles, not social or professional classes.

The reconstruction of the history of the field of the civil service examinations has an immediate bearing on broader areas of Chinese history, specifically intellectual history. One question that has loomed large on the research agenda of Chinese intellectual historians is how the Neo-Confucian movement of the Learning of the Way obtained the support of Chinese elites. In contrast to the conclusions of political and social historians who have attributed its spread to its imposition as state orthodoxy in 1241 or to the appeal of its rejection of an examination degree as the hallmark of the scholar-official,[32] this book explains how Learning of the Way ideology was disseminated in the field of the civil service examinations.

---

32. For the first view, see James T. C. Liu, "How Did a Neo-Confucian School Become the State Orthodoxy?"; and idem, *China Turning Inward*. For the second view, see Bol, *"This Culture of Ours,"* 333–34; and Bossler, *Powerful Relations*, 205. On the involvement of Daoxue proponents in local educational and social welfare programs, see Gardner, *Learning to Be a Sage*, 23–34; de Bary and Chaffee, eds., *Neo-Confucian Education*, esp. the articles in parts II–IV; Hymes, "Lu Chiu-yuan, Academies, and the Problem of the Local Community"; and von Glahn, "Chu Hsi's Community Granary in Theory and Practice."

An example of my use of Bourdieu's concept of the field will illustrate the relevance of this project to the broader history of Chinese intellectual culture. My examination of the relationships between the intellectual field and the examination field demonstrates that the intellectual and political agendas of literati traditions were mediated by the conventions of examination preparation, and that individual traditions' adaptations to such conventions related to their wider social and political success. The examination capital of the Learning of the Way movement was small and suspect in the twelfth century. Learning of the Way teachers and students proved unable to translate their modes of exegesis and of discussing government into sustainable examination capital. At the same time, teachers of the Learning of the Way were accumulating capital in the larger field of literati intellectual culture by writing and printing large numbers of books and establishing schools and shrines. By the mid-thirteenth century, Learning of the Way modes of interpretation dominated the field of the examinations because teachers and candidates successfully converted intellectual capital into examination capital. This conversion, however, placed the intellectual legacy of the Learning of the Way in the hands of different agents and transformed Learning of the Way ideology. Parts III and IV of this book discuss how the examination field functioned as a key location for the adaptation, extension, and contestation of Learning of the Way ideology.

The history of the Learning of the Way in the twelfth and thirteenth centuries as traced in this book also suggests a different answer to the larger question of how the civil service examinations contributed to the continuity (or reproduction) of the imperial order. As the civil service examinations became central to literati status, proponents of political and moral philosophies had to participate in the examination field in order to appeal to literati. Proponents of the Learning of the Way therefore inserted themselves forcefully into this field, albeit reluctantly in some cases (for example, Zhu Xi 朱熹 [1130–1200]).

Examination capital could be converted into political capital. The adoption of Learning of the Way standards by the late Southern Song emperors followed, and subsequently strengthened, the position of authority Daoxue leaders had already obtained in the intellectual and examination fields. For all players, participation in the imperial civil service examinations implied acceptance of the rules not only of the examination field but also of the imperial order of which it was part. The examinations contributed to the imperial order because they led to the court's recognition of standards for literati status negotiated between literati groups and representatives of the imperial state in the examination field.

The history of two twelfth-century intellectual traditions animates the field model of the civil service examinations developed in this study. This book traces how "Yongjia" and Learning of the Way teachers promoted distinctive modes of exegesis and of discussing government in the examination field. It also attempts to understand how other teachers adopted and adapted these modes of scholarship, how students used them, how editors and printer/publishers sold them, and how examiners, court officials, and emperors reacted to them.

The Yongjia tradition and the Learning of the Way movement originated as regional intellectual formations and appeared on an empire-wide scale in the second half of the twelfth century. As the name suggests, the Yongjia tradition was closely associated with Yongjia county (which housed the seat of Wenzhou 溫州 prefecture; in present-day Zhejiang province). In a narrow sense, it refers to those scholars native to Wenzhou prefecture working in a distinctive tradition of exegesis, historical scholarship, and administrative reasoning and advocating a distinctive political reform program. More frequently I use the term "Yongjia" (in quotation marks) in a broader sense to refer to teachers active in nearby prefectures, all located in the East Zhe circuit (Zhedong 浙東; present-day eastern Zhejiang), who shared the methods and vision of the Yongjia teachers in many respects. The term was used in this broader sense in the twelfth century. Even though "East Zhe teachers," a term frequently employed in modern intellectual history, is more accurate, I will follow the suggestion of an anonymous reviewer of this book, and use "Yongjia" throughout to avoid confusion.

The histories of the "Yongjia" and Learning of the Way traditions were closely intertwined. Proponents of both traditions were dismissed from office for their opposition to court policies in the 1190s. Because of their shared experience, these two traditions have frequently been treated as the same. Chapter 1 describes the differences between "Yongjia" scholarship and the Learning of the Way movement. The subsequent chapters, outlined below, recount their divergent histories in the twelfth- and thirteenth-century examination field.

Chapter 2 highlights the relevance of examination writing to literati culture. In examination expositions and policy response essays, the two genres of examination writing discussed in this book, scholar-officials demonstrated competence in two of the areas that defined the scholar-official: textual exegesis and the discussion of government policy. By the mid-twelfth century, private teachers and intellectual formations participating in the examination field focused on training in these two areas. Their curricula were shaped by contemporary conventions that governed both the selection of source materials and the methods of analysis and presentation. The overview of the general characteristics of twelfth-century examination writing in this chapter depicts the conventions that shaped the activity of the "Yongjia" and Learning of the Way teachers in the examination field.

"Yongjia" teachers occupied a central position in the examination field in the second half of the twelfth century. Based on an analysis of the reception of examination writing, Chapter 3 demonstrates that the work of the "Yongjia" teachers shaped examination standards in the last decades of the twelfth century. This chapter attributes the appeal of "Yongjia" teachers to their commitment to a political reform program aimed at restoring the authority of the Song court over all its former territories and the successful translation of this political program into examination writing. Chapter 4 sets out the curricular programs underlying the success of the "Yongjia" teachers. It shows that they taught the skills and the political program admired in their examination writing in a comprehensive curriculum that included courses in composition, institutional history, administrative reasoning, and the textual analysis of the Classics, philosophers, and the histories.

Zhu Xi, the central figure in the Learning of the Way movement in the twelfth century, perceived the dominance of "Yongjia" scholarship in the examination field as a pernicious influence on classical exegesis, historical scholarship, and government, the very activities defining literati status. Beginning in the last decades of the twelfth century, Learning of the Way teachers began to formulate alternative standards for examination writing. By the mid-thirteenth century, Learning of the Way examination curricula had displaced "Yongjia" curricula, and

Learning of the Way texts were officially endorsed as authoritative sources for examination preparation. Chapter 6 analyzes the Learning of the Way curricular offensive. It traces the transformation of Learning of the Way ideology—from an antagonistic stance toward competing forces in the examination field to the reconciliation of Learning of the Way moral philosophy with diverse literati traditions in the more traditional areas of competence tested in the examinations such as history, government, and composition. Chapter 7 discusses the ascendancy of Learning of the Way standards by analyzing examination essays written between the 1180s and the 1270s. It explains the impacts of Learning of the Way discourse on both the exegesis of classical and historical texts and the discussion of issues in administrative policy in examination writing. The gradual ascendancy of Learning of the Way discourse in examination writing paralleled the trajectory of its ideological transformation in examination preparation. By the mid-thirteenth century, examination candidates shed the confrontational language of philosophical debate and adopted the textual legacy of the Learning of the Way, that of Zhu Xi in particular, as canonical texts.

The shifting curricular strategies of Learning of the Way teachers and their growing impact on examination writing corresponded to shifts in court policy regarding the examination curriculum. After a major persecution campaign in the 1190s, the court gradually endorsed Learning of the Way teachers and texts as authoritative sources for students engaged in examination preparation. Chapter 5 discusses the role of the court and the central bureaucracy in the definition of examination standards between the 1130s and 1270s. Government regulations from this period bear evidence of a shift in court policy from curricular ecumenism in the twelfth century to curricular standardization in the last half century of Song dynastic rule. Both positions contrasted sharply with the reformist curricular policy of the last decades of Northern Song rule and attest to the lasting influence of the Southern Song court's early decision to withdraw from a trendsetting to a policing role in the examination field.

# Part I

---

*Prolegomena*

# I

## Intellectual Traditions
## and Teachers

The main protagonists in the twelfth- and thirteenth-century examination field were two intellectual traditions, the Learning of the Way and "Yongjia." Both traditions played a prominent role in the development of the examination curricula; their nature and their relationship remain, however, subject to historiographical controversies.

### The Learning of the Way

DEFINITIONS

"The Learning of the Way" is one of a variety of terms used to refer to the re-envisioning of Confucian teachings that began in the eleventh century. Other names used among scholars of Chinese intellectual history to interpret the tide of new explanations of the Confucian legacy are Neo-Confucianism, Neo-Confucian orthodoxy, Cheng-Zhu orthodoxy, the Learning of the Heart-and-Mind (Xinxue 心學), and the Learning of Principle (Lixue 理學). Besides "the Learning of the Way," other translations of Daoxue such as "Tao School" and "True Way Learning" are also in current use. These expressions are indicative of the different perspectives and methods that scholars have brought to bear on the study of intellectual change in middle-period and late imperial China.

"Neo-Confucianism," the most common locution in twentieth-century European and American accounts of Chinese intellectual history, may have originated as early as the eighteenth century;[1] it has gained further currency through the work of Wm. Theodore de Bary and the series of "Neo-Confucian studies" produced by scholars affiliated with Columbia University. De Bary defines the object of Neo-Confucian studies as a set of terms, ideas, and institutions established and debated by East Asian thinkers since the eleventh century.[2] In response to Buddhist ways of mind-cultivation, these thinkers advocated a view and a method of self-cultivation based on the Confucian tradition.[3] De Bary's approach is characterized by a "humanistic hermeneutics,"[4] which aims to explain Neo-Confucian philosophical problems and to rehabilitate Neo-Confucian values by examining them in the context of the life and thought of individual Neo-Confucian thinkers. De Bary distinguishes the broader tradition of Neo-Confucianism from Neo-Confucian or Cheng-Zhu "orthodoxy," which, in his terminology, refers to the endorsement, by the state or other institutions, of a narrower line of thinkers—in the case of "Cheng-Zhu orthodoxy" centered around the Cheng brothers, Cheng Hao and Cheng Yi, and Zhu Xi.[5]

"The Learning of Principle" or "the Learning of the Heart-and-Mind" are translations of indigenous terms. They have been adopted as general terms for the teachings of Neo-Confucian thinkers, but they also reflect sectarian debates about the core values of the Neo-Confucian tradition. "The Learning of Principle" stresses the emphasis of some thinkers on the notion of an inherent "pattern" or "coherence" (*li* 理) in all things that mirrors the natural moral order. It evokes the need for the continuous investigation of pattern in things and events through self-cultivation. "The Learning of the Mind," on the other hand, embodies the priority assigned to the realization of the pattern inherent in the human mind without the need for gradual training in understanding the coherence in things outside the mind. The con-

---

1. Elman, "Rethinking 'Confucianism' and 'Neo-Confucianism' in Modern Chinese History," 526.

2. De Bary, "The Uses of Confucianism," 549–52.

3. De Bary, *Neo-Confucian Orthodoxy and the Learning of the Mind-and-Heart*, preface, xiv–xvi.

4. I borrow this term from Thomas Wilson; see his "The Indelible Mark of an Overlooked Scholar."

5. De Bary, "The Uses of Confucianism," 548–49.

tested meaning of such native terms gives de Bary reason to defend the appropriateness of the generic western term to emphasize the continuities and the commonalties in Confucian and Neo-Confucian traditions.[6]

Hoyt Tillman advocates the use of the indigenous term "Daoxue" (the Learning of the Way), which he translates as "Tao School." In adopting this Chinese term, Tillman intended to rid American academic discourse on the intellectual history of the Song dynasty of the determinism he associates with the use of "Neo-Confucianism," which Tillman believes is narrowly understood as what later became the "Neo-Confucian orthodoxy." In Tillman's view, de Bary's interpretation of Neo-Confucianism has led to a simplistic evolutionary model of the Confucian renaissance in which Neo-Confucian history is refracted through the thought of Zhu Xi. The history of Neo-Confucianism prior to Zhu Xi was cast in light of his later importance, its later history was similarly read as the history of the influence of Zhu Xi, and the impact of Zhu Xi's contemporaries on both his work and that of his disciples was ignored.

In *Confucian Discourse and Chu Hsi's Ascendancy* (1992), Tillman presented a contextualized reconstruction of the rise to dominance of Zhu Xi's version of the Learning of the Way. This work traces the development of the Learning of the Way from "a concentration upon Tao learning" in the eleventh century to "a fellowship" in the twelfth century to a more rigorously defined school of thought under the impulse of Zhu Xi.[7] *Confucian Discourse* and a programmatic article on the scholarly implications of the adoption of "the Learning of the Way" published in the same year had an immediate effect. Monographs and articles on Song history published in the decade since the publication of Tillman's studies have increasingly adopted Daoxue or various western-language equivalents to refer to the Confucian resurgence.[8] In this study,

---

6. Ibid., 546–47.

7. Tillman, "A New Direction in Confucian Scholarship"; idem, *Confucian Discourse and Chu Hsi's Ascendancy*.

8. Note that the adoption of the new terminology has not led to a consensus on the understanding of its meaning. Benjamin Elman ("Rethinking 'Confucianism' and 'Neo-Confucianism' in Modern Chinese History," 529) endorses Tillman's historical reconstruction of the Learning of the Way, but defines the term differently as the "orthodox, Zhu Xi–oriented trends in classical learning from 1000–1700."

I likewise opt for "the Learning of the Way" in an effort to contribute to the history of the intellectual formation commonly referred to by this name in the last decades of the twelfth century.

The historical trajectory of the Learning of the Way presented in *Confucian Discourse* supports a hypothesis of reductionism. According to this thesis, the Learning of the Way was an intellectual formation characterized by a diversity of opinion and sources of authority in the first century of its existence (ca. 1080–1180). In the last two decades of the twelfth century, Zhu Xi excluded competing sources of authority in an effort to construct a narrow line of the transmission of the Learning of the Way and its principal teachers, texts, and beliefs. This interpretation is based on the assumption that teachers belonging to distinct regional intellectual formations, including the "Yongjia" tradition, were identified as members of a common Learning of the Way fellowship before the 1180s.

Ascribing membership in the Learning of the Way movement is complicated by the fact that the Chinese term carried at least three meanings and connotations in the twelfth century. First, it was a generic word for moral cultivation, here translated as "learning of the Way." In a passage written by Lu Jiuyuan 陸九淵 (1139–93), "His writing approached [the model of] Antiquity and resembled the style of Tuizhi [Han Yu 韓愈 (768–824)] and Zihou [Liu Zongyuan 柳宗元 (773–819)]; his learning of the Way reached profundity and followed up on the meaning of Zi Si 子思 (5th c. BCE) and Mencius 孟子 (372–289 BCE),"[9] "learning of the Way" stands for one type of learning or area of literati distinction. This type of learning could be juxtaposed with other categories of learning, such as literary skill and historical or classical scholarship. As such, it was not the domain of one particular group of literati.

For those who prioritized learning of the Way, moral learning guided all areas of literati activity. This implied both that moral learning had to be expressed in the social and political arena and that other areas of literati distinction had no legitimate independent status. The earliest usage of the term in Song times opposed its disinterestedness and

---

9. Lu Jiuyuan, *Xiangshan ji*, 27.313. Lu was citing praise for his brother Lu Jiuling 陸九齡 (1132–80). Zi Si was the grandson of Confucius.

grounding in classical moral values such as humanity and righteousness to literati activity motivated by personal gain.[10] The term *daoxue* was further used in this sense by Cheng Yi, later hailed as one of the founders of the Learning of the Way fellowship. He noted that several of the men commemorating his brother Cheng Hao after his death noted his excellence in moral cultivation (learning of the Way).[11] One of Cheng Yi's disciples, Yang Shi 楊時 (1053–1135), similarly drew attention to the impact of Cheng Hao's moral cultivation. In Yang Shi's reading, Cheng Yi wrote a biography of his brother to preserve the memory of how his "moral learning and righteous actions" had transformed those around him to the benefit of the human realm as a whole.[12] "Learning of the Way" understood as moral cultivation referred to moral knowledge acquired through the classical tradition as well as to the practice of moral self-cultivation based on the acquisition of that knowledge. For its advocates, moral learning encompassed classical and historical scholarship, literary composition, and political practice.

Second, "Learning of the Way" referred to a group of scholar-officials accused of setting up an exclusive party. Court officials and literati used the term in this pejorative sense throughout the 1180s and 1190s. I use the phrase "Doctrine of the 'True' Way" for *daoxue* as used by its critics, because it conveys the derogatory overtones in their explanations of the meaning and intent of "the Learning of the Way." The first reference to the Doctrine of the "True" Way in court communications dates to 1182 or 1183. In 1183 Censor Chen Jia 陳賈 submitted a memorial in which he argued that Doctrine of the "True" Way discourse had gained popularity as a means for partisans to obtain

---

10. Yü Yingshi (*Zhu Xi de lishi shijie*, 1: 173) traces the earliest usage to Liu Kai 柳開 (947–1000).

11. Tillman, *Confucian Discourse and Chu Hsi's Ascendancy*, 6; idem, "A New Direction in Confucian Scholarship," 460.

12. Yang Shi, *Yang Guishan ji*, 3.52. Hans van Ess (*Von Ch'eng I zu Chu Hsi*, 25n97) cites a passage in which Hu Yin 胡寅 (1098–1156) juxtaposed various areas of literati distinction and similarly associated "learning of the Way" with Cheng Yi. My reading of this passage differs somewhat from van Ess's as he does not make the distinction between the learning of the Way as an area of literati achievement and the Learning of the Way as a movement.

political power.[13] He labeled their interpretation of the Classics hypo-critical and false (*wei* 偽) and charged that their words never matched their actions (for further discussion of this memorial, see Chapter 5). By the 1190s, the "Doctrine of the 'True' Way" was used interchangeably with "False Learning." The politicians behind the False Learning cam-paign of the 1180s and 1190s portrayed the Doctrine of the "True" Way as a tight-knit group of scholar-officials who had claimed an exclusive identity for their members based on a shared interest in the intellectual legacy of the Cheng brothers. Among the means this party used to bond its members, they identified the use of philosophical jargon in speech and writing, distinctive dress and outward behavior, the prioriti-zation of a new set of classical texts, a moralistic critique of the court, and the promotion of party members through the civil service examinations.

The campaign harkened back to government action in the 1130s, 1140s, and 1150s under the regime of Qin Gui 秦檜 (1090–1155) against "Cheng Learning"—contemporary shorthand for the written and orally transmitted legacy of the Cheng brothers and the activities that kept this legacy alive in society and politics. Despite the claims of Qin Gui and his partisans that those practicing Cheng Learning claimed exclu-sive knowledge of the truth and were forming a political faction, the perception that Cheng Learning represented a mode of scholarship that was of general interest to scholar-officials and without a political infra-structure prevailed. The call for action against "Cheng Learning" ended soon after Qin Gui's enemies, the real targets of the campaign, were removed from court. Literati continued to discuss the work and the ideas of the Cheng brothers. This kind of Cheng Learning was, how-ever, not an organized project to transform society and polity. When I use the qualifier "Chengist," I am referring to the nonprogrammatic literati interest in the moral philosophy of the Cheng brothers.

---

13. Zheng Bing 鄭丙 (1121?–94), then minister of personnel, submitted a memorial denouncing "the Doctrine of the 'True' Way" several months before Chen Jia. This memorial cannot be dated precisely, but according to Shu Jingnan's (*Zhu Xi nianpu changbian*, 1: 756) findings, it had to have been written between September 1182 and January 1183. For Chen Jia's memorial, see ibid., 1: 772. In *Zhuzi da zhuan* (525), Shu dates Zheng's memorial to November–December 1182.

The substitution in the 1180s and 1190s of "Doctrine of the 'True' Way," for "Cheng Learning" was an indication that the successive factions dominating the court at the time recognized the intellectual and political threat posed by the radicalization of a group of scholars advocating Cheng Learning. The campaign against the Doctrine of the "True" Way was a response to the challenges posed by Zhu Xi. Through a series of publications, political actions, and educational activities starting in the 1160s and lasting through the 1190s, Zhu Xi presented himself as the leader of the tradition of Cheng Learning and as a vociferous critic of court policy. Zhu Xi institutionalized the Chengist call for the moral reform of society and created an exclusionary identity for the members of the Learning of the Way. Through his actions Cheng Learning was transformed from a mode of learning into a cultural movement (see Chapter 5 for further discussion).

Scholars of Song political and intellectual history have read the government campaign against the Doctrine of the "True" Way as a sign of the existence of a *fellowship* of advocates of Cheng Learning known as the Learning of the Way throughout the course of the twelfth century. Several historians have further pointed to the 1197 blacklist of scholar-officials accused of having had ties to the clique of the Doctrine of the "True" Way as evidence of the inclusive membership of this fellowship in the twelfth century. Similarly, historians proposing a broad definition of the twelfth-century meaning of the Learning of the Way have regarded the diverse blacklist of Confucian scholars as evidence that the Learning of the Way referred to a diverse fellowship of Confucians, from different regional branches.[14] For example, Chen Fuliang 陳傅良 (1137–1203) and Ye Shi 葉適 (1150–1223), both representatives of the Yongjia tradition of scholarship, appeared on the official blacklist. The inclusion of Yongjia teachers as well as scholars affiliated with other regional traditions of scholarship confirms, from this perspective, that a

---

14. Tillman, "A New Direction in Confucian Scholarship," 465; Schirokauer, "Neo-Confucians Under Attack," 184–96. The list referred to in this work, i.e., the list as it now appears in Li Xinchuan's 李心傳 (1166–1243) *Record of the Destiny of the Way (Dao ming lu* 道命錄), may have been a later interpolation; see Hartman, "Bibliographical Notes on Sung Historical Works," 33.

broad understanding of the Learning of the Way best reflects the general twelfth-century use of the term.

The broad definition of the Learning of the Way accords with the court's use of the label in its efforts to rid the bureaucracy of those supportive of Zhu Xi's politics. The dominant court faction cast its web widely, but its indiscriminate application of the label does not mean that a Learning of the Way fellowship with many members existed in fact. The historical reconstruction of the Learning of the Way as a fellowship with a wide-ranging membership needs to move beyond the verification of personal, social, geographical, political, or intellectual ties to address the question of identity. Did those included in the broad definition of the Learning of the Way identify with it? And how did they articulate their identity as members of the Learning of the Way?[15] Chapters 3 and 4 demonstrate that several of those on the blacklist did not identify with the Learning of the Way. The Yongjia teachers seldom used the term *daoxue* in their writing. When they did, they cast it in a critical light and concurred with the official censure of its advocates' exclusivist attitude toward other traditions of scholarship. Chen Fuliang and Ye Shi appeared on the official blacklist because they had written in opposition to the court's purge of Zhu Xi, not because they identified with the Learning of the Way.

"The Learning of the Way" was used in a third sense in the late twelfth century, to refer to a tradition of moral philosophy transmitted through a narrowly defined genealogical line of true transmitters and captured in a new set of canonical texts. This was the meaning that Zhu Xi created for it. Recent studies of Zhu Xi's intellectual development concur that his moral philosophy took on a definite shape during the 1170s.[16] By age fifty, the age by which Confucius claimed to have come to know what

---

15. Tillman (*Confucian Discourse*, 3) intentionally defined "fellowship" rather loosely: "By 'fellowship,' I mean that they had a network of social relations and a sense of community with a shared tradition that distinguished them from other Confucians. They forged personal, political, and intellectual ties in a common effort to reform political culture, revive ethical values, and rectify Confucian learning"; cf. "A New Direction in Confucian Scholarship," 459. Tillman interprets the participation of scholars interested in Cheng Learning in social, political, and intellectual exchanges as the basis for their identification with an ecumenical fellowship.

16. Shu Jingnan, *Zhuzi da zhuan*, 380, and *passim*; Ichiki, *Shu Ki monjin*, 194, and *passim*.

Heaven decreed, Zhu Xi had identified the process and the extension of moral self-cultivation outlined in the eight-step sequence in *The Great Learning* (*Daxue* 大學) as the core of Confucian learning.

In Zhu Xi's reading of the sequence, sociopolitical order depended on "rectifying the mind," "making the will sincere," "extending knowledge," and "investigating things." His thinking focused on the theoretical elaboration of the last two steps in the process of self-cultivation. The significance of moral self-cultivation depended for Zhu Xi on the apprehension of "coherence" (*li*, also translated as "principle"). He adopted Cheng Yi's interpretation of "coherence" as a structuring principle giving shape and purpose to the psycho-physical energy (*qi* 氣), of which everything is made up. Coherence is the normative principle of organization in the universe and dictates the proper path of development in each thing and its proper relationship to other things. Zhu Xi defined self-cultivation as the apprehension of coherence in all things and events. He understood things and events primarily as the human relationships that structure society and the polity and held that apprehension of the universal patterns of coherence in past and present human relationships results in the full realization of the moral obligations inscribed in them.

Zhu Xi's views on how coherence could be apprehended and activated by the human mind differed from those of others engaged in Cheng Learning in the twelfth century. The investigation of things was the first step in the process of self-cultivation and consisted, in Zhu Xi's interpretation, of the mind's engagement with things external to it. Zhu Xi agreed with the Cheng brothers' view that coherence inhered in the mind and that it was the mind's endowment of coherence that allowed it to apprehend coherence in other things. He opposed, on the other hand, the conclusion that the mind was self-reliant and could effect its own enlightenment. In Zhu Xi's moral philosophy, the mind cultivated itself through the continuous probing and realization of the patterns of coherence in events external to itself. Reading was an essential training ground in this process of self-cultivation. Zhu Xi developed a reading method incorporating his vision of reading as the process of realizing coherence (see Chapter 6). Zhu Xi's method proposed a gradual curriculum, in which the key sources for his moral philosophy were studied in a fixed sequence. It began with the Four Books, a new set of classical texts, which Zhu Xi established as the core of the Learning of the Way. Before the first combined edition of all Four Books (*The Great Learning*, *The Analects*, *Mencius*, and *The Doctrine of the Mean*) in 1182, he had published separate editions and commentaries for all four texts between the 1160s and 1180s (Chapter 5).[17]

When Zhu Xi first started using the designation "the Learning of the Way" in the 1170s, he defined it not only by a core message (the cultivation of the mind to apprehend coherence and thus revive the Way, or the order of the universe) and a new classical canon (the Four Books) but also by a genealogy of true transmitters of the Way. Following the model of Cheng Yi, who claimed that his brother Cheng Hao had recovered the Way after centuries in which the moral order of Antiquity had been lost, Zhu Xi claimed that the core of Confucian learning had been revived by Zhou Dunyi 周敦頤 (1017–73), who then transmitted his understanding of the cosmic imperative of moral self-cultivation to Cheng Hao and Cheng Yi. In Zhu Xi's genealogy, the Cheng brothers in turn transmitted this message to Zhang Zai 張載 (1020–78). The historical accuracy of this genealogy is highly questionable,[18] but Zhu Xi's genealogical thinking was an essential contributing factor to both the intellectual coherence of the Learning of the Way and the formation of a Learning of the Way identity.

The genealogy of the Four Masters provided a template for imposing unity on their disparate works and imprinting in the minds of Learning of the Way followers the notion that their work coincided with the Way of the sages of Antiquity. The inclusion of Zhou Dunyi's work, in Zhu Xi's interpretation of it, provided a cosmological foundation for Cheng Learning. Zhang Zai's *Western Inscription* (*Ximing* 西銘) was an influential and controversial text among twelfth-century scholars,

---

17. Shu Jingnan, *Zhuzi da zhuan*, 766. The first combined edition of the Four Books has traditionally been dated to 1190 (Wang Maohong, *Zhu Xi nianpu*, 558). Shu Jingnan (*Zhuzi da zhuan*, 766–70, 814–15) has argued that the *"Four Masters"* Zhu Xi printed in 1190 were not his commentaries on the Four Books. The edition of his commentaries printed in 1192 became the standard edition, but earlier editions were printed in 1182, 1184, and 1186. Sano Kōji traces the term "Four Books" to Zhu Xi and discusses early attempts at publishing Zhu Xi's commentaries on the Four Books in *Shishogaku shi no kenkyū*, chap. 4, esp. 203–9.

18. Tsuchida, "Dōtōron sai kō"; idem, "Shū Tei juju sai kō"; idem, "Sōdai shishōshi jō ni okeru Shū Toni no ichiki"; Wilson, *Genealogy of the Way*, 199–210.

and Zhu Xi's co-optation of it contributed a powerful vision of the immanence of the (hierarchical) sociopolitical order in the cultivation of the self. In the *Western Inscription*, Zhang Zai envisioned the Confucian virtue of "humaneness" as an all-encompassing love linking Heaven and the individual in a parental relationship on a cosmological scale, a love that encompasses all relationships and everything that intervenes between Heaven and the individual. Zhu Xi edited and commented on Zhang Zai's text and promoted it as the articulation of the metaphysical truth central to the Learning of the Way: coherence is one but manifested in different ways in the myriads of things (*li yi fen shu* 理一分殊). The emphasis on the diverse manifestations of coherence in this interpretation modified Zhang Zai's vision of humaneness as the foundation of the unity of all things and turned the *Western Inscription* into a vision of a unified hierarchical moral order in line with Cheng Yi's moral philosophy.[19] In *A Record for Reflection* (*Jinsi lu* 近思錄, 1173), Zhu Xi culled extracts from the works of the Four Masters to elucidate the core concepts and the process of self-cultivation in his version of the Learning of the Way. This anthology, discussed in Chapter 6, was early testimony to Zhu Xi's systematic effort to shape the Learning of the Way into a coherent textual, intellectual, and moral community.

Intellectual genealogy was also a way of laying claim to a position of authority in the transmission of the Learning of the Way. Teachers and disciples inserted themselves within an ongoing chain of transmitters by affiliating themselves with teachers and disciples linked to the main figures in the genealogy. As noted by Thomas Wilson, genealogical discourse also served as a means of exclusion.[20] Zhu Xi's genealogy was intended to exclude those among his contemporaries interested in Cheng Learning who articulated competing interpretations. Genealogical discourse and exclusivism served to define the Learning of the Way

---

19. Zhu Xi was following up on Cheng Yi's praise for this particular piece in Zhang Zai's oeuvre. Cheng Yi also read it as a confirmation of his metaphysical thesis that coherence was one but manifested itself differently in individual things. See Kasoff, *The Thought of Chang Tsai*, 142–43. For Zhu Xi's reading of Zhang Zai's work, see Qian Mu, *Zhuzi xin xuean*, 3: 97–112; Wing-tsit Chan, *Chu Hsi: New Studies*, 297–99; Shu Jingnan, *Zhuzi da zhuan*, 281; and Yü Yingshi, *Zhu Xi de lishi shijie*, 1: 200–218.

20. Wilson, *Genealogy of the Way*, esp. chap. 2.

identity. In the 1190s the public display of an exclusive intellectual gene-alogy spurred the largest campaign against Cheng Learning in history.

In the last three decades of his life, Zhu Xi commonly referred to the tradition of Cheng Learning as "the Learning of the Way." Not only in letters to colleagues, friends, and disciples or in conversations (as they were recorded by disciples) but also in more public genres such as stele inscriptions and prefaces to published editions and commentar-ies, he wrote of the Learning of the Way as a coherent moral philoso-phy transmitted through the unified voices of its founders. Between 1175 and 1195, Zhu Xi was involved in the construction or dedication of a dozen shrines to Zhou Dunyi, Cheng Hao, and Cheng Yi.[21] In in-scriptions written for these shrines, which must have been on public display in them, he highlighted the roles of Zhou and the Cheng broth-ers in the "recovery" of "the Learning of the Way" and their commit-ment to take its further transmission as their foremost responsibility.[22] Likewise, in prefaces to his commentaries on *The Analects* (*Lunyu yaoyi* 論語要義, 1163) and *The Doctrine of the Mean* (*Zhongyong zhangju* 中庸章句, 1189) and in introductions to his editions of the works of the Four Masters (*Taiji tongshu* 太極通書, 1179; *Cheng shi yishu* 程氏遺書, 1168), he underscored the contributions the masters had made to "the Learning of the Way" in terms of the restoration of the true meaning of the Classics and the revival of the moral philosophy that explained their underlying unity.[23] Zhu Xi distinguished the Four Masters from other

---

21. Neskar, "The Cult of Worthies," 225. Neskar's tabulation of the total number of Transmission Shrines (shrines dedicated to the founders of the Learning of the Way) by decade shows a steep increase between the 1170s and 1195 (before the beginning of the False Learning campaign). Of the total number of shrines, Zhu Xi and Zhang Shi were involved in the dedication of more than half. Zhu Xi's disciples were involved in the construction of several other shrines (ibid., 219–20, 228–30). Neskar gives thirteen as the total number of shrines; Meng Shuhui (*Zhu Xi ji qi menren de jiaohua linian yu shijian*, 426–27) gives a figure of twelve.

22. For examples, see the inscriptions to the Shrine for Master Mingdao (Cheng Hao) in the Prefectural School of Jiankang (1176), the Shrine for Master Lianxi (Zhou Dunyi) in the District School of Shaozhou (1183), and the Shrine for the Two Masters Cheng in the District School of Huangzhou (1192) in Zhu Xi, *Zhu Xi ji*, 7.4064–65, 4105–6, 4135–37. Cf. Neskar, "The Cult of Worthies," 234–35.

23. Zhu Xi, *Zhu Xi ji*, 7.3923–25, 3937–38, 3967–70, 3994–96.

eleventh-century teachers sometimes credited by his contemporaries with the revival of the classical tradition:

[A student asked:] "Why is the ascendance of the Learning of the Way dragging on so long during our dynasty?"

The Master [Zhu Xi] replied, "It was after all a gradual process. Beginning with Fan Wenzheng [Fan Zhongyan 范仲淹 (989–1052)], there was already some good discussion. For example, in Shandong [the Jingdong 京東 circuit] there was Sun Mingfu [Sun Fu 孫復 (992–1057)], by the Culai Mountains [Jingdong Circuit] Shi Shoudao [Shi Jie 石介 (1005–45)], and in Huzhou [Liang Zhe 兩浙 circuit] Hu Anding [Hu Yuan 胡瑗 (993–1057)]. Then afterward Master Zhou, Master Cheng [Hao], and Master Zhang appeared. Therefore Master Cheng [Yi] could not forget these gentlemen during his life and respected them all along. . . . However, even though the natural ability of these [early Song] men was great, even though they knew how to honor kingship and censure hegemony, clarify integrity and ban self-interest, they reached only that level and that was it. *They did not see coherence (li), and therefore they were unable to get it.*"[24]

本朝道學之盛, 豈是滾纏?

先生曰: "亦有其漸. 自范文正以來已有好議論, 如山東有孫明復, 徂徠有石守道, 湖州有胡安定, 到後來遂有周子程子張子出. 故程子平生不敢忘此數公, 依舊尊他. . . . 然數人者皆天資高, 知尊王黜霸, 明義去利. 但只是如此便了, 於理未見, 故不得中."

Zhu Xi excluded contemporary students of Cheng Learning from the Learning of the Way on the same grounds. He originally included men like Lü Zuqian 呂祖謙 (1137–81) and Lu Jiuyuan in the group of contemporaries he believed constituted the Learning of the Way in the late twelfth century, but he persistently pointed out incongruities between their scholarship and what he saw as core Learning of the Way beliefs. He took Lü Zuqian to task for his interest in examination teaching and particularly his attention to institutional history and composition (see Chapters 4 and 6). His criticism of Lu Jiuyuan centered on Lu's philosophy of the mind, which attributed the apparent existence of coherence in things to the workings of the human mind. Zhu Xi questioned

---

24. *ZZYL*, 129.3089–90. This passage was recorded by Zheng Kexue 鄭可學 (1152–1212). Zheng's notes reflect his exchanges with Zhu in 1191 (Wing-tsit Chan, *Zhuzi menren*, 340–41).

the loyalty of both men to the legacy of the Cheng brothers because of
their unwillingness to put the investigation of things, defined as the
pursuit of moral coherence in all affairs, first in their teaching.

It was in reaction to Zhu Xi's narrowly construed Learning of the
Way that ruling court factions in the 1180s and 1190s launched the False
Learning campaign through which the label "Doctrine of the 'True'
Way" gained notoriety.[25] Court officials adopted *daoxue*, the term Zhu
Xi had fashioned to bond those interested in Cheng Learning into a
tighter community, but broadened its meaning to implicate many who
were sympathetic to Zhu Xi's call for moral reform but critical of his
efforts to set up an exclusionary Learning of the Way movement. In
1188, Lin Li 林栗 (*js.* 1144), then vice minister of military affairs (*bingbu
shilang* 兵部侍郎), submitted a memorial accusing Zhu Xi of establishing the Learning of the Way as an exclusive intellectual tradition with
the goal of claiming intellectual and political authority:

[Zhu] Xi fundamentally has no scholarship. He just steals the leftovers of
Zhang Zai and Cheng Yi and makes them into empty principles. This he calls
the Learning of the Way. He unrightfully aggrandizes himself. He now has
several dozens of disciples. They mimic attitudes of the Spring and Autumn
and Warring States periods and vainly aspire to the model of Confucius' and
Mencius' successive court invitations.[26]

熹本無學術. 徒竊張載程頤之緒餘, 以爲浮誕宗主. 謂之道學. 妄自推尊.
所至輒攜門生數十人. 習爲春秋戰國之態, 妄希孔孟歷聘之風.

Among those leading the campaign against the Learning of the Way,
Lin Li was probably most familiar with Zhu Xi's scholarship. Prior to

---

25. Other historians have also noted that the use of the term *daoxue* to denote a
group of scholars did not become widespread until the 1180s; see Chaffee, "Chao Ju-yü,
Spurious Learning and Southern Sung Political Culture," 36, 37*n*36. Chaffee adds that
this fact makes the use of it in reference to a Learning of the Way fellowship before the
1180s questionable. Nevertheless, in contrast to the argument outlined above, Chaffee
still accepts Tillman's account of the history of the Learning of the Way in the twelfth
century. Yü Yingshi (*Zhu Xi de lishi shijie*, 2: 314, 345–46) has argued that the creation of
the label "Doctrine of the 'True' Way" in the early 1180s contributed to the formation
and the growth of the Learning of the Way and that it was first targeted at Zhu Xi and
his disciples.

26. Li Xinchuan and Xu Gui, *Jianyan yilai chaoye zaji*, yi ji, 7.617; cf. Shu Jingnan,
*Zhuzi da zhuan*, 647.

his impeachment of Zhu Xi in 1188, Lin had exchanged views with him on *The Changes* (*Yijing* 易經), Zhou Dunyi's *Diagram of the Supreme Ultimate* (*Taijitu* 太極圖), and Zhang Zai's *Western Inscription*. In a private meeting with Zhu Xi ten days before he submitted his memorial, Lin Li repudiated the work of Zhou Dunyi and defended his own commentaries on *The Changes*. He further dismissed Zhang Zai's *Western Inscription*. Lin Li's rejection of Zhang Zai's cosmological explanation of Confucian virtues as an apology for Buddhist theories on the identity of the Buddha (the enlightened) and the phenomenal world undercut Zhu Xi's interpretation of it.

During Lin's courtesy visit, Zhu Xi ridiculed his commentary and summarily rejected his criticisms of the men Zhu Xi regarded as the founding thinkers of the Learning of the Way. This visit confirmed the picture that Lin Li and other court officials had been forming of the Learning of the Way since the late 1170s and early 1180s. The passage from Lin's 1188 letter of impeachment translated above illustrates the perception among a group of court officials that Zhu Xi was transforming Cheng Learning into a Learning of the Way movement. The signs of this transformation were his creation of a coherent moral philosophy, co-optation of eleventh-century sources of authority to this end, suggestion of a line of transmission between the eleventh-century masters and himself, and the formation of a group of disciples committed to the Learning of the Way philosophy and dedicated to its spread at court and in local society.[27]

This third meaning of the Learning of the Way is the one that applies throughout most of this book as it addresses the question of how this radicalized movement of scholars of Cheng Learning positioned themselves in the examination field. In those cases where either the more general meaning or the pejorative meaning is appropriate, I use "learning of the Way" and "the Doctrine of the 'True' Way," respectively. My decision to use "the Learning of the Way" in a sectarian

---

27. Shu Jingnan, *Zhuzi da zhuan*, 631–35. Yü Yingshi (*Zhu Xi de lishi shijie*, 2: 168–73) argues that Lin Li was originally merely Zhu Xi's intellectual opponent and became a political enemy of the Doctrine of the "True" Way as a result of the escalating tensions between the two factions at court, namely, the reformers (which he terms the Learning of the Way camp in a broad sense) and the career bureaucrats.

sense is motivated by the fact that Zhu Xi's definition gained widespread publicity in the last decades of the twelfth century and became the core of thirteenth-century definitions of the Learning of the Way or, as it was rechristened, the Learning of Principle. The application of the narrow definition of the Learning of the Way (referring to an intellectual community centered around Zhou Dunyi, the Cheng brothers, Zhang Zai, and Zhu Xi and their teachings) in writing the intellectual history of the late twelfth and thirteenth centuries need not ignore the broader context that shaped Zhu Xi's definition.

Cheng Learning as a mode of scholarship carried on through the discussion of the works of the Cheng brothers and the questions they raised had existed for half a century before Zhu Xi first began to study the work of the Chengs and some of their disciples. The history of the reception of Cheng Learning in the first half of the twelfth century requires further study.[28] Did the tradition they spearheaded develop from a unitary to a dispersed community as some historians have suggested?[29] What were the defining characteristics of the community the Cheng brothers created around themselves? What tied the community together after dispersal set in?

As recent scholarship has amply demonstrated, Zhu Xi's grand synthesis was shaped by lengthy discussions with contemporary peers and students working on the legacy of Cheng Learning, especially Zhang Shi 張栻 (1133–80), Lü Zuqian, Lu Jiuyuan, Chen Liang 陳亮 (1143–94), Chen Fuliang, and Cai Yuanding 蔡元定 (1135–98). The history of these

---

28. The recent work of Hans van Ess focuses on the role of the Hu family in the transmission of Cheng Learning between the Northern and Southern Song periods. Van Ess argues that Zhu Xi was indebted to the Hu family because they were the first to "canonize" the Four Masters and because the idea of the genealogy of the Way originated with Hu Hong 胡宏 (1105–55). More work remains to be done to place van Ess's findings in the broader context of the transmission of Cheng Learning in the early twelfth century. Even though van Ess typically adopts the broader definition of the Learning of the Way, he concludes his study on the publication history of the Cheng brothers' *Posthumous Scriptures* (*Cheng shi yishu* 程氏遺書) with the question "whether a 'Learning of the Right Way' Movement really existed as such before Chu Hsi" (van Ess, "The Compilation of the Works of the Ch'eng Brothers," 298). For the reasons explained in this introduction, I share this doubt.

29. Ichiki, *Shu Ki monjin*; Tillman, *Confucian Discourse and Chu Hsi's Ascendancy*; idem, "A New Direction in Confucian Scholarship."

scholarly exchanges attests to the continuity of elite interest in Cheng Learning, oftentimes coexisting and intermingled with other modes of scholarship such as Su Learning (Suxue 蘇學, the Learning of Su Shi and the Su family more broadly), Wang Learning (Wangxue 王學, the Learning of Wang Anshi), or Buddhist scholarship. It does not, in my view, substantiate the existence of an ecumenical fellowship of diverse intellectual traditions. The questions of what kind of intellectual formation was represented in Chengist discourse and how this intellectual formation changed shape in the twelfth century depends on the question of identity, or, in twelfth-century terms, intellectual affiliation and moral commitment. In contrast to Zhu Xi, most of his peers (including Lü Zuqian, Lu Jiuyuan, Chen Liang, Chen Fuliang, and Ye Shi) did not articulate an affiliation with "the Learning of the Way."[30] In official communications concerning the civil service examinations before the 1180s, the label "Cheng Learning" rather than "the Doctrine of the 'True' Way" was used.[31] The court then censured a mode of scholarship without associating scholarly discourse with a community of self-identifying followers of the Learning of the Way.

My use of the narrow definition of the Learning of the Way underscores a major transformation in the history of the reception of the legacy of the Cheng brothers. From the term's association with one or two individuals (Cheng Yi and, to a lesser extent, Cheng Hao) in the designation "Cheng Learning" or from its association with a regional tradition in "Luo Learning" (Luoxue 洛學, referring to the Luo River

---

30. This impression is substantiated by a search of the electronic edition of the *Siku quanshu*. Based on a search of the Digital Heritage electronic edition, the term *daoxue* appears only three times in Chen Fuliang's work; only one of the three results is a valid use of the term and can be attributed to Chen Fuliang. A search of Lü Zuqian's work yields 29 hits, but the vast majority of the results are inapplicable or do not reflect Lü's usage of the term as they occur in writings of others about him or in texts of others included in anthologies compiled by him. In contrast, a search of Zhu Xi's works included in the *Siku quanshu* yields 178 results. Zhu Xi's posthumous recognition of Lü Zuqian as a leader of the Learning of the Way is frequently cited as evidence of Lü's central role in it. This evidence can, however, also be read as an attempt on Zhu Xi's part to claim Lü Zuqian's legacy and incorporate it into his Learning of the Way. For an example of Zhu Xi's appropriation of Lü's legacy, see Tillman, "Reflections on Classifying 'Confucian' Lineages," 45–46.

31. *WXTK*, 32.300.

region in Shaanxi and Henan where the Cheng brothers taught), on a par with other modes of scholarship associated with individual thinkers or regional traditions, Chengist scholarship spun off an exclusivist movement in the last two decades of the twelfth century. The court's perception of the ideological and organizational strengths of the burgeoning "Learning of the Way" movement under Zhu Xi's direction helps explain why the ongoing factional disputes between pro-Cheng and anti-Cheng politicians developed into a full-scale campaign, complete with an official blacklist, only in the 1190s (see Chapter 5).

## DISCURSIVE PRACTICES

This study not only defines the Learning of the Way by the tension between inclusiveness and exclusiveness central to the scholarship on it in the past decade but also analyzes Zhu Xi's Learning of the Way movement as a community characterized by a cluster of identity-forming discursive practices—habits of thinking, speaking, writing, and acting that form a systematic whole. As a cluster, these practices facilitated both the delimitation of the Learning of the Way movement in the twelfth century and its transformation in the thirteenth century.

The first of the four discursive practices defining the Learning of the Way community was the recognition of coherence in the textual tradition and unity among its leading teachers. In Zhu Xi's definition, the Learning of the Way represented a moral philosophy based on the fundamental principles of the organization of the cosmos. This moral philosophy was shared by the sage-kings of Antiquity, Confucius and his disciples down to Mencius, and the eleventh-century founders of the Learning of the Way. The words and descriptions of the sages recorded in the Classics, the teachings of Confucius and Mencius, and the writings and records of the Four Masters therefore had to be interpreted as being in agreement. From the thirteenth century on, Zhu Xi's legacy was added to this list as the great synthesis of the Learning of the Way tradition.

The full textual tradition of the Learning of the Way, as reorganized in Zhu Xi's reading program, was a carefully organized introduction to the theory and practice of the cosmic imperative of moral self-cultivation. The recognition of a body of knowledge as a comprehen-

sive and systematic explanation of the human condition and the normative sociopolitical order was fundamental to Learning of the Way identity. It provided the foundation for the individual literatus' ability to ob-
jectify, that is, to interpret, individual experience (for example, that of
reading a textual passage) in terms of a theoretical framework. This
mechanism for converting individual experience into moral philosophy
transformed discrete events in a literatus' life into theorized group identity.[32] It allowed the individual to interpret and celebrate his experience
in the same terms as those articulated by the great teachers as well as
the larger community of followers of the Learning of the Way.

Second, Learning of the Way discourse was characterized by hierarchical authority. The genealogical mode of discourse inserted students
in vertical relationships to figures, past and present, transmitting the
imperative of moral self-cultivation. Learning of the Way discourse in
the twelfth century prioritized personal contact between teacher and
disciple. The records of conversations (*yulu* 語錄), which preserved exchanges between teacher and disciple, symbolized the adoption of a
new mode of discourse. As discussed in Chapters 5 and 6, this new
mode of discourse centered on the disciple's gradual acquisition and
personal articulation of Learning of the Way truths under the ever-
watchful eye of a teacher.

The moral authority accorded the leading thinkers in the genealogy
of the Way and the teachers who followed in their wake was based on a
new image of the teacher. In the recorded conversations and commemorative writings, teachers emerge as transmitters of the Way.[33] The
Learning of the Way teacher was linked to the founders of the Learning
of the Way through an intellectual genealogy. He had acquired a full
understanding of Learning of the Way moral philosophy and demonstrated that in a lifelong commitment to live and teach its core beliefs.
He was indifferent to rank and wealth and put moral principle first in
all matters. He taught without tiring and selected texts and methods
suited to the individual disciple's moral transformation. Disciples testified that the teacher's vocation and devotion to the Learning of the

---

32. Kohn and Roth, *Daoist Identities*, 1–11.

33. On the formation of competing images of the teacher in twelfth-century biographical writing, see De Weerdt, "The Ways of the Teacher."

Way inspired new generations to assume responsibility for the continuity of the transmission of the Learning of the Way.

Commitment, a focal attribute in the ideal image of authoritative figures, was a third feature in the cluster of identity-forming discursive practices. Commitment was expressed in the personal and explicit identification of the individual with the Learning of the Way. Starting in the last decades of the twelfth century, students of the Learning of the Way affirmed their belief in the leading thinkers of the movement, the new textual tradition edited by Zhu Xi, and Learning of the Way philosophical concepts. The individual's affirmation of the truth of the Learning of the Way was more than an illustration of the effect of the interpellation of the subject as a general characteristic of ideologies.[34] According to Louis Althusser, ideologies call on the individual and thus identify the individual as a unique and indispensable contributor to their project. In the case of the Learning of the Way, study was by definition "learning for oneself" (*wei ji zhi xue* 爲己之學). Self-awareness and moral self-cultivation were the only means by which sociopolitical order could be achieved. The individual's understanding of the self (in accordance with Learning of the Way moral philosophy) defined one's membership in the transmission of the Learning of the Way.

Scholars articulated commitment in various ways and in varying degrees. The performance of rites honoring predecessors in the transmission as well as living teachers, participation in the promotion of the transmission through study, teaching, and publishing, and the use of distinctive dress, speech, and behaviors were all read as signs of commitment. Chapters 6 and 7 discuss the assertion of commitment to an intellectual lineage in exegesis as an equally public and controversial act of identification. From the Learning of the Way perspective, the individual's moral transformation necessitated personal reflection on the textual tradition as well as the personal articulation of Learning of the Way philosophical truth.

Commitment to the Learning of the Way was strengthened through the application of exclusivism and stridency in intellectual and political

---

34. Althusser saw the interpellation of the subject as a structural feature of all ideologies; see his "Ideology and Ideological State Apparatuses," 134–36. See also Eagleton, *Ideology*, 48, 142–46.

activity. Opposition to heterodoxy (Buddhism and Daoism) and criticism of rival traditions of scholarship were part of the process of moral cultivation. Zhu Xi included opposition to Buddhism and Daoism in his first textbook on the process of moral cultivation, *A Record for Reflection*. As attested by the recorded conversations, he trained his disciples in criticism of other Confucian traditions (see Chapters 6 and 7). Followers of the Learning of the Way were taught to defend its exclusive claims to truth.

Zhu Xi also engaged his students in criticism of contemporary politics. He considered moral self-cultivation the basis of social and political order and saw its absence as the most pressing problem in Song politics. The cultivation of the emperor's mind, in particular, was the foundation for the rectification of officialdom, the restoration of social order, and, ultimately, the repossession of the northern homeland of the Song dynasty. The Song's escalating military and financial problems after the loss of the north in 1127 were the subject matter of a constant flow of memorials from Zhu Xi harshly criticizing conditions in the inner court and urging Emperors Xiaozong 孝宗 (r. 1163–89), Guangzong 光宗 (r. 1190–94), and Ningzong 寧宗 (r. 1195–1224) to take moral self-cultivation seriously.[35] His impeachment of Tang Zhongyou 唐仲友 (1136–88), discussed in Chapter 5, is another example of the Learning of the Way's uncompromising insistence on moral principle in government. Both his political enemies and those sympathetic to the call for moral reformation attributed the persecution of the Learning of the Way to Zhu Xi's uncompromising and exclusivist attitude. Zhu Xi continued to defend the Learning of the Way's exclusive claims to truth despite political persecution. This last element in the cluster, combined with the others, created a strong sense of identification with the movement as well as a missionary zeal among its members.

These general characteristics of Learning of the Way discourse in the twelfth century were reflected in examination writing from the same period. Surviving examination essays from the 1180s and 1190s written by Learning of the Way diehards who defended its message in the face of the False Learning campaign combined exposition of the theory of

---

35. Zhu Xi submitted such memorials in 1163, 1180, 1182, 1188, and 1194 (Shu Jingnan, *Zhuzi da zhuan*, 199–200, 416, 468–70, 635–38, 711, 911–12).

coherence and invocation of the genealogy of its transmission with explicit personal affirmation of its truth and refutation of alternative modes of scholarship (Chapter 6). Changes in examination writing advocating the Learning of the Way in the thirteenth century illustrate the transformation of the movement as it gained official approval. In thirteenth-century Learning of the Way discourse, stridency and exclusivism gave way to conciliation. The co-optation of different modes of scholarship supplemented the cluster of discursive practices that had characterized Learning of the Way discourse in its first, antagonistic phase of development.

Analysis of the Learning of the Way as a rhetorical community thus makes it possible to trace the historical trajectory of Learning of the Way ideology. The definition of the Learning of the Way tradition through a cluster of discursive practices also helps explain why the tension between inclusiveness and exclusiveness or between combativeness and conciliation was never resolved. The exclusivist and combative discursive practices that characterized the early Learning of the Way community remained part of the tradition. As shown in the internal criticism of the adoption of the Learning of the Way canon in examination curricula (Chapter 7), this residue persisted in the potential for the reformation of the Learning of the Way from official ideology into a movement for moral renewal.[36]

### From Yongjia to "Yongjia"

The Yongjia tradition was named after Yongjia county. In the twelfth century, Yongjia was a prosperous county in the East Zhe circuit and housed the prefectural seat of Wenzhou prefecture, an area that now occupies the southern part of the Zhejiang coast. Wenzhou was then, as now, a manufacturing and commercial center. After the withdrawal of the Song court to the south in 1127, Wenzhou's population tripled to an estimated 910,000 in the 1170s and 1180s. Besides rice cultivation, the population was employed in shipbuilding, the manufacture of lacquer and porcelain goods, and paper and silk production. Situated on the

---

36. For an example of the periodic revival of moral reformism in the Learning of the Way tradition, see Bol, "Neo-Confucianism and Local Society."

border with Fujian and south of the capital of Lin'an, Wenzhou was a gathering place for southern merchants traveling to and from the capital region. Its port also attracted overseas traders, most of whom engaged in trade with Japan and Southeast Asia.[37]

Wenzhou's wealth translated into a vibrant cultural scene. Apart from the diversions that accompanied the development of urban culture in twelfth- and thirteenth-century Song China, Wenzhou boasted an especially large and prominent intellectual elite.[38] The presence of educated men and the empire-wide prominence of some of them are illustrated by the success of Wenzhou men in the civil service examinations under the Southern Song and the competitiveness of the local examinations. Wenzhou was the prefecture with the largest number of graduates in Liang Zhe circuit and ranked second in the total number of *jinshi* graduates in Southern Song territories, ceding place only to Fuzhou prefecture in Fujian.[39] The competitiveness of the Wenzhou examinations was twice as intense as in most other Song prefectures. This was borne out in an imperial decree issued in 1156 stipulating that the prefectural ratio of graduates to candidates was 1 to 100, except for the prefectures of Wenzhou, Wuzhou 婺州 (East Zhe, now Zhejiang Province), and Taizhou 台州 (East Zhe), in which the quota was set at 1 to 200.[40]

In the minds of twelfth-century scholars, Yongjia symbolized the phenomenal examination success of Wenzhou. Of the four districts that made up Wenzhou prefecture, it produced the largest number of *jinshi* graduates. Yongjia was therefore one of the most successful counties in the Song territories in terms of numbers of *jinshi* graduates. Its reputation was further enhanced by the top rankings in the palace examinations achieved by several Wenzhou *jinshi* between the 1150s and the 1170s and their subsequent high-level official appointments (Chapter 4).

---

37. Zhou Mengjiang, *Ye Shi yu Yongjia xuepai*, chap. 1.

38. Gu Hongyi (*Jiaoyu zhengce yu Songdai Liang Zhe jiaoyu*, 328–29) notes that of the 1,839 scholars listed in *The Cases of Song and Yuan Scholarship (Song Yuan xuean* 宋元學案), 120 came from Wenzhou. Wenzhou thus ranked third among the prefectures in Liang Zhe, the region with the highest number of scholars.

39. My description of the success of Wenzhou prefecture is based on Oka, "Nan Sō ki Onshū no meizoku to kakyo"; cf. Chaffee, *The Thorny Gates*, appendix 3, 196–202.

40. Chaffee, *The Thorny Gates*, 125, 155.

The reputation of Yongjia scholarship in the twelfth century was the product of the examination success of Wenzhou prefecture and the commercialization of examination preparation. The stereotypical Yongjia scholar was the examination teacher who limited his teaching to the basic knowledge and skills tested on the civil service examinations. This negative stereotype was part of a larger critique of the commodification of the examinations among twelfth-century critics. In the eyes of critics like Zhu Xi, the examinations had become a means to an end; they were manipulated by teachers and students to ensure students' success and future careers. Because examination preparation courses were available to those who could afford them, the examinations no longer served to select the most worthy for government service and moral leadership.

This critique captured a trend in literati life. Examination teaching became an attractive business as a result of the combined effect of supply and demand. Increases in the numbers of prefectural examinees as well as of students preparing for the examinations, which was even larger but for which we have no estimates, created a demand for instruction. This demand was met in part by the expansion of commercial printing and private teaching. Teachers of successful students gained large followings; several of them reportedly taught hundreds of students.[41] Commercial printers were eager to publish their works. The increase in the number of teachers was spurred not only by the increased demand but also by the abundant supply of unoccupied scholars eager to engage in teaching, preferably as a temporary money-making venture. This trend was especially evident in Wenzhou, where there was both a large number of candidates preparing for the prefectural examinations and an abundant supply of educated men unable to obtain or maintain government positions, some of whom had examination degrees.

Chen Fuliang and Ye Shi, the two Yongjia teachers discussed in this book, were examples of this trend. They taught to support themselves. After they passed the departmental examinations in the 1170s, their reputations spread beyond Wenzhou. Commercial printers collected their examination writings, and printed editions of their examination

---

41. For examples beyond Wenzhou prefecture, see Liu Hsiang-kwang, "Yinshua yu kaoshi"; and Liang Gengyao, "Nan Song jiaoxue hangye xingsheng de beijing."

policy response essays were, according to official and nonofficial accounts, bestsellers among examination candidates (Chapter 4).

To a certain extent, the stereotype of the Yongjia teacher as an examination tutor held true. Yongjia scholarship reflected the emphasis on classical exegesis, historical studies, administrative reasoning, and prose composition skills that lay at the core of the Yongjia teachers' examination preparation curricula. The stereotype was flawed insofar as it underestimated the intellectual and political agenda that inspired Yongjia examination teaching and the scholarship produced by Yongjia teachers more broadly.

In contrast to the copious studies on Neo-Confucianism or the Learning of the Way in its narrow definition, little scholarship has been devoted to the development of the Yongjia tradition. With the exception of Ye Shi, whose political thought and sharp critique of Zhu Xi's genealogy of the Way exerted a strong appeal on modern intellectual historians, indepth scholarship on individual Yongjia intellectuals or the intellectual history of Yongjia as a whole is lacking.[42] This study is not intended to fill this gap; it is, rather, a first step toward the reconstruction of interactions between regional intellectual traditions in Song political culture. Chapters 3 and 4 analyze the "Yongjia" position on exegesis and government as it was reflected in examination essays and manuals. These chapters and the discussion of Zhu Xi's critique of "Yongjia" scholarship in Chapter 6 underscore the correspondence between "Yongjia" institutional history and both classical scholarship and examination teaching.

Yongjia scholarship was associated with a group of teachers active in Yongjia and the other counties in Wenzhou in the twelfth century. Its close association with a group of intellectuals was reflected in such

---

42. In English there are only the two book-length studies on Ye Shi by Winston Lo and Niu Pu. Zhou Mengjiang's work offers a survey of the history of Yongjia scholarship, but the bulk of his work is devoted to Ye Shi's life and intellectual career. Oka Motoshi has written a series of articles on the social and intellectual history of Wenzhou. See also Bol, "Reconceptualizing the Nation"; and Chu Ping-tzu's and Kondō Kazunari's works. On the occasion of the 860th anniversary of Chen Fuliang's birth, a commemorative volume was published based on a conference held in 1997. This title, *Chen Fuliang danchen babai liushi zhounian jinian ji*, was unavailable to me. I thank Song Jaeyoon for bringing it to my attention and sharing those chapters he copied. The dissertation by Lo Wing Kwai, "Chen Fuliang yanjiu," is currently unavailable to the public.

contemporary references as "the scholars of Yongjia" (*Yongjia zhuru*
永嘉諸儒) and "the Yongjia gentlemen" (*Yongjia zhugong* 永嘉諸公).[43]
Even before the emergence of Chen Fuliang and Ye Shi on the intellectual scene in the 1160s and 1170s, Zhou Xingji 周行己 (1067–?), Zheng
Boxiong 鄭伯熊 (1127–81), and Xue Jixuan 薛季宣 (1134–73) had already established reputations in classical scholarship, institutional history, and governance. Chen Fuliang and Ye Shi carried on this tradition
of scholarship, which was characterized by a common administrative
agenda, the analysis of institutional history as an aid in formulating policy proposals, and an interest in the main literati traditions of the eleventh century, including the legacy of the Cheng brothers and the Su
family. The Yongjia tradition was further cemented by teacher-disciple
relationships between its main representatives. Zheng Boxiong and Xue
Jixuan taught Chen Fuliang, who in turn taught Ye Shi.

Unlike the Learning of the Way, the Yongjia tradition was not defined by a body of texts or a genealogy of teachers. Chen Fuliang and
Ye Shi specifically objected to genealogical discourse in the definition
of literati traditions of learning. In their view, genealogies established
divisions among scholars and introduced partiality into scholarly discourse. Similarly, teacher-student relationships were not genealogical.
They conceived of relationships between teacher and students in horizontal terms, with students referring to teachers as friends or brothers
and not masters or fathers.[44]

Rather than defining the Yongjia tradition by mere geographical location, position in the examination field, a body of texts, or an intellectual genealogy, I see it as a mode of scholarship, a method of historical
and policy analysis and textual exegesis, developed by three successive
generations of teachers in Wenzhou. The primary characteristic of
Yongjia scholarship and teaching was a preoccupation with current
affairs.[45] Chen Fuliang identified four broad topics as important: the

---

43. *ZYYL*, 86.2207, 97.2480, 107.2660, 123.2962, 136.3250.
44. Oka, "Nan Sō ki no chiiki shakai ni okeru 'yū'"; De Weerdt, "The Ways of the
Teacher."
45. My discussion of the main attributes of Yongjia scholarship is based primarily on
the commemorative writings Chen Fuliang dedicated to his teachers Zheng Boxiong
and Xue Jixuan. See De Weerdt, "The Ways of the Teacher."

restoration of imperial power (in reaction to the monopolization of power by individuals and factions at court), border affairs and military planning, supernumerary troops and officials, and taxation. The Yong-jia teachers did more than share a common set of concerns in these areas; they developed a set of proposals to address the problems they identified. They formulated an administrative agenda that amounted to a call for a general scaling back of central government activity. They envisioned a smaller government, fielding a smaller army and a more effective bureaucracy and requiring lower tax revenues. In their view, a reduced government was essential to the restoration of Song power and to the well-being of local society.

Second, Yongjia scholarship was characterized by a commitment to realize its administrative agenda. This commitment entailed an ir-reverent and independent attitude. Independence, or impartiality, was a principle of government for Yongjia teachers. It expressed their be-lief that factional affiliations should be avoided in policy debates and decisions. This attitude was amply demonstrated in their criticism of both the emperor and court factions, on one hand, and the exclusivist intellectual politics of the Learning of the Way, on the other hand (Chapter 3).

Third, in Yongjia scholarship administrative action was grounded in broad scholarship. The Yongjia teacher read widely and published on all manner of subjects. The teacher's consulting of a variety of texts re-flected the Yongjia principle of impartiality—just as the emperor was to consult all scholarly opinion, the teacher was to read all scholarly writ-ing. In response to critics who perceived this emphasis as a sign of the lack of a unified vision and integrated philosophy, Yongjia teachers de-fended broad and inclusive scholarship as the basis for the analysis of the Song's military, tax, and bureaucratic problems and the foundation for its political program.

Fourth, Yongjia scholarship critically incorporated Cheng Learning. The Yongjia teachers demonstrated in their intellectual exchanges and publishing record a clear interest in Cheng Learning. Xue Jixuan wrote commentaries on those classical texts that were at the core of Chengist moral philosophy, *The Great Learning*, *The Doctrine of the Mean*, and *The Analects*. He defended the value of these texts against those who found fault with their use in Chengist discourse. Chen Fuliang cited Xue's response to a comment from Military Affairs Commissioner Wang Yan 王炎 (1137–1218):

They [the high officials] cannot reform their minds and correct the beginnings to complete the task of dynastic restoration. They only chase after achievement and profit and boast to deceive the public. Today they commonly say that *The Doctrine of the Mean* and *The Great Learning* are old expressions, and they hate to listen to them.

不能格心正始以建中興之業. 徒僥倖功利夸言以眩聽. 今俗皆曰: 中庸大學陳編厭聞.

The Yongjia teachers shared the Chengist view that the moral reform of the emperor and high officialdom were essential to the reinvigoration of the Song state. They also agreed that the classical texts the Cheng brothers had selected to propagate the message of moral cultivation contributed to this end.

The Yongjia teacher was, on the other hand, not an advocate of the exclusive truth of Cheng Learning. For the Yongjia teacher, Cheng Learning was but one of many schools in the intellectual legacy of the past that provided answers to contemporary issues. There were historical connections between Cheng Learning and Yongjia scholarship. Zhou Xingji had studied with Cheng Yi; Xue Jixuan had a meeting during his youth with the hermit Yuan Gai 袁溉, a student of the Cheng brothers.[46] These connections demonstrate an interest in Cheng Learning on the part of the Yongjia teachers, but they did not claim a line of transmission in Cheng Learning. The association with Cheng Learning did not provide an identity for the Yongjia teacher as it did for the teacher of the Learning of the Way. As indicated in the Introduction, I use the term "Yongjia" (in quotation marks) to refer to teachers active outside Wenzhou prefecture whom contemporaries associated with Yongjia scholarship because of shared interests in Cheng Learning, institutional history, government, and composition.

---

46. Chen Fuliang, *Zhizhai ji*, 51.2a.

## Scholarship and Movement

The strong sense of identification fostered in Learning of the Way discursive practices turned it into an intellectual formation of a different type than the regional traditions such as Yongjia. The Learning of the Way became a movement in the twelfth century. Its members were self-identified and advocated, under its flag and in an organized and systematic manner, a radical reorientation of Song society and the polity in line with Learning of the Way moral philosophy. The Yongjia tradition made no coordinated effort to promote its political and intellectual agenda. It was a mode of scholarship, a way of historical and policy analysis and textual exegesis. There was a political vision associated with it (Chapters 3 and 4), but that vision was not based on universal claims; it did not generate a canon or proselytizing institutions. Its aim was to transform the way in which scholars (as aspiring officials) and officials thought about government. Yongjia administrative thought was based on historical and policy analysis and textual exegesis broadly conceived; its teachers did not envisage a transformation of society as a whole.[47]

The difference between the Yongjia tradition as a mode of scholarship and the Learning of the Way as a movement is illustrated in the ways in which the discourses of the two schools diverged. The Yongjia preoccupation with current affairs and the development of a political program contrasts with the Learning of the Way's development and transmission of a coherent moral philosophy. Yongjia emphasized the art of argumentation in political writing and the teaching of that skill; the Learning of the Way, the personal affirmation of its truth (Chapter 4). Broad scholarship, the critical evaluation of Cheng Learning, horizontal relationships between teachers and students, and the claim of impartiality and a refusal to construct an exclusive intellectual or political identity in the Yongjia tradition are matched against the centrality of

---

47. In *Zhu Xi de lishi shijie*, Yü Yingshi sees all Song Confucians as fighting for social and political reform. I am not convinced that we can equate a commitment to political reform with a commitment to social reform. The Learning of the Way had a comprehensive reform of society and polity (moral reform) in mind; Yongjia was more narrowly focused on administrative reform (social and economic reform). For a thorough critique of Yü's approach and conclusions, see Hartman, "Zhu Xi and His World."

a new canon of classical texts and the writings of the Four Masters, the vertical relationships of intellectual genealogy, and stridency and exclusivism in intellectual and political activity in the Learning of the Way.

These differences have significant implications for the way in which we analyze the relationship between the Learning of the Way and Yongjia scholarship. The Yongjia teachers have frequently been included in broad definitions of the Learning of the Way by Chinese literati in the imperial period as well as by modern scholars. The differences between the Learning of the Way movement and the Yongjia teachers, particularly the refusal of many of the latter to identify with the Learning of the Way, suggests that they represented distinct intellectual and political formations, with a different organizational nature, even though they shared intellectual and political interests.

With regard to their mutual relationship, the intellectual history of the twelfth- and thirteenth-century examination field recounted in the following chapters pursues two lines of argument. First, the histories of Yongjia scholarship and the Learning of the Way movement demonstrate that intellectual formations were defined by their interaction in the examination field. The contrasts outlined above were the result of actions in the examination field. The curricular strategies of the Yongjia teachers and the cluster of discursive practices that characterized the Learning of the Way were shaped in part by teachers' and students' rejection or modification of alternative modes of examination learning.

Second, the examination field functioned as a barometer measuring and forecasting the authority of intellectual formations. Examination manuals and essays reveal the shifts in the interpretation of intellectual traditions among scholars and officials. The intellectual history of the examination field allows us to explore the interpretation of intellectual traditions not only among leading thinkers and their disciples but also among virtually unknown literati. The changes in the cluster of Learning of the Way discursive practices, for example, were indicative of the change in its status from the ideology of a radical minority in the twelfth century to that of an official ideology invoked by the mainstream by the mid-thirteenth century.

# 2

# Examination Expositions
# and Policy Response Essays in
# Literati Culture

As explained in the Introduction, during the Southern Song period, the second session of each level of the civil service examinations required candidates to compose one exposition (*lun*), and the third session, three policy response essays (*cè*). The requirements in the first session of the examinations varied, depending on the candidate's track. Those candidates opting for the poetry track were tested on their ability to compose poetry in two genres: regulated poem (*shi*) and poetic exposition (*fu*). Those opting for the Classics track were examined on the classic of their choice as well as *The Analects* and *Mencius*.

My analysis of twelfth- and thirteenth-century Song examination preparation is based exclusively on an investigation of expositions and policy response essays, the manuals that were produced to train students in these two genres, and discussions of conventions and criteria governing exposition and policy response writing in government and literati circles. My decision to limit the scope of the analysis to only two of the five examination genres is based on two considerations. First, examination expositions and policy response essays have survived in much greater numbers than have examination poetry and essays on the meaning of the Classics for the Southern Song period. Expositions, policy questions, and response essays can be found in several anthologies and in the collected writings of a relatively large number of twelfth- and thirteenth-century Song literati. By contrast, few examples of examination poetry collections or essays on the meaning of the Classics are extant, too few to allow for the reconstruction of long-term developments.[1] Second, the preservation of examination expositions and policy response essays in individual authors' collected writings suggests that Song literati attached particular significance to them. Their significance derived from the decisive impact of these genres in the higher-level examinations and their centrality in literati culture broadly speaking. Expositions and policy response essays were expressions of competence in two of the areas that defined the scholar-official: textual exegesis and the discussion of government policy.

As did contemporary observers, modern historians have downplayed the import of the last two sessions on the overall result of the examinations.[2] In the twelfth century, candidates and examiners shared the assumption that the grades on the first session determined the final grade at the prefectural level of the examinations. According to Wu Cong 吳琮 (twelfth c.), papers in the second and third sessions at the lowest level of the examinations counted only when no decision could be made on the basis of the first session.[3] Wu Cong's observation confirmed the more general perception among Song literati that the burden of grading so many candidates made it impossible for examiners to give each paper due consideration. The magistrates and provincial officials

---

1. The three examination essays on *The Spring and Autumn Annals* in Lin Xiyi's 林希逸 (ca. 1210–ca. 1273) collected works (*Zhuxi yan zhai shiyi gao xuji*, 8.1a–9a) are, to my knowledge, the only example of the inclusion of examination essays on the meaning of the Classics in a Southern Song collection. Pu Yanguang's article on Song essays on the meaning of the Classics is based primarily on *Standards for the Study of the Exposition* (*Lunxue shengchi* 論學繩尺). Although examination essays on the meaning of the Classics and examination expositions certainly overlapped, they should not be seen as the same thing. Northern Song essays on the meaning of the Classics have been better preserved. Seventeen essays appear in Liu Anjie's 劉安節 (1068–1116) collected writings, *Liu Zuoshi ji* 劉左史集. Two essays on the meaning of the Classics were included in Lü Zuqian's anthology (*Song*) *Wen jian* 宋文鑑. A dozen or so Song essays appear in Ming and Qing dynasty collections of essays on the meaning of the Classics. See *Jingyi mofan* 經義模範; and Yu Changcheng 俞長城 (*js.* 1685), *Keyi tang yibaiershi mingjia zhiyi* 可儀堂一百二十名家制義. I was unable to consult Yu's *Song qi ming jia jingyi* 宋七名家經義. Liu Chenweng's 劉辰翁 (1231–94) *Xuxi's Collection of Poems on the Four Seasons* (*Xuxi sijing shiji* 須溪四景詩集) is the only surviving collection of examination poetry.

2. See, e.g., Elman, *A Cultural History of Civil Examinations*, 26.

3. *Lun jue*, 3a.

serving as examiners allegedly checked the expositions and policy response essays only when in doubt about the quality and the ranking of the papers submitted in the first session. Examiners read the expositions and policy response essays of those examinees whose first-session papers were borderline. They also graded the papers of the second and third sessions to determine the appropriate ranking of those candidates whose first-session papers received the same grade.

Whereas the ability to compose poetry and knowledge of the Classics came first in the evaluation of examination candidates at the local level, the ability to write prose expositions and policy response essays took pride of place in the departmental and palace examinations. According to Wu Cong, candidates were judged primarily by their responses in the last two sessions at the departmental examinations.[4] This meant that a successful answer to an exposition or policy question could in and of itself be sufficient to make the cut. Average grades in all three sessions did not necessarily lead to success, but distinction in either the second or third session could ensure a place on the list of departmental examination graduates.

The emphasis on expository writing was even more apparent in the palace examinations. After 1070, the highest level of the examinations consisted of only one session and featured a policy response question. The ranking of the cream of the examinees was thus based on their ability to discuss the policy issue raised by the emperor or, more commonly, his court advisers. The decision in 1070 to change the focus of the palace examinations from poetry to policy essays was part of a larger trend. The abolition of regulated poetry and the poetic exposition as the main criteria of evaluation at the highest level of the examinations reflected the widely shared belief that the use of poetry as an evaluative tool in civil service recruitment was questionable. From 1070 on, the policy response replaced poetry as the hallmark of the holder of the *jinshi* degree.[5]

Twelfth-century literati debate revolved around the interpretation of classical, historical, and philosophical texts and the discussion of

---

4. Ibid., 2b.
5. Araki Toshikazu, *Sōdai kakyo seido kenkyū*, 299.

Song dynastic history and current affairs. The ability to participate in exegetical and political discussions was a defining characteristic of the scholar. It mattered in officialdom and in private exchanges among scholars. Exegesis had traditionally been closely associated with officialdom. The quotation and interpretation of passages from the Classics and previous dynastic histories characterized political discourse in oral and written form at court and in bureaucratic communication more generally. The testing of these skills in the civil service examinations underscored their connection with the state. Political discourse at all levels of the bureaucracy connected the exegesis of classical and historical texts with the discussion of Song policy. Court debates, memorials, and reports bear witness to the fact that Song officials were well informed about Song historical precedent and the current political climate. The currency of the history of early Song politics and information about state affairs in officialdom spilled over into the examination field. Students requested instruction in current affairs in addition to the Classics and the history of the Han and Tang dynasties and bought encyclopedias and anthologies that reproduced recent official documents (a violation of publishing laws; see Chapter 4).[6]

The expansion of the literati class under Song rule, as attested by the growing numbers of prefectural examination candidates, ensured that the discussion of policy cut across official and private arenas. As an examination degree, or simply participation in the examinations, became the hallmark of the scholar and a license to exert local leadership, the discussion of policy joined exegesis and the exchange of poetry as activities defining literati status. The inclusion of political argument in the self-definition of the scholarly elite was manifested in the pride taken in examination policy questions and response essays. The validation of political argument in the examination field further created conditions for the scholars' creative adaptation of examination genres. Literati pursued the policy question and response outside the immediate context of examination preparation and participation in the private discussion of contemporary policy (see Chapter 7).

---

6. On the dissemination of archival documents, see De Weerdt, "Byways"; and idem, "What Did Su Che See in the North?"

## Exegesis and Examination Writing

The application of exegetical skills was required in both the examination exposition and the policy response. Examination exposition topics were quotations from classical, philosophical, historical, or, less frequently, literary texts. Candidates were required to demonstrate their familiarity with the textual tradition by identifying the source of the passage and evaluated on their ability to formulate a coherent interpretation of the meaning of the passage and its immediate context. Exegetical skills were also tested in policy response essays. The discussion of classical and historical cases was an essential rhetorical step in the two main types of policy questions, questions on the Classics and the histories and questions on contemporary affairs. Furthermore, questions on the Classics and the histories were specifically designed to test candidates' ability to explicate classical and historical texts in relationship to one another and to engage them in exegetical debates that spanned the history of textual commentary.

Examination standards affected exegesis in two ways. First, the selection of topics shaped the focus of exegetical effort. The preference for particular Classics and histories, or chapters from them, channeled literati attention. Both the court and examination teachers had the power to so direct literati attention. The following chapters explain why and how the authority to define the focus of exegetical effort shifted from the court in the eleventh century to teachers in the twelfth century. Second, the rhetorical conventions of examination genres molded the ways in which students read and applied texts. Politicians and teachers eager to change the ways in which literati interpreted the classical and historical heritage therefore vied to reform the structural requirements of examination genres. Both "Yongjia" and Learning of the Way teachers transformed exegesis through examination teaching. Their efforts need to be understood against the background of the twelfth-century examination conventions sketched below.

Standards for examination writing were shaped by government regulations. At frequent intervals during the eleventh and twelfth centuries, the Song government posted examination regulations. These rules, discussed in more detail in Chapter 5, dealt primarily with administrative aspects. Instructions concerning evaluation were limited to setting the length of examination papers, stipulating proper procedures for filling out blank sheets, and listing the violations that led to immediate disqualification, such as the use of taboo characters and plagiarism. The official regulations did not specify the structural requirements of each genre to be tested, nor did they identify a core curriculum of examination texts or a list of banned books.[7]

The authors of the official regulations and their audience nevertheless believed that structural requirements guided the composition of papers in each of the genres tested on the examinations. In an 1171 report on the departmental examinations, Liu Zheng 留正 (1129–1206), then imperial recorder (*qiju sheren* 起居舍人), wrote: "As for the state's selection of scholars, [the papers tested in] each of the three sessions have their structural requirements (*tizhi* 體制). Therefore, we call to be selected 'according with the norms' (*hege* 合格)."[8] Here Liu Zheng was confirming a theoretical assumption common among twelfth-century Chinese literati. At the core of each genre of prose and poetry lay a set of characteristics that defined the makeup of the written text. Liu Zheng evidently felt a need to confirm this assumption explicitly because examination papers from the 1160s demonstrated that the core characteristics of the examination genres were subject to significant change.

The authors of the official regulations assumed that most texts in the bibliographic categories of the Classics, the histories, and the philosophers were legitimate sources for examination questions and answers. The concerns they expressed about the use of Buddhist writings and the works of contemporary authors suggest that they felt that some categories of texts should be excluded. However, because in the 1130s the Song government had committed itself to the "Great Impartiality" policy (Chapter 5), official regulations did not prescribe writing and curricular standards. Examiners and students turned to professional teachers and private printers for guidance.

---

7. For a more detailed discussion of examination regulations, see De Weerdt, "The Composition of Examination Standards," 16–23.

8. *SHY, XJ*, 4.41a.

## CURRICULAR STANDARDS

Teachers collected examination papers and authored mock examination essays and used them in class to teach students the sources and types of questions found in recent examinations. Through careful analyses of such texts, teachers taught current conventions in examination writing. Private printers, in collaboration with teachers, printed collections of examinations papers. Such collections often featured annotations and, in some cases, evaluations by the examiners or renowned teachers. Among the large repertoire of examination cribs discussed in Chapter 4, anthologies of examination essays, especially those gathered from graduates after the examinations, were considered the most reliable guide to examiners' questions and evaluation criteria. One extant example of such an anthology, *Standards for the Study of the Exposition* (*Lunxue shengchi* 論學繩尺, late 1260s or early 1270s), allows us to gauge curricular and writing standards for examination expositions in the second half of the twelfth century.

Little is known about the compilers. Wei Tianying 魏天應, a *jinshi* degree–holder from Jian'an 建安, selected the essays. Wei was a disciple of Xie Fangde 謝枋得 (1226–89), an examination teacher of some renown, who was also involved in the business of anthologizing (Chapter 4). Lin Zichang 林子長, an instructor at the Lin'an Prefectural School, annotated the volume.[9] At least 28 percent of the expositions selected by Wei Tianying were from departmental examinations, and at least 44 percent were taken from internal and preliminary Imperial College (Taixue 太學) examinations. He also included a few expositions from local examinations, local and departmental avoidance examinations (*caoshi* 漕試, *bieyuanshi* 別院試), examinations at the school for the imperial family (*zongxue* 宗學), and local schools (see Table 2 in Appendix B).[10] The essays of those who ranked first in the departmental

---

9. Following the gloss on *jingxue* 京學 in Nakajima, "*Sōshi*" "*Senkyoshi*" *yakuchū*, 1: 10.

10. The edition from the 1330s at the Seikadō in Tokyo mentions the type of examination for which the essay had been written as well as the ranking the authors of the expositions attained. The great majority gained first place in the relevant examination. Both Zhu Shangshu (*Song ren zongji xu lu*, 366–72) and Zhang Haiou and Sun Yaobin ("*Lunxue shengchi* yu Nan Song luntiwen") fail to mention this edition and list fifteenth-

examinations were the most prestigious models. As indicators of the examination standards at the highest center of learning in the capital, the essays of Imperial College students also commanded special attention.[11]

The essays span the last century of the Southern Song dynasty. The earliest essay was written around 1154, the last few date from 1268; there are essays from all intervening decades.[12] The preponderance of essays date from the last decades of the twelfth century and from the 1250s and the 1260s (see Appendix A and Table 3 in Appendix B). The relatively large number of essays written before 1200 may reflect a preference for the works of Chen Fuliang; the even larger percentage of essays from the 1250s and the 1260s[13] shows the preoccupation with the latest trends in examination writing. The range in the dates of the expositions and in the types of examinations for which they were written makes *Standards for the Study of the Exposition* an excellent source for an investigation of curricular and writing standards.

Table 4 in Appendix B charts the sources of the topics of the expositions in *Standards for the Study of the Exposition.* Two other anthologies printed in the thirteenth century are included for comparison. Table 4B charts the sources of the topics of those essays in *Standards for the Study of the Exposition* that can be dated to the period 1150–1200. The essays included in *Chen Fuliang, the Founding Father of the Exposition (Zhizhai*

---

century editions as the earliest editions. For the Imperial College preliminary examination and the avoidance examinations, see Chaffee, *The Thorny Gates,* 103, 108.

11. For other examples of the prestige of the Imperial College in examination preparation, see Chapter 4. Imperial College papers may also have been attractive because many tried to take the college entrance examinations to enhance their chances for examination success; see Chaffee, *The Thorny Gates,* 104–5.

12. Zhang Haiou and Sun Yaobin ("Lunxue shengchi luntiwen yu Nan Song luntiwen," 95) write that the earliest essay dates from 1132. They identify the author of one of the essays, Chen Shizhong 陳時中, as a man who passed the examinations in 1132. Most likely they are referring to Chen Shizhong 陳時仲, who is listed as a successful candidate in a local gazetteer (Liang Kejia, comp., *Chunxi Sanshan zhi,* 28.11b). The fact that the name of the 1132 graduate is written differently combined with the fact that there were other examination graduates named Chen Shizhong 陳時中 makes this identification problematic. In general, I have not assigned dates to essays for which there are multiple plausible authors.

13. The number of essays from the 1250s and the 1260s is higher than indicated in the table. The similarities between the essays dated to the 1250s and the 1260s and many of the undated essays suggest that many undated essays were also written at that time.

*lunzu* 止齋論組) were also written during the second half of the twelfth century. These tables reveal a remarkable homogeneity both in terms of the sources used and also in the share of each source in the total number of questions. *The History of the Han Dynasty* (*Hanshu* 漢書) was by far the most popular source for exposition topics. Historical subjects, predominantly from *The History of the Former Han*, *History of the Later Han* (*Hou Hanshu* 後漢書), and *The New History of the Tang* (*Xin Tangshu* 新唐書), account for one-third to one-half the total.[14] *Mencius* ranked second, followed by *The Analects*, *Instructions of Master Yang* (*Yangzi Fayan* 揚子法言),[15] and *Xunzi* (荀子).

This selection bears witness to the authoritative position of Ancient Prose (古文 *guwen*) in the twelfth and thirteenth centuries.[16] In the eighth and ninth centuries, the first Ancient Prose masters to design a genealogy of the Way of Antiquity, Han Yu and Pi Rixiu 皮日休 (ca. 834–ca. 883), had listed Confucius, Mencius, Xunzi, Yang Xiong, and Wang Tong 王通 (584–618), the alleged author of *Explaining Centrality* (*Zhongshuo* 中説), as the legitimate transmitters of civilization.[17] Tenth- and eleventh-century proponents of Ancient Prose also gave priority to the works of these masters in their teaching and added Han Yu to the list.[18]

Twelfth-century masters of Ancient Prose adopted the standard histories of the Han and Tang dynasties as the major sources for their historical studies.[19] The priority assigned these histories became a matter of concern to the government. In 1185, Ni Si 倪思 (1147–1220), then an Erudite Scholar at the Imperial College, reported that topics for expositions on history at the time were selected solely from the histories of the Han and Tang dynasties. He objected:

---

14. For the publication history of the dynastic histories in the Song period, see Ozaki, *Seishi Sō Gen han no kenkyū*.

15. Yang Xiong 揚熊 (53 BCE–18 CE) finished the *Fayan* in 12 CE.

16. For a further discussion of the history of Ancient Prose writing in twelfth- and thirteenth-century examination preparation, see Chapters 4 and 6.

17. He Jipeng, *Tang Song guwen xin tan*, "Tang Song guwen yundong zhong de wentong guan," 253–63.

18. Ibid., 264–71, 282–83.

19. For discussions of the conflict between Chen Liang and Zhu Xi on this matter, see Tillman, *Utilitarian Confucianism*, 134–52; idem, *Confucian Discourse and Chu Hsi's Ascendancy*, 169–78; and idem, "Ch'en Liang on Statecraft."

As for the Three Kingdoms, the Six Dynasties, and the Five Periods, people see them as periods of decline, they despise them and are ashamed to talk about them. However, the advantages and disadvantages of their recruitment, the efficiency and inefficiency of their defense, the tight or loose nature of their policy making, the skillfulness and ineptness of their planning, as well as their methods of handling soldiers and the people, and the records about the conditions of defeat and victory—as mistakes of the past and warnings for the future, all of these are useful. Students should investigate all these matters. . . . I beg that the examiners be ordered to select from various histories without constraints when designing topics.[20]

至若三國六朝五代, 則以爲非盛世事. 鄙而恥談. 然其進取之得失, 守禦之當否, 籌策之疏密, 計慮之工拙, 與夫兵民區處之方, 形勢成敗之迹, 前事之失, 後事之戒, 不爲無補. 皆學者所宜講. . . . 乞申敕考官課題命題雜出諸史, 無所拘忌.

Questions based on the Classics, on which many famous earlier expositions were written,[21] are remarkably absent from these anthologies. The prominence of the Classics in the first session of the examinations accounts for this. In the first session, candidates in the Classics track were tested on their classic of specialization; topics for poetic expositions and regulated poems often came from the Classics as well. Yet, the possibility of duplication in sessions one and two was not regarded as problematic in the case of *Mencius* and *The Analects*. Candidates in the Classics track had to write one essay on *Mencius* and one on *The Analects* in the first session of the examinations. This discrepancy attests to the prominent position *The Analects* and *Mencius* held in twelfth-century literati circles. The inclusion of *The Analects* and *Mencius* in the Four Books,

---

20. *SHY, XJ*, 5.7b–8a.

21. Among Su Shi's six expositions for the special examination of 1061, one dealt with a passage from *Gongyang's Commentary on the Spring and Autumn Annals* (*Chunqiu Gongyang zhuan* 春秋公羊傳); another was on a passage from *The Book of Songs*; see Jin Zheng, *Keju zhidu yu Zhongguo wenhua*, 126. He wrote one of his most famous essays, "The Most Compassionate Penalties and Rewards" ("Xingshang zhonghou zhi zhi lun" 刑賞忠厚之至論), in reply to a question on a passage from *The Book of Documents* during the departmental examinations of 1057 (ibid., 115). *The Changes, The Book of Rites,* and *The Book of Documents* were regular sources for exposition topics in the palace examinations between 978 and 1063. For the topics of the palace examinations for these years, see the list in *SHY, XJ*, 7. See also Ning, *Bei Song jinshi ke kaoshi neirong zhi yanbian*, appendix, table 2, 183–85.

the core curriculum of the Learning of the Way, was a confirmation of the ranking these books already enjoyed in the Ancient Prose tradition (Chapter 6).[22]

## Writing Standards

Model examination expositions not only taught students the main sources for examination questions but also familiarized readers with the rhetorical characteristics of the exposition genre. Teachers, editors, and printers alerted twelfth-century students to the typical layout of the examination exposition in lectures and in annotated anthologies such as *Standards for the Study of the Exposition*. By the 1150s, the sequence of the main parts of the examination exposition had become standardized. Students were expected to apply the typical format in the examinations, even though these requirements were not specified in examination regulations. Standardization affected not only the structural layout of the genre but also its argumentative strategy.[23] Argumentation in parallel format had become standard according to twelfth-century observers. These two practices shaped the "Yongjia" teachers' innovations in the

---

22. See Chapters 5 and 6. Learning of the Way critics were more upset about Learning of the Way uses of *The Doctrine of the Mean* (*Zhongyong* 中庸) and *The Great Learning*. In memorials during the Campaign Against False Learning, Learning of the Way critics urged scholars to "take Confucius and Mencius as their teachers and to study the Six Classics, the philosophers, and the histories" and not to "devote themselves entirely to false teachings from 'recorded conversations' and to *The Doctrine of the Mean* and *The Great Learning* in order to give literary expression to their wrong views" (*SHY, XJ*, 5.17b; *WXTK*, 5.302b; see also *XZZTJ*, 729). For a discussion of the campaign, see Chapter 5.

23. Despite the similarities, a crucial difference exists between the regulated examination expositions from the twelfth and thirteenth centuries and later eight-legged essays. The difference lies in the second component of the rhetoric of the examination exposition, the argumentative strategy. Parallelism defined the argumentative strategy of the eight-legged essay after the early fifteenth century, whereas it was only one of many argumentative strategies used in regulated expositions; see Elman, *A Cultural History of Civil Examinations*, 380–403. According to Zhu Shangshu (*Songdai keju yu wenxue kaolun*, 210–32), the standardization of layout affected all examination genres with the exception of the policy response essay, whose structure he considers dependent on randomly ordered questions and thus inherently resistant to standardization. I discuss standardization in the policy response below.

teaching of composition and the Learning of the Way critique and reform of examination conventions.

According to the eighteenth-century editors of *The Complete Collection
of Books in the Four Bibliographic Categories* (*Siku quanshu* 四庫全書), there
was an unmistakable trend toward formalization in the examination expositions written after the establishment of the Southern Song dynasty
in 1127.

> In the beginning people did not yet stick to established formats. Su Shi's "The
> Most Compassionate Penalties and Rewards," for instance, manifested an in-
> novative structure.[24] People did not yet value sticking to the details of the
> head-neck-heart-belly-waist-tail format. After the move to the south, people
> endeavored to adhere to a more elaborate format, and the structural regula-
> tions became more rigid. The examiners held to fixed formats in their expecta-
> tions of others, and everybody followed their fixed formats to comply. There-
> upon the dogmas of "double bolts" [explaining words or phrases in a parallel
> structure] and "three doors" [developing three lines of argument] arose.[25] Con-
> sequently, there were separate norms for examination essays. Even those with
> unlimited and extraordinary talent could not overcome this. It must be because
> of this reason that this collection has "standards" ["cord" (*sheng*) and "foot"
> (*chi*) are both units for measuring things] in the title!
>
>   . . . The structural rules for the sections called "broaching the topic" (*poti*),
> "following up on the topic" (*jieti*) "minor discussion" (*xiaojiang*), "the main
> discussion" (*dajiang*), "going back to the topic" (*ruti*), and "tracing the origins
> of the topic" (*yuanti*) are in fact the beginnings of the eight legs (*babi*). Thus
> one can see wherefrom the writing for the imperial examinations originates.[26]

---

24. See note 21 above. For translations of this essay, see Margouliès, *Le kou-wen
chinois*, 271–74; and Shi Shun Liu, *Chinese Classical Prose*, 232–35.

25. For examples, see *Lun jue*, 25a–28a. Although the "double bolts" can metonymi-
cally stand for parallel prose and the eight-legged essay, as a rhetorical technique it was
also very popular among Ancient Prose writers. Han Yu's name is commonly associated with the technique in Southern Song anthologies. See *LXSC*, *passim*; and Suzuki,
"Hakkobun no enkaku oyobi keishiki," 699. This is an illustration of the blurred lines
between parallel and Ancient Prose writing. In many cases the difference in the application of parallelism (fairly strict vs. ad hoc) indicates what style the author is adopting;
other stylistic features provide further indicators. For an example of the metonymical
use of the "double bolts" and similar terminology, see Gu Yanwu, *Rizhilu jishi*, 2: 16.21a;
cited in Yin Gonghong, *Pianwen*, 158.

26. *SKTY*, 38.4162.

其始尚不拘成格. 如蘇軾 "刑賞忠厚之至論" 自出. 未嘗屑屑於頭項心腹
腰尾之式. 南渡以後講求漸密程式漸嚴. 試官執定格以待人. 人亦循其定
格以求合. 於是雙關三扇之説興, 而場屋之作遂別有軌度. 雖有縱橫奇緯
之才, 亦不得而越. 此編以繩尺爲名, 其以是歟!
　其破題接題小講大講入題原題式實後來八比之濫觴. 亦足以見制舉
之文源流所自出焉.

The head-neck-heart-belly-waist-tail format refers to subdivisions of
essays that had become standard by the second half of the twelfth century. The examination exposition comprised six subdivisions, arranged
in a fixed sequence. The first three subdivisions were relatively short.
The author introduced the main argument succinctly in the opening
lines, or *poti* (an introduction of two to three lines). An outline of the
main subarguments, of equal or slightly greater length, followed the
summary statement of the argument in the continuation (*jieti* or *chengti*
承題). The author could opt to continue the introductory part with a
more detailed abstract of the exposition in the minor discussion, or
*xiaojiang*. Together the mandatory one-paragraph statement of the argument in the *poti* and *jieti*, and the optional clarification of the argument in the minor discussion formed the introductory part of the exposition (*maozi* 冒子). The main segment of the text (main discussion, or
*jiangti*) was usually preceded by the identification of the source of the
question in the citation, or *yuanti*. In rare cases, the citation followed the
main discussion. The citation then linked the main argument to the conclusion (*jieti*), instead of marking the end of the introduction and the
beginning of the main body of the exposition.[27]

The eighteenth-century editor singled out the "double bolts" and the
"three doors" (or "three leaves") as signs of the formalization of exposition writing after 1127. He did so for good reasons. The double bolts
and three doors reflected the trend toward parallel argumentation in
regulated examination expositions.[28] A piece by Huang Huai 黄槐, who

---

27. This structural division may go back to similar organizational schemata for regulated (examination) poetry during the Tang. See the arguments by Mao Qiling 毛奇齡
(1623–1716) and Qian Daxin 錢大昕 (1728–1804) in Liang Zhangju 梁章鉅 (1775–1849),
*Zhiyi conghua*, 1.6b; see also Suzuki, "Hakkobun no enkaku oyobi keishiki," 697; and Jin
Zheng, *Keju zhidu yu Zhongguo wenhua*, 61. The Siku editors see the eight legs as an extension of this subdivision, as do these literary historians.

28. Cf. Zhu Xi's criticism cited in Chapter 6.

obtained the *jinshi* degree in 1154, illustrates both the standardized format of and the use of parallel argumentation in regulated expositions of the twelfth century. It further underscores the authority accorded the Ancient Prose tradition in examination preparation by the mid-twelfth century.

"[If] the knowledgeable man [also] does what causes him no trouble" [topic of the exposition]

The exposition:

All affairs in the world are accomplished by the good planning of knowledgeable men, and they are defeated by the busy meddling of knowledgeable men.

Therefore, that which completes the affairs of the world is knowledge, but that which destroys the affairs of the world is equally knowledge.

When the rights and wrongs, the benefits and harms, of the affairs of the world have not yet become manifest, who else but the knowledgeable man can discern these things? If he follows the set momentum of the rights and wrongs and the benefits and harms, and handles affairs on the basis of the necessary development of the rights and wrongs and the benefits and harms, then what harm could there be in the morally superior man's use of knowledge? *On the other hand,* if one estimates knowledge too highly, breaks moral standards to pursue fame and utilizes techniques to handle affairs, one will bring confusion to what can be easily done in the world. Only then will there be things in the world that do not accord with their natural principles. When things do not accord with their principles, one will cause harm by relying on benefit at a time when one has to consider the rights and wrongs and the benefits and harms. The origins of busy meddling really come from this.

"[If] the knowledgeable man [also] does what causes him no trouble," one should be excited by Mencius' arguments [citing the topic from *Mencius*].[29]

智者行其所無事 [*topic*]
　　論曰：
　　天下之事成於智者之善謀，
　　而敗於智者之多事. [*poti*]
　　故成天下之事者智也.
　　而敗天下之事者亦智也. [*jieti*]
　　天下之事是非利害之未形, 非有智者孰能辨之? 因其是非利害之定勢
而處之以是非利害之當, 然則君子於智何惡之有!

---

29. *LXSC*, 2.89a–93b; Legge, *The Chinese Classics*, vol. 2, *The Works of Mencius*, 331.

惟夫智者過而矜之裂道以徇名任術以處事, 取天下之所安行者而畀
之膠膠擾擾之地. 天下之事始有不循其理者矣. 不循其理而從事於是非
利害之際, 將以利之適以害之. 多事之原實基於此. [*xiaojiang*]
　　智者行其所無事. 宜有激於孟子之論也. [*yuanti*]

The topic the examiner assigned in the departmental examinations of 1154 was adapted from a line in *Mencius*. In a speech reported in Chapter IVB, Mencius debates the topic of "knowledge" (*zhi*) and "knowledgeable men" (*zhizhe*).[30] Even though the passage covers only Mencius' response and not the context of the debate, we can imagine Mencius responding to Mohist arguments. The two main themes of Mencius' speech are "knowledge" and "inherent character" (*gu* 故), concepts central to Mohist philosophy. Mencius argues that knowledge is ambivalent. He objects to knowledge that ignores and distorts the natural disposition of things. Mencius' interpretation of knowledge was predicated on his views on human nature and development and, more generally, the nature and development of all things. In other dialogues and speeches recorded in *Mencius*, the master maintained that all human beings and things possess inherent capacities, which, when properly nurtured, determine the course of natural and moral development.

In the passage Huang Huai was asked to consider for his examination exposition, Mencius argues that when discussing the nature of things (*xing* 性), we must pursue their inherent character (*gu*). Mencius maintains that it is easy to pursue the inherent character of things. That is what Yu did in directing the waters that had flooded the Chinese plains in Antiquity. In Mencius' rendering, Yu made the waters flow in their natural direction. His work was easy and successful because he did not interfere with the natural course of the waters. The examiner lifted the topic for the examination exposition from the line, "If *knowledgeable men do* also *what causes them no trouble* [as Yu did], then their knowledge is great indeed."

In the *poti* and *jieti*, the first two sentences in the translation, Huang set out the polarity that guides the argumentation of the essay: using knowledge is both necessary and dangerous. In the one-paragraph

---

30. For translations, see Legge, *The Chinese Classics*, vol. 2, *The Works of Mencius*, 331; and D. C. Lau, *Mencius*, 133.

summary, Huang Huai indicated that the exposition pursues two lines of argument. On one hand, all affairs in the world depend on knowledgeable men for their completion. On the other hand, knowledge is also a potential source of destruction for all things. The author worked out both sides of the proposition in the main body of the text (see Fig. 1).

In the minor discussion quoted in full above, Huang Huai offered a theoretical explanation of the underlying cause for the different uses of knowledge. Knowledge of the irreprehensible kind acts on opportunities to move things in the direction of their preordained course of development. Knowledgeable men who put their knowledge to positive use monitor their surroundings. They determine the rightness (*shifei*) and beneficialness (*lihai*) of the course of affairs and act to steer affairs in the direction of their normative paths of development. Knowledgeable men of the reprehensible kind, on the other hand, determine rightness and beneficialness without reference to the normative paths of development. For this reason, their actions are based on partial interest (*li* 利) and cannot be undertaken with the smoothness that characterized Yu's guiding of the waters.

Huang Huai's exposition on the need to monitor the set course (*dingshi*) of affairs and his use of rightness and beneficialness as the criteria for monitoring events are reminiscent of the political philosophy of the Yongjia teachers, discussed in more detail in Chapter 3. In their lectures and course materials, the Yongjia teachers taught students preparing for the examinations to evaluate circumstances and to calculate the benefits and harms resulting from different courses of action. Huang Huai's argument differs from the more overtly utilitarian approach to calculating benefit and harm in typical Yongjia expositions, however. Beneficialness in Huang's exposition is predicated on the normative pattern of development of each affair. In this perspective, calculations of benefit and harm cannot determine the course of action because the inherent patterns of development of individual things are just one of several aspects factored into such calculations.

Huang Huai's arguments differed from those of another twelfth-century school of thought. Huang Huai's reference to the normative patterns of development (*li* 理) of individual things calls to mind the

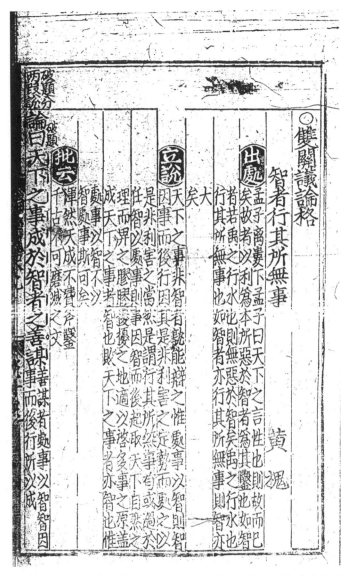

Fig. 1 The beginning of Huang Huai's examination exposition in *Standards for the Study of the Exposition* (1270s). This page demonstrates the use of front commentary (including the full citation of the essay topic, an essay abstract, and a critical appreciation), top-margin commentary, intralinear commentary (in small characters), and interlinear commentary, along with punctuation and stress marks in thirteenth-century examination anthologies (Yuan edition, 1335–40, held at the Seikadō bunko; microfilm of this edition at the Harvard-Yenching Library).

interpretations of this same concept developed by the teachers associated with the Learning of the Way. *Li* (pattern, coherence) was the cornerstone of the philosophy developed by four generations of teachers after the mid-eleventh century and synthesized from the 1150s on by Zhu Xi. Teachers associated with the Learning of the Way defined *li* as the moral coherence of the world. *Li* occupied a central position in Neo-Confucian moral philosophy because it expressed both the belief in the operation of an absolute moral force in the cosmos and the conviction that the operating principles of this moral force were inherently accessible to the human mind. Moral self-cultivation, of which Neo-Confucian philosophers developed several models, was the path each human being had to follow to gain insight into and act upon the universal principles of morality embedded in the cosmos and in their own minds. Fundamental Confucian moral values, including humaneness (*ren* 仁), filial commitment (*xiao* 孝), loyalty (*zhong* 忠), and wisdom (*zhi* 智), were redefined as forms of the universal principle of moral coherence.

Huang Huai did not accord *li* the explanatory power it had in the work of his Neo-Confucian contemporaries. In his exposition, he interpreted it as the natural patterns of development (*ziran zhi li*) of individual affairs. By "affairs," he meant administrative interventions, and the examples he marshaled in the discussion of the positive use of knowledge (main discussion) were the social welfare policies associated with benevolent rulers in Antiquity. These rulers taught their people the techniques of agriculture, the manufacture of clothing, and the construction of boats because these three activities satisfied natural needs. These rulers also invented weapons, seals, and city walls to protect their people from devious and violent behavior. In Huang's assessment, their educational policies and protective measures were proof of their ability to steer public affairs in the right and beneficial direction.

Several of Huang Huai's examples were identical to those Han Yu used in "Finding the Origins of the Way" ("Yuan dao" 原道).[31] This eight-century Ancient Prose essay, and Han's prose oeuvre more generally, had gained a wide following in the twelfth century among teachers preparing students for the examinations. The "Yongjia" teachers advo-

---

31. For a discussion of this essay, see Hartman, *Han Yü and the T'ang Search for Unity*, 145–62.

cated analysis of the works of Han Yu and Liu Zongyuan as models of Ancient Prose writing, but they went beyond Tang models. In the decades following the examinations of 1154, "Yongjia" teachers modified the eighth-century models of Ancient Prose and promoted more contemporary models.

Huang Huai's prose was modeled on the older eighth-century Ancient Prose models. In his comments on Huang's exposition, Lin Zichang pointed out Huang's indebtedness to Liu Zongyuan's prose.[32] Despite their criticisms of parallelism, eighth-century Ancient Prose authors did not totally abandon the parallel structures that were the hallmark of the writing style practiced at the courts of the Six Dynasties and the Tang dynasty. Rather, they applied parallelism flexibly, as in the philosophical prose written before the second century CE. In Ancient Prose, parallelism was not the governing rhetorical pattern; it was simply one of many devices used to heighten the effect of the author's argumentation.[33]

In the prose of Han Yu and Liu Zongyuan, rhetorical flexibility was matched by a comparable relaxation in diction. Liu Zongyuan's prose was characterized by concision, but not at the cost of clarity. The language of Ancient Prose was straightforward and shorn of dense allegorical phrases. The sentences were laced together with grammatical particles to clarify the relationships between the polysemous characters. The following passage from the main discussion in Huang's exposition illustrates the clear-cut diction and the use of parallelism in Ancient Prose:

Are there really no affairs in the world? If one does not perform, there are no benefits. If one does not act, there are no accomplishments. So, there has never been a time when there were no affairs in the world. Therefore, to get

---

32. Lines A1–B1 and A4–B4 in the passage translated below were modeled on the opening lines of Liu Zongyuan's "On Feudalism" ("Fengjian lun" 封建論):

> Does the cosmos really have no beginning? I cannot find out.
> Do human beings really have a beginning? I cannot find out.
> 天地果無初乎? 吾不得而知之也.
> 生人果有初乎? 吾不得而知之也.

For a brief discussion of this essay, see Chen Jo-shui, *Liu Tsung-Yüan and Intellectual Change*, 96.

33. On the use of parallelism in Ancient Prose, see Yin Gonghong, *Pianwen*, 161–63.

rid of wisdom on the grounds that there are no affairs—this will prevent one from completing things that can be achieved in the world.

Are there really affairs in the world? Those who do things destroy them. Those who hold on to things lose them. So, there has never been a time in the world when there were things. Therefore, to use wisdom on the grounds that there are affairs—this will open up all sorts of disasters in the world.

What should a noble person faced with this situation do?

A.

天下果無事乎?
不爲則不利;
不行則不成.
天下蓋未始無事也.
故以無事而去智, 則不足以立天下可成之功.

B.

天下果有事乎?
爲者敗之;
執者失之.
天下蓋未始有事也.
故以有事而任智, 則適以啓天下紛紛之禍.

C.

君子於此當何如哉?

In this passage at the very beginning of the main discussion, Huang Huai used parallel structures to describe a dilemma in the relationship between knowledge and affairs. Does knowledge come first, and does it create affairs to act upon for itself? Or, do affairs present themselves *a priori* and activate the proper kind of knowledge? Parallelism works at two levels in this passage. Lines one, four and five in sections A and B match syntactically and semantically. Lines two and three within each section are parallel, and these two sequences also correspond structurally with each other in sections A and B. The syntactic parallelism in the opening passage of the main discussion presents in condensed format the two lines of argumentation Huang pursued in the remainder of the text. Huang first argued that affairs should precede the use of knowledge and cited actions of the rulers of Antiquity to illustrate this line of argument. In the second part of the main discussion, he supplied evidence for his argument that knowledge wielded improperly leads to catastrophes. Based on the cases of the strategists and militarists of the

Warring States period, Huang concluded that disaster ensues when knowledge not guided by the course of affairs dictates action.

In sum, Huang Huai's 1154 exposition exemplified the standards for examination prose in the mid-twelfth century. Its layout embodied the ideal structural format described in contemporary writing manuals.[34] His prose demonstrated the influence of Ancient Prose on examination writing and validated the contemporary critique of the role of parallel argumentation in classical and philosophical discourse.[35]

In contrast to exposition topics, which focused on an isolated passage, policy questions on the Classics and the histories tested students' broader familiarity with classical, philosophical, and historical texts and the commentarial tradition. Policy questions juxtaposed passages from one or several sources and asked students to explain obscure and contradictory statements. The following question illustrates this technique. The examiner, Lü Zuqian, asked students to discuss a series of conflicting statements on the significance of "humaneness" (*ren* 仁) and "sagehood" (*sheng* 聖) from *The Analects, Mencius,* and *The Rites of Zhou* (*Zhouli* 周禮):

As for the discussion of sages and humane people in the Confucian school, even Confucius did not presume to be up to it. The way of the sages and the humane people is all-encompassing; other people are not up to it. However, when the Duke of Zhou [ca. eleventh c. BCE] compiled *The Rites of Zhou*, he listed humaneness and sagehood between wisdom and rightness on one hand and unbiasedness and balance on the other. He instructed people by listing these terms next to one another without ranking them. Even Confucius did not think it appropriate to apply these terms to himself, but the Duke of Zhou held such expectations for all scholars of the empire. Could it be that Confucius and the Duke were talking about different kinds of morality?

---

34. *The Art of the Exposition* (*Lun jue* 論訣) is a collection of tips on exposition writing citing the work of twelfth-century teachers. Several of the teachers discussed the structural format exemplified in Huang's exposition.

35. The essays of Yang Wanli 楊萬里 (1127–1206) provide further examples of this technique. Nine out of his ten essays were structured on a polarity between two phrases that differ only slightly; see the examination expositions in his collected works, *Chengzhai ji*, 90.

When Zi Gong [ca. fifth c. BCE;[36] Duanmu Si 端木賜] asked about "giving extensively and providing for the multitude," Confucius replied, 'How could such action qualify someone as just a humane person? Such a person must be a sage!'[37] Does this mean that there is a qualitative difference between humaneness and sagehood?

Allegedly, Confucius seldom spoke of humaneness.[38] However, the number of his statements and responses to questions touching upon it is innumerable. How could it be that he seldom spoke of it?

As for sagehood, it is clear from the start that it cannot be said that Confucius seldom talked about it. However, there are fewer passages in *The Analects* about sagehood than about humaneness. How come?

Confucius acknowledged Bo Yi [ca. eleventh c. BCE] as a humane man but ranked him as a wise person.[39] Mencius acknowledged Bo Yi as a humane man but proceeded to rank him as a sage.[40] What is the explanation for this?

These are all questions that I have not yet seen through. I hope you can explain them and report back to me.[41]

問孔門之論聖與仁. 雖夫子, 有所不敢居. 其道至大固非它人之所能與也. 而周公之制周禮列仁聖於智義中和之間. 並舉以教人而無所輕重. 夫子猶不敢以此自居, 而周公乃以此待天下之學者, 周孔豈二道邪?

子貢問博施濟眾而孔子對以何事於仁必也聖乎. 是仁與聖果有優劣耶? 仁之一字號爲夫子所罕言. 然其立言其答問及於仁者不可縷數. 安在其爲罕言耶?

至於聖初非夫子之所罕言. 而載於論語反不若言仁之多. 抑又何也?

夫子許伯夷以仁, 止目之以賢. 孟子許伯夷以仁, 遽目之以聖. 其說復安在耶? 此皆疑而未喻者. 願明以告我.

---

36. For more biographical information on Zi Gong, see Brooks and Brooks, *The Original Analects*, 290.

37. *Lunyu*, VI:30; D. C. Lau, *The Analects*, 85.

38. *Lunyu*, IX:1; D. C. Lau, *The Analects*, 96.

39. *Lunyu*, VII:15; D. C. Lau, *The Analects*, 87–88. Bo Yi was held up as a model of both brotherly love and dynastic loyalty. He was the eldest son of the Lord of Guzhu. His father designated his second son, Shu Qi, as heir. At his father's death, Shu Qi deferred to Bo Yi, but the latter would not accept the fief. When King Wu of Zhou put an end to Shang rule, Bo Yi and Shu Qi starved themselves out of loyalty to the Shang king.

40. *Mengzi*, VB:1; D. C. Lau, *The Works of Mencius*, 149.

41. Lü Zuqian, *Donglai ji,* waiji, 2.15b–16b. Cf. idem, *Lü Zuqian quanji,* vol. 1, waiji, 2.639–40.

This question fell under the contemporary rubric of "questions about the Classics" (see Table 5 in Appendix B). This designation reveals the connection between this type of question and the larger context of twelfth-century intellectual culture. "Doubts about the Classics" is a more literal rendering of what I translated as "questions about the Classics." The critical investigation of inconsistencies (in word usage or argument) within a given classic and research into discrepancies among classical texts was a trend in the classical scholarship of the eleventh and twelfth centuries.[42] In the work of some exegetes, the trend of critically analyzing the linguistic and semantic consistency of the classical tradition led to the questioning of the authenticity of canonical texts.

In twelfth-century policy questions, however, the juxtaposition of conflicting passages usually did not lead to the radical questioning of the Classics. Rather, this technique of questioning was used to test candidates' broad familiarity with the textual tradition, their philological skills, and their ability to reconcile contradictions within the Classics. As we shall see in Chapter 6, the emphasis on the abilities to reduce contradictions to paradoxes and to interpret the Classics as a coherent moral philosophy characterized policy questions authored by Learning of the Way teachers.

Policy questions on the textual tradition bore the traces of intellectual debate in more explicit ways. Questions about the Classics and histories not only were designed to force students to struggle with conflicting statements in the original sources but also could require the discussion of disagreements in later commentaries and scholarship (Table B5, subject 11). The relative merit of Han, Tang, and Song interpreters of the classical tradition was a bone of contention that divided scholars in the twelfth century but was increasingly resolved in favor of the intellectual lineage of the Learning of the Way in the thirteenth century. Chapter 7 analyzes the debate between the Masters of "Yongjia" and the advocates of the Learning of the Way on the relative merit of Song interpreters of the classical tradition in policy questions.

---

42. Ye Guoliang, *Song ren yijing gaijing kao, passim*. The first appendix in Ye's work (pp. 177–203) provides a chronological overview of publications doubting or changing classical texts. For a more narrowly focused study of doubting and changing classical texts in the early Song period, see Feng Xiaoting, *Song chu jingxue fazhan shulun*, pt. I.

## Government and Examination Writing

Policy response essays and expositions were often linked in the minds of twelfth-century students preparing for the examinations. In both the exposition and the policy response sessions, students were tested on their ability to answer questions from the philosophers and the histories, texts that did not appear on the official lists of the Classics promulgated in the eleventh and twelfth centuries. The exposition and policy response sessions were frequently discussed together because of their curricular similarities and the different requirements of the Classics session. Policy questions drew from a broad range of textual sources including the Classics, the dynastic histories, contemporary official documents, the philosophers, and the collected writings of major Tang and Song intellectuals. Occasionally, and in violation of court edicts banning such sources, examiners used nonwritten materials such as unidentified sources of opinion.

In contrast, candidates in the "meaning of the Classics" track were tested only on their familiarity with the classic of their choice and associated commentarial traditions in their first session; noncanonical sources were typically not used in the questions or the answers. The list of the Classics was subject to some alteration, but it was strictly defined at any given point in time. During the Southern Song dynasty, the curriculum consisted of *The Book of Songs* (*Shijing* 詩經), *The Changes*, *The Book of Documents* (*Shujing* 書經), *The Rites of Zhou*, *The Book of Rites* (*Liji* 禮記), *The Spring and Autumn Annals* (*Chunqiu* 春秋), *The Analects*, and *Mencius*.[43]

The differences between the exposition and the policy response were predicated on the different uses of historical evidence and philosophical arguments in each genre. In the exposition, students were asked to distill a general truth from the event referred to in the question. Expositions were by definition concerned with the underlying principles of events and ideas. According to the description of the genre in *The Literary Mind and the Carving of the Dragons* (*Wen xin diao long* 文心雕龍), a treatise on literature that had shaped Chinese literary theory

---

43. Yuan Zheng, *Songdai jiaoyu*, 50. The list had been subject to change in the last decades of the Northern Song. See ibid., 29, 34, 38–39.

since its first appearance ca. 500 CE, expositions "encompass a multitude of explanations and investigate one underlying principle."[44] The demonstration of one general truth informed the layout and the rhetoric of examination expositions. In Huang Huai's exposition, the general truth was announced in the opening sentence of the essay. Each subdivision of his exposition contributed to the thesis that knowledge based on the understanding of the proper course of events is required for the successful completion of affairs. Expositions thus tested the student's ability to discover philosophical truths in the textual tradition. Evidence from the histories and the philosophers provided the material for theoretical debate.

Policy questions asked students to perform the reverse mental process: concretization rather than abstraction was the main goal of the exercise. A policy question confronted students with a list of events and asked them to apply lessons from these events to contemporary administrative and cultural questions. In theory, the application of cultural knowledge and the formulation of concrete proposals and solutions were the goals of the policy response exercise.

The Song government left not only the sources but also the range of topics for policy questions undefined. Teachers at government and private schools developed courses and manuals to introduce students to the main topics and sources used in policy questions. Zeng Jian 曾堅, who was active as a teacher in the mid- to late thirteenth century, briefed his students on the range of questions they could expect in his *Secret Tricks for Responding to Policy Questions* (*Dace mijue* 答策秘訣),[45] a survey of policy question topics and a digest of lines of argument for tackling them. Zeng Jian's hints were based on an extensive reading of contemporary examination essays. He quoted policy response essays by forty-one authors; nine of the twenty-one men who can be identified

---

44. Liu Xie, *Wen xin diao long*, 4.18, 327; for a translation of this passage and the full chapter on the exposition, translated as "treatise," see Shih, *The Literary Mind and the Carving of the Dragons*, 101–8. This passage is on p. 102. On the early history of the exposition, see Kinney, *The Art of the Han Essay*.

45. An extant copy of *Dace mijue* is appended to a Yuan anthology of policy essays, *The Mirror of Peace: A Collection of Policy Essays* (*Taiping jinjing ce* 太平金鏡策), preserved at the National Palace Museum Library in Taibei. For more information on editions, see the Appendix.

obtained their *jinshi* degrees in the second half of the twelfth century; the remainder received theirs in 1202 and 1205. This manual thus provides an excellent index of the range of subjects in twelfth-century policy questions (see Table B5 and Appendix A).[46]

Zeng Jian's list of topics and subtopics coincides roughly with the subjects covered in twelfth-century questions preserved in the collected works of individual authors. The omission of sections on economic and financial matters, such as monetary policy, the budgets of the central and local governments, or the state monopolies on salt, tea, and liquor in Zeng's manual is the only notable difference from other contemporary sources.[47] The inclusion of sections on the learning of human nature (*xingxue* 性學) and the transmission of the Way indicates how concepts central to the Learning of the Way found their way into examination manuals from the late twelfth century on.[48]

Zeng Jian's repertoire of topics was, broadly speaking, shared across the political and intellectual spectrum and reflected fundamental characteristics of Chinese political culture. The preparations for the examinations trained Song dynasty elites in imperial political culture in three major areas. The first was the proper relationship between sovereign and official servant. Candidates were asked to articulate the virtues of the imperial form of government as well as the standards to which the exercise of imperial power should conform (Zeng's category 1). Students were also asked to reflect on the attributes of the official and his responsibilities in assisting the emperor in the exercise of imperial power (categories 2, 6, and 10). Second, examination candidates were expected to demonstrate versatility in the textual tradition that formed the basis for contemporary political discourse. Policy questions on the Classics, the histories, literature, and scholarship in general measured students' exposure to classical texts as well as their ability to contribute to contemporary debates on the interpretation and value of these sources (categories 7 and 11). Finally, policy questions trained examina-

---

46. Wang Yinglin (*Yuhai*, 201.3a–11b) cited similar lists of topics compiled by scholars preparing for the more demanding polymaths examination.

47. Compare the examination encyclopedias discussed in Chapter 4.

48. See Chapter 7 for a discussion of twelfth- and thirteenth-century questions on the genealogy of the way.

tion candidates in the business of imperial government, its organization, functions, and procedures. The topics covered in this broad category correspond to the operation of the Six Ministries, the main central government agencies: the Ministry of Rites (sacrifices, palace buildings, rituals and music, education and examinations, disasters, and astronomical observations, etc.; categories 3, 5, 9, and 12), the Ministry of Punishments (Xingbu 刑部; legal affairs; category 1), the Ministry of Revenue (Hubu 戶部; land taxes; category 3—and commercial taxes and monopolies not mentioned by Zeng Jian), the Ministry of Works (Gongbu 工部; water conservation; category 3), the Ministry of Personnel (Libu 吏部; recruitment and promotion, category 5), and the Ministry of Military Affairs (Bingbu 兵部; military organization, military strategy, and reunification; categories 3 and 8). These topics figured on the curricula of teachers and examiners of different persuasions and reappeared in manuals and questions in the thirteenth century. The specific questions asked under these broad topical categories and the methodologies required in answering them were subject to variation and change.

Essay questions were subdivided into two main categories, those on the Classics and the histories (*jingshi* 經史), discussed in the previous section, and those on contemporary affairs (*shiwu* 時務).[49] The second category of policy question, questions on government business, followed a format similar to that of the questions on classical and historical texts. Teachers and examiners typically listed different approaches to governmental organization and administrative practice and invited candidates to evaluate their pros and cons. They cited cases drawn from different periods of Chinese history and asked students to take sides in policy debates, past or contemporary. Such exercises in argumentation tested the ability of examinees as prospective officials to formulate and defend plans in policy debates. By listing competing interpretations,

---

49. Song Qi 宋祁 (998–1061) proposed that the three policy essays required in the first session of the departmental examinations consist of one essay on the Classics and the histories, and two on contemporary government. Emperor Renzong implemented this proposal in 1044, but, following negative reports, agreed to rescind it the following year. The regulations issued in 1045 required three essays in the last session but did not specify their contents (Araki Toshikazu, *Sōdai kakyo seido kenkyū*, 96).

policy questions involved teachers, examiners, and examination candidates in controversies about a wide range of issues concerning government and scholarship.

The high esteem for policy response essays in the Song dynasty is readily apparent from their place in Song records. Many Song scholars preserved their policy response essays and had them included in their collected writings; in contrast, almost no examination poems and relatively few essays on the Classics and expositions were left for posterity.[50] Biographies of Song figures quote passages from the subject's departmental, palace, or decree examination essay to prove his moral stature or administrative insight.[51] Intellectuals of different orientations also compiled lists of questions for use in the classroom or as puzzles for acquaintances or random readers.[52]

The prestige of the genre among examiners and examinees derived from its association with a tradition of high officials who had exploited the genre to give frank advice to their sovereign. Foremost among these was the Han intellectual Dong Zhongshu 董仲舒 (179?–104? BCE). His essays on the way of government, formulated in a response to the questions posed by Emperor Wu 武帝 (r. 141–87 BCE) on the occasion of a decree examination, demonstrated the direct impact examination writing could have on the central government. Dong Zhongshu's answers to the emperor's questions resulted in the emperor's promotion of Dong's interpretation of Confucian learning.[53]

Under the Song dynasty, the policy response was the last genre to be added to the list of genres tested in the *jinshi* examinations. After re-

---

50. Most collected writings from the Southern Song period do not include examination expositions. The collected writings of Lu Jiuyuan, Yang Wanli, Fang Fengchen 方逢辰 (1221–91), and Wei Liaoweng 魏了翁 (1178–1237) are the rare exceptions. See also note 1 to this chapter.

51. Examples abound. See, e.g., the biography of Chen Fuliang, in *SS*, 423.12634–36; and that of Wang Mai in Chen Fuliang, *Zhizhai xiansheng wenji*, 52.8b.

52. The policy questions of many Song officials were included in their collected writings. For examples, see the questions by Zhu Xi, Chen Fuliang, Lü Zuqian, Zhen Dexiu 眞德秀 (1178–1235), Cheng Bi 程珌 (1164–1242), and Wu Yong 吳泳 (*js.* 1208) discussed in this chapter and in Chapter 7.

53. For a brief description of the three essays and their impact on the Han court, see Twitchett and Loewe, eds., *The Cambridge History of China*, 1: 710–13, 753–56. According to *The Cambridge History*, the examination took place in either 140 or 134 BCE.

peated requests, Emperor Renzong agreed to add policy questions to the *jinshi* examinations in the mid-1020s.[54] The valorization of the policy response in the Song civil service examinations started with the Qingli 慶曆 Reforms in 1043. Fan Zhongyan and Ouyang Xiu proposed that the policy response session be the first in the series of three sessions and requested that the results for the first session be decisive in the ultimate outcome. Students who failed this part of the examinations would not be allowed to sit for the subsequent sessions, which featured expositions and poetic genres.[55] Fan's emphasis on candidates' ability to discuss government shaped the history of the Song examinations. The arrangement was canceled the next year, but policy response essays, henceforth tested in the third and final session of the examinations, became a crucial factor determining examination results and scholars' reputations. According to fellow reformer Song Qi 宋祁 (998–1061), the demotion of the session on poetic composition to third place accelerated a trend among scholars to value policy over poetry.[56]

Following in the footsteps of the reformers of the Qingli period, Wang Anshi implemented a series of measures to enhance the weight of the policy response in the civil service examinations. Most important among these was the establishment of the policy response as the only genre to be tested in the palace examinations. Henceforth, the discussion of contemporary government determined the ultimate outcome of the examinations.[57]

The policy response essays required examinees to respond to a long list of questions. The questions tended to be relatively short at the beginning of the Song Dynasty, when the examiners condensed their

---

54. The poetic genres (*shi* 詩 and *fu* 賦), expositions (*lun* 論), and essays on the Classics (*moyi* 墨義) had been tested in the *jinshi* examinations since the beginning of the dynasty (Araki Toshikazu, *Sōdai kakyo seido kenkyū*, 94, 372–74). For a more detailed overview of early requests for the incorporation of policy essays in the examinations and the history of their official adoption, see Ning, *Bei Song jinshi ke kaoshi neirong zhi yanbian*, 104–6.

55. Araki Toshikazu, *Sōdai kakyo seido kenkyū*, 377–80.

56. Ibid., 374–80.

57. Wang Anshi did away with the poetry session in the *jinshi* examinations and increased the number of policy essays to be written in the departmental exams from three to five. (Araki Toshikazu, *Sōdai kakyo seido kenkyū*, 298–99; Ning, "Songdai gongju dianshi ce yu zhengju," 147).

questions to ten to twenty lines and students copied them out in full on their answer sheets. By the 1190s when policy questions from the first half of the eleventh century were upheld as models of concision,[58] the questions had lengthened considerably. Examiners had become accustomed to writing long treatises. Although of a later date, the questions collected in *Standards for the Study of the Policy Response* (*Cexue Shengchi* 策學繩尺), a compilation of nineteen school and civil service examination essays from the last decades of the Southern Song (Chapter 7), illustrate a trend starting in the twelfth century. Instead of a hundred characters or so, the number usual in the first half of the eleventh century, the length of the questions in this collection range from 536 to 1,373 characters, and average 833. In the twelfth and thirteenth centuries, candidates were no longer asked to copy down the questions; they were printed and distributed to each examinee.[59]

The student's response to the policy question resembled the memorial, the official medium for the communication of administrative problems and proposals. Official regulations set the length for the examination policy response at 500 characters in the early Northern Song period. Wang Anshi's 1070 regulations stipulated that the essay be no longer than 1,000 characters. The examination regulations issued in 1145 did not mention the required length, but the re-endorsement of the regulations of the Yuanyou period (1086–94) suggests that a maximum length of 700 characters was in effect.[60]

Mirroring the license that examiners took in the formulation of policy questions, however, candidates felt free to write as much as they deemed appropriate. The nineteen essays in *Standards for the Study of the Policy Response*, written for local examinations and Imperial College tests, averaged 1,467 characters and ranged from 822 to 2,205 characters long. Students attached even greater weight to their essays for the palace and decree examinations and took even more liberties with them. The ex-

---

58. Zhu Xi, "Xuexiao gongju siyi" (Private opinion on schools and selection through examinations) in *Zhu Xi ji*, 69.3639–40.

59. Questions were first printed in 1008 for display in the examination compound; later, printed copies were distributed to all candidates (Araki Toshikazu, *Sōdai kakyo seido kenkyū*, 338).

60. *SHY, XJ*, 4.21b–22a, 28b. For the number in the Yuanyou period, see Araki Toshikazu, *Sōdai kakyo seido kenkyū*, 95.

pectations for the policy response essays in these examinations were higher. Han models were about 2,000 characters long; during the Song period, the minimum length for essays in the decree examinations was set at 3,000 characters.[61] Yet, candidates exceeded the minimum at a rate comparable to that in their essays for the lower-level examinations. Zhou Nan's 周南 (1159–1213) policy response for the palace examination of 1190, discussed in Chapter 3, was 8,061 characters long.

The disregard for the requirements regarding length brings out the similarities between the policy response and genres, such as the memorial, reserved for the discussion of policy by professional officials. At 10,651 characters, Yao Mian's 姚勉 (1216–62) essay for the palace examination of 1253 is one of several surviving examples of policy response essays that took on the dimensions of a "ten-thousand word memorial," the stereotypical memorial proposing major reform policies.[62] Due to the formal resemblance of essays on administrative issues to memorials, the policy response essays of several Southern Song examination candidates were ranked with the official proposals they formulated after obtaining their *jinshi* degrees. Wang Mai's 王邁 (1185–1248) 1217 palace examination essay, in 7,553 characters, and his essay for the 1235 decree examination, of comparable size, were included in the memorials section in his collected works.[63] In a postface to another examination candidate's palace examination essay, Li Maoying 李昴英 (1201–57) wrote that there was no difference between Xu Guangwen's 許廣文 essay and a memorial.[64] The relevance of examination policy writing to the factional politics of the twelfth and thirteenth centuries is discussed in Chapters 3, 5, and 6.

---

61. Araki Toshikazu, *Sōdai kakyo seido kenkyū*, 408.
62. Yao Mian, *Xuebo ji*, 7.1a–33a.
63. Wang Mai, *Quxuan ji*, 1.1a–23b, 23b–46a.
64. Li Maoying, *Wenxi ji*, 4.2b.

# Part II

---

## *The "Yongjia" Teachers in the Examination Field*

# 3

## The "Yongjia" Teachers' Standards for Examination Success (ca. 1150–ca. 1200)

In the second half of the twelfth century, Yongjia county gained empire-wide renown for producing large numbers of examination graduates. Its graduation success and the presence in Yongjia of a tradition of teachers specializing in examination-oriented instruction contributed to its reputation as a center for examination preparation. Due to this reputation, Yongjia county lent its name not only to the teachers active there and the scholarship produced within the county but also to teachers in Wenzhou prefecture, of which it was the seat, and, as shown in Chapter 4, in nearby Wuzhou prefecture. The term "Yongjia" in quotation marks refers to this broader meaning of the place-name.

This chapter analyzes the reasons behind the commercial success of Yongjia collections of policy response essays and examination expositions. It attributes the ascendancy of the Yongjia teachers in the examination field of the 1170s to the effective translation of their approach to the administrative and military problems of the Song court. Both in exegetical and in policy response essays, the Yongjia teachers addressed the problems of overextension and factionalism, which they perceived as the major problems of the Southern Song court. Through the discussion of administrative theory in expositions and the proposal of concrete measures in policy response essays, they promoted a political program advocating both a central government retreat and the reassertion of imperial power.

*Standards for
Policy Response Essays*

## AWAITING RECEPTION AND
## PRESENTED SCROLLS

According to both official and private sources, two collections of policy response essays were especially popular among examination candidates in the 1190s. In a letter sent to Ye Shi in 1191, Zhu Xi mentioned that Ye's *Presented Scrolls* (*Jin juan* 進卷) was circulating widely among examination candidates. Wu Ziliang 吳子良 (1197–1256), writing in the first half of the thirteenth century, testified that examination candidates vied to study Chen Fuliang's *Awaiting Reception* (*Dai yu ji* 待遇集) as soon as it appeared in the 1170s.[1] Official sources confirm the popularity of these two collections. In a memorial to the court in 1196, the minister of personnel, Ye Zhu 葉翥 (*js.* 1154), reported: "There are Ye Shi's *Presented Scrolls* and Chen Fuliang's *Awaiting Reception*. Scholars recite the essays in these two collections. Every time they make use of them [in the examinations], it has an immediate positive effect for them."[2] Ye Zhu wrote this report after supervising the departmental examinations of 1196. He proposed that the woodblocks for these two collections be destroyed throughout the empire.

The destruction order seems to have been effective. Contemporary witnesses commented on the ban. Zhu Xi supported it and lauded the government's decision in private correspondence and conversations.[3] The ban suggests that printed editions of both works circulated between the 1170s and the 1190s, but from the early thirteenth century on they were no longer available as independent collections. Not a single copy remains today. Chen Fuliang's collection was lost; only the preface

---

1. Wu Ziliang, *Lin xia ou tan*, 4.42.

2. *WXTK*, 5.302b. Schirokauer ("Neo-Confucians Under Attack," 180) mistakenly dates this memorial to 1197. This memorial also appears in *SHY, XJ*, 5.17b–18a; the passage on the ban is omitted. Cf. Kondō Kazunari, "Yō Teki no kai kan," 58.

3. *ZZYL*, 123.2967. For comments in letters, see Niu, "Confucian Statecraft in Song China," 100–101.

remains.[4] Ye Shi's compilation was included in his collected works in the thirteenth century and reconstituted from various sources in 1448.[5]

Both collections were written in the 1170s. Chen Fuliang wrote his essays around 1170, when he entered the Imperial College and before he passed the departmental examinations of 1172. Ye Shi's essays probably date to the mid-1170s.[6] In these collections the two presented their views on all matters relating to central administration and elite culture. Ye Shi's *Presented Scrolls*, for example, consisted of fifty essays covering roughly the same topics as Zeng Jian's list of policy response subjects (see Table 5 in Appendix B). Ye Shi included essays on the body politic (the primacy of the virtue of the ruler), the business of imperial government (financial administration, personnel management, and, especially, foreign policy and military organization), and literati learning (the Classics, philosophers, and histories, as well as important historical figures).

Collections of this nature were not uncommon. From the Tang period on, ambitious students compiled samples of their work and sent them to officials in the hope of gaining their support and recommendation. Tang students presented scrolls of their works before the regular *jinshi* examinations. Self-promotion through such means was standard in the seventh and eighth centuries. The practice fell into disuse after the first Song emperors implemented regulations to prevent favoritism

---

4. The preface is included in a Southern Song Ancient Prose anthology preserved in the Palace Museum Library in Taipei (Liu Zhensun, *Gujin wenzhang zheng yin*, houji, 14.2b–3b). For further discussion of this anthology, see Chapter 6.

5. Kondō Kazunari, "Yō Teki no kai kan," 58; *YSJ*, bieji, preface, 1.631; postface, 871–73.

6. Ye Shi's biographers disagree on the date for his collection and place the collection's conception between his early years (1170s) and the call for recommendations for the special decree examination of 1184. Zhang Yide (*Ye Shi pingzhuan*, 61, 103, 365) proposes 1178, suggesting that Ye Shi worked on this after he had succeeded in the *jinshi* examinations and while he was observing the mourning period for his mother. Zhou Xuewu (*Ye Shuixin xiansheng nianpu*, 7011–12) concludes that the question cannot be resolved based on the available evidence. Those scholars, cited in Zhou's work, arguing for a date before 1178 point to a report submitted in 1177, which stated that collections of "presented scrolls" (the required set of policy response essays) received at court were the work of men who would gain prominence (*WXTK*, 33.316). Ye Shi may have written his collection for the earlier round of submissions.

and maximize fairness in the *jinshi* examinations.[7] Contact between examiners and students before the examinations was prohibited. The names of the examiners were not announced until shortly before the examinations, and they were then immediately escorted to the examination halls and locked inside until the examinations began. The government also took measures to enforce equitable evaluation procedures within the examination compounds. Clerks, for example, copied all examination papers to ensure that examiners would not recognize the calligraphy of the candidates.

The major exception to the stricter regulation of networking activities before the examinations was the decree examination, held throughout the Song dynasty. Unlike the triennial *jinshi* examinations, decree examinations were held by imperial fiat only. But even though an imperial decree was necessary, decree examinations were held regularly in the twelfth century. Between the 1130s and 1160s decrees calling for candidates were issued every three years.[8] The major difference between the regular *jinshi* examinations and the decree examinations was thus not the frequency and regularity of the examinations but their difficulty. Throughout the Southern Song period, only one candidate passed the decree examination for the "wise and honest, those able to speak straightforwardly and remonstrate vigorously" (*xianliang fangzheng neng zhi yan ji jian* 賢良方正能直言極諫), one of the most common types of decree examinations held during the Song period.

The "wise and honest" examination track was first established in 178 BCE by the Han government to recruit exceptional talents. In the twelfth century, the exceptionally talented could bypass the various stages of the regular examinations if they qualified for the "wise and honest" examination, which was held at the capital only. In order to qualify for the examination, the candidate had to obtain a recommendation from a high-ranking official. The recommendation had to be accompanied by a series of fifty essays in the format of the policy

---

7. For the impact of the early Song measures as well as the continued use of essay collections in pre-examination networking, see Azuma, "Kōkan yori mita Hoku Sō shoki kobun undō ni tsuite"; Takatsu, "Song chu xingjuan kao"; and Zhu Shangshu, *Songdai keju yu wenxue kaolun*, 340–62.

8. Wang Deyi, *Songdai xianliang fangzheng ke ji cike kao*, 10–11.

response. The recommender submitted the candidate's collection of essays to the court. Only after the essays received a passing grade from selected court officials in the Two Departments (the Secretariat-Chancellery), and in some cases from the emperor himself, was the candidate qualified to take the examinations. The examination consisted of two sessions of six expositions and one policy response essay. Those who passed were directly eligible for court positions.[9]

Very few candidates were allowed to take the decree examinations, and even fewer graduated. Chen Fuliang and Ye Shi may have thought of the decree examination as a way to gain recognition and a court position quickly. The title and the preface of Chen Fuliang's *Awaiting Reception* suggest that he wrote it with the same goal as other candidates for the decree examination, namely, to gain the recognition and support of those in influential government positions. In *Presented Scrolls*, Ye Shi declared openly that he was taking advantage of imperial decrees encouraging high officials to promote commoners of exceptional wisdom and men willing to speak without fear about the most pressing issues in government.

Chen Fuliang and Ye Shi may have been inspired in this undertaking by the success in 1171 of Li Gou 李垕, the only candidate to pass the "wise and honest" examination during the Southern Song period.[10] There was more enthusiasm in the 1170s for the "wise and honest" track as a handful of candidates were recommended for it after years in which no one answered the call to submit an essay collection. Ye Shi's bid was not successful. His collection, however, became a bestseller. Why did this collection of policy essays become a hit among students preparing for the regular *jinshi* examinations in the 1180s and 1190s? Why was it banned by the court in the late 1190s?

The success of these two collections of policy essays was related to the examination success of teachers and students from Yongjia county, Chen's and Ye's hometown. As noted in Chapter 1, from the 1130s to the end of the twelfth century, Yongjia was one of the most successful counties in Song territory as far as the number of *jinshi* graduates was

---

9. Ibid., 1–31.
10. Ibid., table, 56; Kondō Kazunari, "Yō Teki no kai kan," 57.

concerned.[11] In 1172, the year of Chen Fuliang's graduation, ten candidates from Yongjia passed. The Yongjia graduates took ten of the nineteen *jinshi* titles awarded to candidates from the four counties in Wenzhou prefecture, and Wenzhou graduates accounted for 4.9 percent of the titles awarded that year. In 1178, the year of Ye Shi's graduation, the results were very similar. There were nine Yongjia graduates out of a total of twenty for Wenzhou prefecture as a whole, and Wenzhou took 4.8 percent of the total number of titles in 1178.

The quota of graduates for Wenzhou prefecture had been raised from thirteen to eighteen in 1156 because of the success of students from this region and the sharp competition among them. The actual number of graduates was frequently higher than the new quota. In 1160 twenty-eight and in 1163 twenty-six men from Wenzhou obtained the highest degree, a trend that persisted throughout the 1190s. The large numbers of graduates from Wenzhou prefecture, and Yongjia county in particular, can be attributed in part to the use of alternative examination routes. Apart from the local qualifying examinations, candidates could become eligible for the departmental examinations by enrolling in the Imperial College in the capital. Candidates successful in the less competitive Imperial College examinations could thus avoid the competition for the few slots in their native place. Chen Fuliang along with three other men from Wenzhou followed this route in 1172. Ye Shi avoided the local competition by taking the avoidance examinations given to the sons, relatives, and associates of traveling officials. Ye Shi was able to take the Liang Zhe regional avoidance examination in 1177 after Hanlin Academician Zhou Bida 周必大 (1126–1204) recognized him as an exceptional young talent. Zhou Bida made him his disciple and used this connection to petition that Ye Shi be allowed to participate in the examination.[12]

Not only the numbers but also the ranking and subsequent careers of Wenzhou candidates must have seemed impressive to their contemporaries. Wenzhou graduates took first place in the palace examinations of 1157 and 1163. In 1172, Chen Fuliang ranked in the top tier of the five-

---

11. My description of the success of Wenzhou prefecture is based on Oka, "Nan Sō ki Onshū no meizoku to kakyo"; cf. Chaffee, *The Thorny Gates*, appendix 3, 196–202.

12. Zhou Mengjiang, *Ye Shi yu Yongjia xuepai*, 160–61.

tiered list of graduates, and one of his students, Cai Youxue 蔡幼學 (1154–1217), took first place in the departmental examinations. In 1178 Ye Shi took second place in the palace examinations. The examination success of Wenzhou prefecture translated into central government positions. Chen Fuliang held the posts of drafting official of the Secretariat (*Zhongshu sheren* 中書舍人) and Chancellery imperial recorder (*qiju lang* 起居郎) and was college registrar (*Taixue lu* 太學錄) for a short period of time. Ye Shi held several central government positions, including positions in the top schools of the Song empire as college supervisor (*Taixue zheng* 太學正) and vice director of education (*guozi siye* 國子司業). Examiners for the departmental examinations were selected from among high-ranking officials, and in these positions as well Wenzhou men were well represented. Two of Chen Fuliang's fellow 1172 graduates from Wenzhou became examiners in subsequent examinations. Oka Motoshi, in his research on the social and political history of Wenzhou elites, has demonstrated that men from Wenzhou served as examiners in all but two of the twenty departmental examinations held between 1142 and 1199.[13]

The success of Yongjia proved to be a catalyst in the development of examination teaching. The sharp competition among students and the decreasing likelihood of employment in the bureaucracy through the examination route in the twelfth century spurred the specialization in examination teaching. Students flocked to reputable teachers; several of them reportedly taught hundreds of students. Commercial printers were eager to publish the work of examination teachers for those students unable to study with them in person. For candidates engaged in years of preparation for the examinations or for unemployed graduates, examination teaching was an attractive career option.

Chen Fuliang and Ye Shi were examples of this trend. Both men were born to less well off families. Their fathers were village teachers teaching basic literacy skills and elementary education, a position that garnered modest respect and pay.[14] Chen and Ye started teaching at an early age to fund their further education and had already established reputations as examination teachers before passing the *jinshi* examinations. Chen

---

13. Oka, "Nan Sō ki Onshū no meizoku to kakyo," 17.
14. Zhou Yuwen, *Songdai ertong de shenghuo yu jiaoyu*, 137–47.

Fuliang continued to teach after 1172 while waiting for official assignment.[15] Ye Shi studied with Chen Fuliang in the 1160s and later commented on the extraordinary appeal of Chen's examination curriculum in Yongjia.[16] Even Emperor Guangzong asked Chen about the "several hundreds of students" studying with him.[17]

Ye Shi took up teaching at the age of sixteen. While teaching, he continued to study with other scholars who shared the scholarly interests of Chen Fuliang. He visited Chen's teacher Xue Jixuan in 1169, and in 1175 he called on Lü Zuqian.[18] Lü taught a curriculum similar to Chen's, and the two had become close friends when they met at the Imperial College in the early 1170s. The compatibility of their interests was confirmed in the 1172 examination success of Chen Fuliang and Cai Youxue; Lü Zuqian was examiner for the departmental examinations that year. Ye Shi was thus familiar with the thinking and teaching of the leading teachers of Yongjia. He continued to teach in Yongjia, with a brief hiatus from 1173 to 1174, until the departmental examinations of 1178. Like Chen Fuliang in the early 1170s, Ye Shi had begun to establish a reputation as an examination teacher in Yongjia in the mid-1170s.

Besides the examination success of Yongjia county and the high profile of Wenzhou men in central government and the administration of the departmental examinations, other reasons explain the appeal of their collections to examination candidates. First, the collections covered many of the topics that might appear in the examinations. Ye Shi's *Presented Scrolls* could be used as an introduction to the writing of policy response essays on any topic, with the added benefit that all the essays were written by a recent second-place *jinshi* graduate. Whatever Ye Shi's intent for the collection may have been, twelfth-century official and private accounts of its popularity among students preparing for the *jinshi* examinations indicate that his policy essays were swiftly adopted for general use in the regular examinations. Students saw little difference between Ye Shi's decree examination essays and those required in the

---

15. Zhou Mengjiang, "*Song shi* 'Chen Fuliang zhuan' buzheng," 29.

16. See the section "Chen Fuliang: The Founding Father of the Exposition" below; Niu, "Confucian Statecraft in Song China," 65–66.

17. Chen Zhenbo, "*Yongjia xiansheng Bamian feng* xin tanxi," 44.

18. Niu, "Confucian Statecraft in Song China," 66.

last session of the *jinshi* examinations. The adoption of *Presented Scrolls* for general examination use is further attested in a fifteenth-century edition of Ye's collected works. In this edition *Presented Scrolls* and *External Draft* (*Wai gao* 外稿), another series of forty policy essays Ye Shi wrote around 1185, were gathered under the title *Standards for the Policy Response Session* (*Cechang biaozhun ji* 策場標準集).[19]

The appeal of the Yongjia tradition as embodied in the teaching and publications of Chen Fuliang and Ye Shi between the 1170s and the 1190s further demonstrates the relevance of current affairs in twelfth-century examination preparation. Two major issues on the agenda were the question of the restoration of Song authority over the northern territories lost to the Jurchen Jin dynasty and the dispute over the threat of factionalism caused by the rise of the Learning of the Way movement. The approach of the Yongjia teachers to these questions clarifies why their policy response essays were held up as the standard among examination candidates in the last three decades of the twelfth century. Their stance on these issues also explains why their collections were both proscribed by the government and criticized by Learning of the Way leaders in the late 1190s.

## FACING THE EXTERNAL ENEMY

The restoration of the political authority of the Song dynasty and the recovery of the northern territories, lost after 1127 and officially recognized as Jin territory in the Song-Jin treaties of 1141 and 1161, were a rallying cry among Yongjia teachers. Ye Shi, in his policy response essay for the palace examinations of 1178, wrote:

To take revenge on the enemy is the foremost duty of the empire. To gain back our former territories is the highest responsibility of the empire. . . . If Your Highness wishes to teach the empire the duty of being a son of the people, let them rest their heads on their lances [expressing their determination not to sleep peacefully until they have taken revenge] and head north without consideration for personal safety. How could the people of the empire object![20]

---

19. Kondō Kazunari, "Yō Teki no kai kan," 58.
20. *YSJ*, bieji, 9.754; also cited in Zhou Mengjiang, *Ye Shi yu Yongjia xuepai*, 198; and Niu, "Confucian Statecraft in Song China," 36.

復仇天下之大義也. 環故境土天下之尊名也. . . . 陛下若欲教天下以為人
子之義, 使枕戈北首, 慮不顧身. 天下之人其又何辭!

Even though the patriotic wording of this passage may seem to suggest otherwise, Ye's advocacy of the reclamation of the northern territories was not an immediate call to arms. In *Presented Scrolls* and subsequent work, Ye developed a plan for domestic reform calling for adjustments to Song institutions and administrative policy intended to prepare the Song empire for the "foremost duty" of reclaiming the northern territories. He argued that the survival of the Song court and its capacity to recover the northern territories depended on the mobilization of popular support. The mobilization of popular support required an overall scaling back of central government activity. In *Presented Scrolls*, in his palace essay, and throughout the remainder of his career, Ye Shi advocated tax reductions and cuts in government spending mainly through troop reductions and downsizing of the bureaucracy.[21]

Ye Shi's emphasis on institutional and policy changes and his interest in the financial and fiscal implications of all policies were characteristic of Yongjia scholarship. Chen Fuliang and his teacher Xue Jixuan studied institutional history in an effort to develop an answer to the geopolitical predicament of the Song empire. Chen's *Awaiting Reception* no longer survives, but policy questions he asked of his own students suggest that his collection of policy essays carried the same message of institutional and policy reform with an eye toward the recovery of the north.

Several of Chen's questions were concerned with issues of military organization. The following question about the history of the navy is representative of Chen's and later Ye's thinking about institutional reform and illustrates the methodology for which Yongjia teachers were famous:

Nowadays, if our waterways come under attack, defense has to be tight. Whenever there is an immediate threat, we are still sending militia fleets to

---

21. Niu, "Confucian Statecraft in Song China," 32–41; Bol, "Reconceptualizing the Nation," 17–24. Kondō Kazunari ("Yō Teki no kai kan," 60–64) discusses Ye Shi's views of the relationship between Chinese and barbarian existence and discusses *Presented Scrolls* in this regard. In *The Life and Thought of Yeh Shih*, Winston Lo analyzes Ye Shi's administrative thinking along the same lines but focuses on his *External Draft*.

assist the imperial forces. Sailors and merchants are not regularly preparing for mobilization, and the prefectures and counties are pressed and unable to provide logistical support. Unexpected inspections reveal that the organization is usually unclear. We have occasionally suffered from this, and some therefore want to station forces from Ezhu [Wuchang county, located on the middle Yangzi River, in present-day Hubei province] along the Yangzi River, from Jingmen [located on the middle Yangzi River] to Yangzhou [located in the lower Yangzi region]. They also want to station forces from Xubu Garrison [located near the mouth of the Yangzi on the eastern border between the Jin and Southern Song empires] along the coastline from Wu [the lower Yangzi region, now Jiangsu province] to Min [the southeast coast, now Fujian province] so that all troops will be linked. All forces will fall under the same military administration so that big and small, far and close, will be connected. However, if we investigate Zhou and Han history, examine Chu and Yue [states located in the south during the Warring States period], and check the records for the Liang [502–57] and Tang dynasties, there are no precedents for this. Nevertheless, if we build large numbers of warships and station troops all along the riverbank and the coast and spend lots of public money, how will we be able to support this? If we tax the people for all of this, they will not be able to tolerate it. . . . I ask you all to investigate carefully and explain this. Do not simply say that morality will provide protection! How can that be useful? [22]

方今江海要擊, 其備嚴矣. 間者有卒然之警. 猶調民艦以佐王旅. 漁賈無擬發之常; 州縣有乏興之遽. 一時趣督, 往往條理未彰. 或被其患. 伊欲以鄂渚之戍施之沿江, 自荊達揚, 許浦之戍推之沿海, 自吳達閩. 聯次比伍. 輯以軍政, 使之大小相維, 遠近相及. 而稽之周漢, 參之楚越, 按之梁唐之間, 靡有成憲. 且夫治船置卒, 多糜官錢, 胡能贍之? 一切科民則有不忍. . . . 幸諸君察察陳之. 毋徒曰道德藩籬! 將安用此?

Chen asked the examinees to reflect on contemporary policy issues related to the restoration of Song authority that remained relevant throughout the remainder of Song history. Should the Song government expand and restructure its navy? Where should the fleets and marines be stationed? Along the Huai River or along the Yangzi River? Should the navy be strengthened along the coastline in defense against the Jurchens (and later the Mongols), or in preparation for naval attacks on the northern territories? How would the armed forces and the fleets

---

22. Chen Fuliang, *Zhizhai ji*, 43.11b–14a; idem, *Chen Fuliang xiansheng wenji*, 33.550.

be coordinated? And, how would the state pay for the expansion of the navy?[23]

He asked examinees to solve these questions by applying historical precedent. The conclusion, quoted above, of the question was preceded by a long list of detailed questions about the organization of the navy during times of imperial unity under the early Zhou, Han, and Tang dynasties, as well as during periods of north-south division, in the Warring States and Six Dynasties periods. Chen went through the historical record and presented pieces of evidence on the absence or presence, the organization, and the funding of naval forces in different time periods. His questioning style demonstrates that answering policy questions required a careful investigation of the history of the relevant institutions and policies.

The scope of relevant history in Chen's work covers all time periods, from the Three Dynasties described in the Classics to the Song period. The periods of real or idealized imperial power under the Zhou, Han, and Tang dynasties were traditionally preferred in historical studies, but Chen included the short-lived states of the Six Dynasties period. The experience of the southern dynasties, such as the Liang dynasty mentioned in the quotation, was particularly relevant since the geopolitical and military challenges the southern regimes had faced between the third and the sixth centuries were similar to those facing the Song government in the twelfth century.

History, in Chen's interpretation, also included recent history. Chen Fuliang was noted among his contemporaries for his knowledge of early Song history.[24] He studied, taught, and wrote about early Song history with the same belief in the contemporary relevance of past administrative experience. His reading of the historical record of the first two centuries of Song rule convinced him that Song institutional history provided solutions to contemporary fiscal and military policy questions.

An eminent example of the comprehensive scope of the study of institutional history as well as the practical orientation of historical study

---

23. For the continued relevance of these questions, see Huang Kuanchong's study of the thirteenth-century court debates about war and peace in south China, "Wan Song chaochen dui guoshi de zhengyi."

24. Niu, "Confucian Statecraft in Song China," 55.

in Chen's work is his monograph on military organization, *Military Organization Throughout the Ages* (*Lidai bingzhi* 歷代兵制).[25] In eight chapters, it surveys the history of military institutions chronologically and without omissions from the Zhou through the Song Dynasties. Chen's work was the first of its kind. In the chapter on Song military institutions and policy, he argued that the armies of the first two emperors were relatively small, numbering around two hundred thousand men. Even so, thanks to their tight organization, these armies had been capable of conquering and consolidating the Chinese territories, which had been divided for almost one century. Starting around the turn of the eleventh century, the size of the armed forces gradually increased until they reached 1,410,000 by mid-century.[26] Soldiers were burdened with grain transport, corvée, and other miscellaneous tasks. The expansion had left the army ill-prepared for combat, and it had begun to constitute a major drain of state revenues.

Chen's interpretation of the history of early Song military institutions exemplifies not only the use of institutional history, including Song precedent, in Yongjia scholarship but also its larger reform project. Chen's description of early Song military institutions and their effectiveness served to underscore his call for troop reductions and cuts in military spending. In the policy question quoted above, he used the history of the navy in previous dynasties to suggest the same argument. Either the existence of a naval force was unattested in the records of Antiquity and the Han and Tang dynasties, or the numbers listed were very small. Chen asked his examinees to evaluate contemporary proposals to further increase the size of the armed forces critically. For Chen, as for Ye Shi later, the escalating costs of military expenditures were a key problem in the financial administration of the Song state. In *Military Organization Throughout the Ages*, his policy response essay of 1172, and throughout the policy questions he set his students, Chen argued

---

25. On questions regarding Chen Fuliang's relationship to this title, see Xu Gui and Zhou Mengjiang, "Chen Fuliang de zhuzuo ji qi shigong sixiang shulue," 12.

26. Chen Fuliang, *Lidai bingzhi*, 8.3b, 6b. Chen's estimate is only slightly higher than the total number of imperial forces (*jinbing* 禁兵) cited for the mid-eleventh century in *The Dynastic History of the Song* (*SS*, 187.4576). Lü Zuqian's numbers in *Xiangshuo* are lower; see Chapter 4.

that tax increases were no longer an option.[27] Like his teacher Xue Jixuan, his disciple Ye Shi, Zheng Boqian 鄭伯謙 (12th c.; also of Wenzhou), and others following the Yongjia line of inquiry, Chen was convinced that a smaller government—fielding a smaller army and a more effective bureaucracy and requiring less taxation—was essential to the restoration of Song power.[28] Concrete proposals for the organization and the operation of this new government were to be based on creative investigations into institutional history.

The concluding hortatory sentences in Chen Fuliang's question on the use of naval forces hint at the relationship between Yongjia scholarship and the emerging movement of the Learning of the Way. In the second half of the twelfth century, the Song court grew increasingly anxious about the political agenda of a movement it referred to as "the Doctrine of the 'True' Way." The conflict between proponents of the Learning of the Way and opposing court factions, and particularly the relevance of the examination field in understanding this conflict, is discussed in more detail in Chapter 5. The question of the legitimacy of the Learning of the Way is significant here insofar as the stance of the Yongjia teachers on this question may have contributed to both the high sales of their policy collections and their subsequent proscription.

## Mediating Internal Political and Intellectual Debates

Chen Fuliang, in his instructions to the examinees, warned against applying the argument of morality to his set of questions about the use of naval forces. In the twelfth century, the argument that the military and financial problems of the Song government could be solved only if the emperor and his subjects turned their attention to moral self-cultivation

---

27. In his policy essay of 1172, Chen called for the abolition of the Privy Purse, a fund originally established for relief in emergencies including warfare but increasingly used for court expenses; see Niu, "Confucian Statecraft in Song China," 59. For Xiaozong's use of the Privy Purse, see Gong Wei Ai, "Emperor Hsiao-Tsung and the Consolidation of Southern Sung China," pt. II, 54–57.

28. Zheng Boqian's *Book of Statecraft for Grand Peace* (*Taiping jingguo zhi shu* 太平經國之書, late 12th–early 13th c.) similarly advocated downsizing of the bureaucracy and tax cuts; see Song Jaeyoon, "Tensions and Balance."

provided an alternative to the search for institutional and policy change. Zhu Xi, who presented himself as the leader of the Learning of the Way, shared with the Yongjia teachers the desire for the recovery of the northern territories. He opposed the peace policy adopted by the Song government after the Song loss to the Jin in 1161. Like the Yongjia teachers, he called for domestic reform; reform was needed to rebuild the strength of the Song state in preparation for the eventual recovery of the north. Zhu Xi's reform proposals focused, however, on the reform of the mentality of the emperor and society at large. In his view, institutional and policy changes were irrelevant as long as the minds of the emperor and his entourage remained obscured by human desires and bad habits. As Chapters 6 and 7 and demonstrate, the Learning of the Way alternative to domestic reform gained increasing influence in the examination field during the second half of the twelfth century.

The attitude of the Yongjia teacher toward the Learning of the Way has been a matter of controversy in recent scholarship. Hoyt Tillman has argued that Chen Fuliang, Ye Shi, and other Yongjia scholars were members of the fellowship of the Learning of the Way. In Tillman's study of the development of the Learning of the Way in the twelfth and thirteenth centuries, Yongjia scholars like Ye Shi grew critical of the Learning of the Way and distanced themselves from it only in the late twelfth century when Zhu Xi claimed exclusive leadership.[29] Others have questioned this interpretation and have suggested that the scholarship of Chen Fuliang and Ye Shi was so fundamentally different from that of the core figures in the Learning of the Way that they never qualified for membership.[30]

-------

29. Tillman, *Confucian Discourse and Chu Hsi's Ascendancy*, 139, 143, 231, 247, 257, and *passim*; Niu, "Confucian Statecraft in Song China," 26, 95–96, 99, and *passim*.

30. Chu Ping-tzu, "Tradition Building and Cultural Competition in Southern Song China," 367–443. Chu further suggests that the horizontal relationships among Yongjia scholars were different from the vertical ties that obtained between teachers and students in the Learning of the Way. The term "fellowship" then would also obscure the different kind of ties that bound intellectuals together during the twelfth century. In secondary research focusing on the Yongjia teachers, they are typically portrayed as a school of thought separate from the Learning of the Way; see, e.g., Xu Gui and Zhou Mengjiang, "Chen Fuliang de zhuzuo ji qi shigong sixiang shulue," 8.

The controversy is largely the result of the ambivalent attitude of the Yongjia teachers toward those who identified themselves as Learning of the Way advocates. On one hand, they were critical of the priority assigned to interpretations based on moral philosophy in discussions of history and government. Yongjia scholars were also skeptical of Learning of the Way advocates' exclusivist claims to truth (see Chapter 7). On the other hand, the Yongjia scholars were sympathetic to Learning of the Way reform efforts and defended its cause against accusations of factionalism.

Like Chen Fuliang, Ye Shi expressed his skepticism about the priority assigned to the moral theory of the mind in discussions about government and literati culture. In *Presented Scrolls*, he repeatedly charged contemporaries who claimed to be engaged in the realization of the Way with concentrating on the internal life of the mind and neglecting external affairs:

Those who are engaged in the Way today devote their efforts to drawing on the internal to govern the external, but in the social and political world of the relationships between sovereign and official, father and son, among brothers and friends, and between husband and wife, they often cannot make the internal and the external match. If they are preserving the mind to gain confidence in the self but cannot make the internal and external match, then how can they realize the Way? Therefore, if they cannot understand the reason why these three things [the supreme standard of government (discussed in *The Book of Documents*),[31] *The Great Learning*, and *The Doctrine of the Mean*] must be put in practice, and if they merely embellish their theories to entertain themselves, then how do they deserve the acclaim of practicing the Way?[32]

今之爲道者務出內以治外也. 然而於君臣父子兄弟朋友夫婦常患其不合也. 守其心以自信, 或不合焉, 則道何以成? 於是三者或不能知其所當施之意. 而徒飾其說以自好, 則何以爲行道之功?

---

31. For the passage in the *Book of Documents* discussing *huangji* 皇極, here translated as "the supreme standard of government," see Legge, *The Chinese Classics*, vol. 3, *The Shoo King*, 328–33. For an interpretation of the cultural relevance of "The Great Plan" chapter, in which the concept of *huangji* first appeared, see Nylan, *The Five "Confucian Classics*," 136–67, esp. 139–41. For Ye Shi's interpretation of the term in *Presented Scrolls*, see Yü Yingshi, *Zhu Xi de lishi shijie*, 2: 566–68. Note that Yü mistakenly dates this work to 1194.

32. *YSJ*, bieji, 7.727.

So, as for those who have no experience in social and political affairs, their words do not match reality, and as for those who do not investigate material things, their Way does not have a transformative effect. Their theories are lofty, but they are far removed from reality. This is truly impermissible![33]

則無驗於事者, 其言不合. 無考於器者, 其道不化. 論高而實違. 是又不可也.

Undoubtedly, by "those who are engaged in the Way today" Ye Shi was referring to Learning of the Way scholars. In the same essay, he first pointed out the contributions made in recent generations to the Confucian tradition. He argued that the transmission of the Way of Antiquity had been discontinued after the fall of the Zhou dynasty and that it had been recovered only recently, when scholars redirected their attention to the underlying meaning of the classical tradition as a whole rather than the exegesis of disparate texts. Not only was this argument identical to that of the central figures in the Learning of the Way (the Cheng brothers and later Zhu Xi), but Ye Shi also identified the central concepts of the Learning of the Way as key elements in the recovered Way of Antiquity. The cultivation of the mind, the dialectic between the mind's extension into the underlying patterns of all things on one hand and its activation of the innate patterns of human nature on the other, were in Ye's view crucial steps in the recovery of the full meaning of the Confucian tradition.[34]

However, after paying tribute to the contributions of Learning of the Way advocates, Ye discussed what he perceived as a fundamental weakness in the scholarship of those scholars claiming to restore the Way of Antiquity. His critique in *Presented Scrolls* of the inward focus of Learning of the Way scholarship and its dismissive attitude toward the study of external things presaged his discontent with it in later work.[35] *Presented Scrolls* could thus be read by Ye's contemporaries as espousing an alternative to the emerging movement of the Learning of the Way.

---

33. Ibid., 5.694–95.

34. Ibid., 7.726–27.

35. I therefore disagree with Tillman and Niu Pu that Ye Shi remained part of the Learning of the Way fellowship until the late 1190s. Ye's criticisms of the Learning of the Way can be traced to the 1170s; see Chu Ping-tzu, "Tradition Building and Cultural Competition in Southern Song China," 399–431.

Ye's alternative included the best part of the Learning of the Way, but it was more comprehensive in scope. For Ye, the Way was ultimately the Way of government, and in order to engage in the Way of government, historical studies and policy analyses, as the study of external things, were an essential complement to moral philosophy.

The ambivalent attitude of Ye Shi and other Yongjia scholars continued into the 1190s. In letters and comments to others, the Yongjia scholars and the leading figure in the Learning of the Way, Zhu Xi, discussed their philosophical differences. At the same time, throughout the 1180s and the 1190s, the Yongjia scholars defended the Learning of the Way in court disputes about the factionalist tendencies of the movement. Ye Shi submitted a letter supporting Zhu Xi following accusations to the court in 1188 that Zhu had set up the Learning of the Way for political self-promotion. In 1194 Chen Fuliang wrote Emperor Ningzong 寧宗 (r. 1195–1224) asking him to reverse Zhu Xi's dismissal from office.[36] These daring acts have been interpreted as evidence for the membership of the Yongjia teachers in a larger Learning of the Way fellowship. The wording of the Yongjia defense of the Learning of the Way against accusations of factionalism indicates, however, that the Yongjia teachers were staking out a third position, a position that would allow them to mediate the escalating conflict between the court factions of successive councilors and self-identified advocates of the Learning of the Way.

In their defense of the Learning of the Way, the Yongjia scholars did not identify as members of that movement.[37] They did not write apologies for its moral philosophy and its critique of the current government. Several of the criticisms launched against the Learning of the Way in the 1180s and 1190s were foreshadowed in Ye Shi's evaluation of the movement, translated above, in *Presented Scrolls*. The Yongjia scholars acknowledged the validity of some of the criticisms but argued that the accusations were excessive and divisive. In a policy response written for the palace examination of 1190, Zhou Nan, a student of Ye Shi, identified the escalation of the conflict between court officials and the leader-

---

36. Niu, "Confucian Statecraft in Song China," 95.

37. For a strong argument against the inclusion of Chen Fuliang in the Learning of the Way, see Zhou Mengjiang, "*Song shi* 'Chen Fuliang zhuan' buzheng," 31–34.

ship of the Learning of the Way as a top issue requiring speedy resolution. This issue was tearing the government apart at a time when it could ill afford divisions. Zhou Nan pointed out that accusations of factionalism against the Learning of the Way were in effect creating the conditions for factionalism:

Those who slander others have no idea what "the Learning of the Way" means. They use it to refer to everybody who does not go along with others in their evil deeds. And so, they give this name to all who disagree with them. . . . Those who came up with this interpretation of the Learning of the Way say of those they are slandering that their techniques of mind-cultivation are difficult to understand, their skills one-sided, and that they resent peace and like to cause trouble. . . . Ever since the label of the Learning of the Way was invented, those without convictions degraded themselves by following suit; those who feared trouble disgraced themselves by seeking to comply; but those who stood in the middle and did not lean in either direction never wavered. Even if the career patterns [their leaving and returning to government] of those [who remained neutral] sometimes coincide [with Learning of the Way followers], how could this cause separatism? They socialize and are unbiased in their likes and dislikes. This must be because they are unwilling to comply with others' slanderous accusations. Those who slander others then say: "We cut off our relationships with the Learning of the Way, but others are socializing with their followers. We consider the Learning of the Way devious and treacherous, but others are claiming that they are without reproach." This is what factionalizes the followers of the Learning of the Way.

. . . Even though their talents can be said to be one-sided, if they can still be put to use, then they should be received and brought to completion. Based on the facts, those who today are labeled as factionalist Learning of the Way scholars are actually people who should be recruited according to the supreme standard of government.[38] How is it possible that the empire's talented and knowledgeable scholars are cast aside and that mediocre men are recruited?[39]

彼譖人者謾不知道學爲何事. 意以爲凡不與人同流合汗者皆是也. . . . 彼爲道學之論者曰: 心術暗也; 才具偏也; 惡靜而喜生事也. . . . 自道學之名既立, 無志者自貶以遷就; 畏或者迎合以自汗; 而中立不倚之人, 則未嘗顧也. 彼其出處偶同, 則何害於私? 相往來, 好惡不偏, 必不肯隨人毀譽.

---

38. See note 30 above.
39. Zhou Nan, *Shanfang ji*, 7.10a–11b.

彼譖人者則又曰: 吾方絕道學, 而彼則與之交通. 吾方以道學爲邪佞, 而
彼則頌言其無過行. 是黨道學之人也.

　. . . 謂其才雖有偏, 而終有可用, 則亦當收拾而成就之者也. 若以實而
論, 則今之所謂朋黨道學之士是乃皇極之所取用之人也. 今奈何廢棄天
下有才有智之士, 取世之所爲庸人?

Zhou Nan's criticism of the crescendo of accusations against the
Learning of the Way, which I discuss further in Chapter 5, targeted the
accusers' indiscriminate use of the label. All sorts of scholars, including
those who "stood in the middle," like the Yongjia scholars, came under
suspicion. Zhou Nan adopted a middle position in line with the stance
of Ye Shi and Chen Fuliang, his teachers in Yongjia. He acknowledged
that the Learning of the Way counted many talented and capable men
among its followers. Granted, their interests could be focused narrowly
on issues of personal morality, as both their accusers and his Yongjia
teachers had pointed out, but this preoccupation should not disqualify
them for office.

Zhou Nan's defense of the Learning of the Way was predicated on
the principle of equitable and unbiased recruitment, a principle that he
saw embedded in the supreme standard of government. The supreme
standard of government, which was central both in Ye Shi's *Presented
Scrolls* and in Zhou Nan's policy response, referred to the public-
spiritedness of the sovereign. According to the "Great Plan" chapter of
*The Book of Documents*, the sovereign maximizes the power of his state
by inducing all to contribute their individual talents and by listening to
all of his subjects with an open mind.[40] Accordingly, Zhou Nan urged
the emperor to use a fair standard in evaluating the achievements of
those accused of membership in the Learning of the Way and to stop
recruiting sycophants. Zhou Nan's emphasis on public-spiritedness exposed the court's use of "great impartiality" as its guiding principle
(Chapter 5).[41] He criticized the court's adoption of *huangji* (glossed as
"great centrality") as the embodiment of its program of moderation,
not because of its inherent meaning but because of the court's failure to
practice its creed. The contrast between Zhou Nan's defense of the
Learning of the Way and contemporary policy response essays and ex-

---

40. Nylan, *The Five "Confucian Classics,"* 140–41.
41. Cf. Shu Jingnan, *Zhuzi da zhuan*, 714; and Yü Yingshi, *Zhu Xi de lishi shijie*, 2: 552–58.

positions written by self-identified students of the Learning of the Way further underscores the mediating position adopted by Yongjia scholars. Zhou Nan did not identify himself as a believer in the Learning of the Way, nor did he give an apology for its beliefs in his policy response. As demonstrated in the examination expositions and policy response essays discussed in Chapter 7, many others did do this, and this contributed to the ire and suspicion of the dominant court factions in the 1190s.

The Yongjia scholars' approach to two of the major challenges confronting the Song court and Song society at large contributed to their authority in the examination field, but because of that approach they also incurred the disapproval of both the court and Learning of the Way advocates. The Yongjia reform project was aimed at the restoration of Song authority over the northern territories, a goal widely shared among Southern Song elites and one that excited young examination students.[42] The Yongjia reform project called for a central government retreat. Yongjia arguments for tax and spending cuts and troop reductions appealed to local elites, who were becoming increasingly interested in managing their local communities.[43] The transition from the Northern to the Southern Song coincided with a change in the orientation of elite strategies from the capital to local society. In the second half of the twelfth century, Yongjia arguments to increase local wealth and to enhance the power of the local elite in managing that wealth fell on sympathetic ears among examination candidates from elite families.

The position of the Yongjia scholars on the spread of the Learning of the Way movement likewise endeared them to a broad spectrum of the literate elite. Chen Fuliang and Ye Shi recognized the contributions

---

42. For the political activities of Imperial College students, see Wang Jianqiu, *Songdai Taixue yu Taixuesheng*, esp. chaps. 7 and 8. Yü Yingshi (*Zhu Xi de lishi shijie*, 2: 475, 481) argues that this position expressed the collective identity of "the Chinese people." He provides evidence that pro-war advocates like Zhu Xi attributed a pro-war stance to the soldiers and the common people; whether the pro-war sentiment was indeed so widespread among the population at large that it became part of a collective identity is highly questionable.

43. For a more elaborate discussion of the Yongjia curriculum as an attempt to reshape imperial politics in local elite terms, see De Weerdt, "The Empire-Wide Significance of Local Intellectual Traditions," esp. the conclusion.

advocates of the Learning of the Way were making to contemporary intellectual and political debates. Their own scholarship was considered to be compatible with that of the Learning of the Way by many of its followers. At the same time, the Yongjia scholars warned against the exclusivist tendencies of the movement and sought to safeguard a more traditional curriculum, emphasizing the rigorous study of the historical record, classical philology, and composition. The preservation of a more traditional curriculum and the application of history to contemporary problems pleased those critical of the application of arguments based on contemporary moral philosophy to questions about government.

*Awaiting Reception* and *Presented Scrolls* were both bestsellers and banned books for the same reason. The reform proposals and the attempts at mediation of their authors were not well received by the principal protagonists in the factional struggles of the 1190s. The Yongjia plan for the recovery of the north implied criticism of the current foreign and domestic policies of the court factions in control during most of the 1180s and 1190s (Chapter 5). The Yongjia strategy of self-strengthening in preparation for recovery was presented as an alternative to these court factions' pro-peace policy.

Some of the Yongjia policies in the area of military preparedness and fiscal conservatism were voiced by others at court. This was especially the case during Xiaozong's reign. Emperor Xiaozong adopted a more aggressive stance toward the Jin court and was convinced of the need for reductions in expenditures. Even though his measures conflicted with those proposed by the Yongjia teachers (the emperor relied heavily on court attendants and thus sabotaged policy debates among officials and scholars; he diverted public funds to the Privy Purse and thus acted against injunctions for fiscal responsibility; his aggressive policy toward the Jin court jeopardized the long-term plan for empire-wide domestic reform and military rebuilding in preparation for future recovery of the north), his interest in reform attracted those with well-thought-out plans for the future. However, it aroused opposition among those who favored the more conservative diplomatic and military agenda of the Gaozong reign.[44] The Yongjia scholars' defense of

---

44. Gong Wei Ai has written extensively on the politics of Emperor Xiaozong and his court; see esp. both parts of "Emperor Hsiao-Tsung and the Consolidation of

the Learning of the Way provided further indication of their dissenting attitude toward the status quo. As Zhou Nan's comments suggest, opposing court factions preferred to see Yongjia attempts at conciliation as proof of their affiliation with the Learning of the Way.

The Yongjia attempt to mediate the conflict between the court and the Learning of the Way was also rejected by the leadership of the movement. Zhu Xi supported the government ban on *Awaiting Reception* and *Presented Scrolls*, a ban that simultaneously targeted his own work. Zhu Xi hereby forcefully underlined his disapproval of the Yongjia emphasis on institutional history and policy change in domestic reform. He perceived "Yongjia" examination writing as the major source for the neglect of personal morality among elites and therefore incompatible with the Learning of the Way. Zhu Xi was not particularly grateful for the help Yongjia scholars offered in the 1190s. His lack of sympathy for their fate in the False Learning campaigns of the mid-1190s (Chapter 5) probably arose from the fact that the Yongjia scholars did not speak as Learning of the Way advocates in their letters of defense, as well as the fact that Yongjia policy writing had been more successful as a standard in the examination field than Learning of the Way moral philosophy.

## Standards for Expositions

### CHEN FULIANG: THE FOUNDING
### FATHER OF THE EXPOSITION

Wei Tianying included only one piece each for the majority of the 131 authors anthologized in *Standards for the Study of the Exposition*. In the cases of some better-known authors like Lü Zuqian, whose examination teaching is discussed in Chapter 4, he selected two or three pieces. But at ten pieces, Chen Fuliang was by far the most popular exposition writer. One critic wrote, "Zhizhai's [i.e., Chen Fuliang's] expositions

---

Southern Sung China." Zhu Shangshu (*Songdai keju yu wenxue kaolun*, 436) underscores Emperor Xiaozong's reaching out to scholars critical of the status quo in explaining the appeal of Chen Fuliang's examination writing. For the opposition to Xiaozong's more aggressive policies, see Yü Yingshi, *Zhu Xi de lishi shijie*, vol. 1, chap. 7, and vol. 2, chaps. 10 and 11.

have established the foundations of the exposition." He added that the particular piece to which this comment was attached had functioned as a model for all students who had excelled on the examinations held at the Imperial College.[45]

The interest in Chen's work among examination candidates and teachers is further attested by the numerous publications anthologizing his model examination essays and by other types of examination guides attributed to him. Chen Fuliang's oeuvre includes several titles in the category of examination manuals. He gained especial acclaim for his work in the genre of the examination exposition. Chen was the first teacher to compile a ranked list of top expositions. This list, *Expositions Register (Lun ge* 論格), arranged according to rhetorical strategies, may have provided the matrix for subsequent anthologies of examination expositions such as *Standards for the Study of the Exposition.*[46] Partly because of his analyses of examination expositions and partly because of his numerous model examination expositions, Chen gained the epithet "founding father of the exposition." Collections of Chen's expositions were circulating in the twelfth and thirteenth centuries and were republished in subsequent centuries. Extant are *Awesome Expositions (Aolun* 奧論)[47] and Fang Fengchen's 方逢辰 annotated edition, *Chen Fuliang, the Founding Father of the Exposition (Zhizhai lunzu* 止齋論祖).[48] The latter contains thirty-nine examination expositions.[49]

---

45. *LXSC*, 6.25a, 30b.

46. Ibid., 5.74b.

47. Zhou Mengjiang, *Ye Shi yu Yongjia xuepai*, 93. See the Appendix for a discussion of *Awesome Expositions from Ten Masters, with Notes (Shi xiansheng aolun zhu* 十先生奧論註). This collection includes individual collections of "awesome expositions" from Lü Zuqian, Chen Fuliang, and others.

48. According to the 1268 preface by Fu Canzhi 傅參之, this new edition was based on earlier printed anthologies of Chen Fuliang's expositions. The attribution of the commentary to Fang Fengchen may have been spurious; see Zhu Shangshu, *Song ren bieji xu lu*, 2: 1070.

49. There were also nonregulated expositions. In contrast to examination expositions, these were not confined to a particular passage but address more general subjects, a particular institution, or a particular historical figure. The difference is very clear, for instance, in the expositions preserved in Yang Wanli's collected works. Whereas the "examination expositions" (*chengshi lun* 程式論) deal with passages from the histories and the philosophers, the nonregulated expositions, which do not stick to some of the structural conventions that applied to examination expositions, freely discuss the

In his mid-twenties, years before he obtained the *jinshi* degree in 1172, Chen Fuliang was already attracting large numbers of examination candidates. Several hundred students gathered at the Tea Garden Temple near the southern gate of the Wenzhou prefectural seat in Yongjia to attend his courses.[50] Chen Fuliang benefited from Yongjia's reputation in the first half of the twelfth century as a center of learning. Chen was in fact a native of Rui'an 瑞安, another county in Wenzhou. In the minds of his contemporaries, however, he was associated with Yongjia, since references to Yongjia conjured up the success of Wenzhou prefecture as a whole. In his work Chen carried on the tradition of scholarship promoted by the senior Wenzhou teachers, Zhou Xingji, Zheng Boxiong, and Xue Jixuan.[51] The historical approach and the emphasis on government in Chen Fuliang's expositions were in keeping with the Yongjia teachers' research in institutional history. Yet, Chen took examination preparation in a new direction. Consequently, the reputation of Yongjia as a center for examination preparation climaxed during the 1160s and 1170s.

A contemporary biographer, Ye Shi, noted that Chen Fuliang outshone "the older teachers who transmitted the older examination learning."[52] Chen's appeal was such that students signed up to take classes with Chen even though they were studying with other teachers in the region. Ye Shi attributed Chen's appeal to the extraordinary interpretations and the novel argumentation his teaching encouraged. What made Chen Fuliang's interpretations and diction so contagious?

In the case of Chen's examination expositions, the answer lies in the discussion of principles of government based on the dramatic presentation of historical events. Chen Fuliang developed the psychological states of his subjects and staged their intentions. Through his dramatic

---

thought of Han Yu and similar broader subjects; see Yang Wanli, *Chengzhai ji*, 84 (for the nonregulated expositions) and 90 (for the examination expositions).

50. *SB*, 103; Sun Qutian, *Chen Wenjie gong Fuliang nianpu*, 3b–4a. A collection titled *The Southern City Wall Collection* (*Nancheng ji* 南城集) probably contained examination materials he wrote during his years of teaching at the Tea Garden Temple (Xu Gui and Zhou Mengjiang, "Chen Fuliang de zhuzuo ji qi shigong sixiang shulue," 25*n*15).

51. Zhou Mengjiang, *Ye Shi yu Yongjia xuepai*, 48, 60–61, 72.

52. Sun Qutian, *Chen Wenjie gong Fuliang nianpu*, 3b, citing Ye Shi's epitaph for Chen Fuliang in *YSJ*, 16.298.

contextualization, Chen created effective case studies as illustrations of principles of administrative management. Under the first subdivision in this section I discuss the connection forged in Chen's work between the explanation of principles of government and the psychological interpretation of history. The second subdivision accounts for the novelty contemporaries admired in his writing.

## PRINCIPLES OF GOVERNMENT

Expositions were by definition concerned with the principles underlying events and things. They were theoretical, but the focus of theoretical endeavor varied. Government was the central field in Cheng Fuliang's examination expositions. He interpreted historical and philosophical texts as illustrations of principles in political philosophy. The themes covered in Chen Fuliang's expositions in *Standards for the Study of the Exposition* included the complementary nature of the relationship between emperor and minister; the selection of men with a blemished career over those with an untarnished record; the priority of merit over geographical background in judging careers; the official's obligation to fully inform the ruler; the establishment of institutions and policies that meet the needs of the time; the role of fate in government; and the attitude of sage-rulers toward the people, historical change, and the task of governing the world. This list of themes in the essays included in *Standards for the Study of the Exposition* focuses on the responsibilities of the ruler and those of central civil and military officials. This implies that for Chen Fuliang the principles that needed to be investigated in exegesis concerned the respective responsibilities of rulers and officials and the relationship between them.

The focus on the ruler-official relationship in examination expositions fit into the larger Yongjia call for shared governance and the restoration of imperial power. Yongjia scholars called on the emperor to share power with the entire bureaucracy and the scholarly elite and believed that this was possible only if the emperor focused on governing. By taking charge, the emperor could avoid channeling imperial power into the hands of individuals and factions at court. The restoration of imperial power consisted, in the view of the Yongjia teacher, of the emperor's active involvement in government and his consultation of broad sectors of the official and scholarly elite. The clarification of principles of administrative management was intended as an aid to the restoration of imperial power. The language of political theory that encompassed these principles of government was not only illustrated in Chen Fuliang's examination expositions but also taught in examination preparation manuals (see Chapter 4).

In Chen Fuliang's examination expositions, the clarification of principles of government was typically laid out in three steps. According to Chen Fuliang's guidelines for writing expositions, the first step was imagining the original context of the quotation embedded in the exposition topic. In his instructions on the art of writing expositions, Chen Fuliang emphasized the primary importance of "understanding the topic" (*renti* 認題): "Of all the guidelines for composing expositions, none is more essential than fully comprehending the meaning of the topic. Thus, after seeing the topic, one should consider in detail its provenance and the context."[53] Exposition questions contained short references to the histories, the philosophers, the Classics, and, occasionally, literary collections. The topic usually consisted of a single phrase of about four to eight characters.[54] The typical format of twelfth-century topics from classical sources and from the histories was direct quotation. The following topics were, for example, lifted verbatim from *Mencius*: "The knowledgeable person does what causes him no trouble," "Confucius never went beyond reasonable limits," and "In the 'Completion of War' [in *The Book of Documents*], I select only two or three passages."[55] Typical historical topics cited a phrase from *The History of the Han Dynasty*: "Emperor Wen, the Filial, almost arrived at having a judicial system that he did not have to put to use," "Emperor Wen, the Filial, liked expositions on punishments and institutions," or "In government care about how to practice government energetically."[56]

---

53. *Lun jue*, 7a.

54. There are only a handful of exceptions to this rule. All but one date from the Northern Song period. See the exposition topics for the years between 978 and 1063 preserved in *SHY*, *XJ*, 7.

55. *LXSC*, 2.89a–93b, 6.25a–30a, 6.1a–7b. The topics appear, respectively, in *Mengzi* IVB:26, Legge, *The Chinese Classics*, vol. 2, *The Works of Mencius*, 331; IVB:10, Legge, 321; and VIIB:3, Legge, 479.

56. *LXSC*, 7.65a–69b (Ban Gu, *Han shu*, 4.134), 70a–75a (*Han shu*, 88.3592), 4.95a–100a or 7.55a–59a (*Han shu*, 88.3608).

Understanding the context in Chen's expositions meant not so much citing and explaining the lines preceding and following the quotation in the topic as creating a sense of the historical circumstances in which the event had occurred. Chen's opening sentence in response to the topic "In government pay attention to energetic implementation" reads as follows: "One day some scholar meets his ruler unexpectedly. Since he has not yet seen his ways of accepting and rejecting, how then should he advise him?" 論曰:士有一旦卒然遇其君者. 未見其趨舍之方, 則亦何以告之? Chen did not address the question of what "energetic implementation" might mean. Nor did he consider how this piece of advice related to the next sentence of the passage from which the quote was taken, "At that time the emperor was just fond of literary writing." Instead, the whole essay focused on the question why the old man Shengong 申公 (2nd c. BCE) made the statement about energetic implementation on being summoned to court.[57]

Chen's re-creation of the circumstances caught the attention of his contemporaries. Feng Yi 馮椅 (*js*. 1193), a contemporary critic, commented on Chen's opening sentence: "This sentence is like drawing a painting in speech. His 'broaching the meaning'[58] is rough." Feng Yi, himself a successful author and teacher of examination writing, noted two key aspects in Chen Fuliang's examination writing. Chen's ability to paint a scene and blow life into the world behind the text exerted great appeal. Chen achieved the effect of visual representation by twisting current conventions of examination writing. His opening sentence sketched a problem: How should Shengong advise his ruler when he had no knowledge of his ruler's policy inclinations? The questions that Shengong must have confronted in Chen Fuliang's historical imagination paved the way for the essay's arguments. Chen's opening sentence drew the reader into the scene and into the rest of his exposition, yet Feng Yi felt that the introduction lacked refinement. Chen's introduc-

---

57. *LXSC*, 7.55a–59a; Ban Gu, *Han shu*, 88.3608.

58. *Poyi* 破義 refers to the first section of the exposition, more commonly called *poti* 破題, in which the author breaks up the topic into (usually two) segments to be elaborated in the rest of the essay. See the discussion of the structural layout of the exposition in Chapter 2.

tion broke with the expectation of the introductory line as the summary statement of the argument.

The historical reconstruction of the context of the quotation prepared the reader for the analysis of the intentions and motivations of historical actors associated with the story, the second step in Chen's examination expositions. After evoking the difficulties facing the emperor's adviser, Chen Fuliang proceeded to outline several reasons for Shengong's reaction. He argued that historical circumstances and Shengong's personal situation provided the rationale for his sentence of advice, "In government pay attention to energetic implementation."

First, he explained that the meeting occurred at the very beginning of Wudi's reign 武帝 (141–87 BCE) when the thrust of his administration was not yet clear.[59] In the main discussion he staged the uncertainties that, in his imagination, must have crossed Shengong's mind. He alerted his readers that most of the things commonly associated with Wudi's rule, such as his aggressive foreign policy and his unceasing pursuit of longevity therapies, lay in the future. Then, having shorn his readers of their preconceptions, he presented Wudi's mixed record up till the day of the meeting. The emperor was young. He had shown signs of a willingness to listen to the good recommendations of wise ministers, but he had also been prone to the suspect advice of courtiers and magicians. Chen argued that, given the lack of clear-cut information about Wudi's early years in government and, taking into account Shengong's personal situation—as an old man living in the woods he was suddenly summoned to the emperor's court—Shengong showed great foresight by withholding judgment and urging the emperor to act. This way, he would have had grounds to praise or criticize the emperor's administration later on.

Chen Fuliang's predilection for situational and psychological explanations also surfaced in an essay entitled "Employing Failures Is Better Than Employing Successful Men."[60] The topic quoted a statement attributed to Tang Taizong 太宗 (r. 626–49) in the *New History of the Tang*. Emperor Taizong made this statement in reference to General Li Jing

---

59. This paragraph corresponded to the section called "tracing the origins of the topic" (*yuanti* 原題).

60. *LXSC*, 5.99a–104b; Ouyang Xiu and Song Qi, *Xin Tang shu*, 93.3812.

李靖 (571–649), who, after having been condemned to death and sub-sequently exonerated and reinstated, beat a band of brigands. Chen dis-cussed the careers of several generals and concluded that success breeds arrogance and leads to disaster. He further showed how men who had made mistakes and needed to redeem themselves used more circumspection and proved beneficial to their sovereign. Instead of us-ing compressed quotations and a long list of historical examples, as Huang Huai did in his 1154 essay quoted in Chapter 2, Chen Fuliang restricted his evidence to a few cases. In each case he used psychologi-cal arguments to explain failure and success. The Song commentator was particularly struck by Chen's deployment of the following piece of evidence:

In the past a man was taking a walk at night. He saw a "sleeping stone" [a large stone] and thought it was a tiger lying down. He drew his bow and shot at it. As soon as he let go, the arrow vanished in the stone. When he looked at it, he realized it was a stone. He stepped back and shot again. Then the arrow jumped up and left no trace. Now, the shooting was the same, but the result was different. Why is this? When he saw the stone as a tiger, there was an inexpressible fear in his heart; when he saw the stone as a stone, he resumed his calm, became devoid of awe, and looked upon it as a game. Thus, when those with merit handle affairs, they see the stone as a stone; when those with mistakes handle affairs, they see the stone as a tiger. As for the sovereign's employing people, if he can get those with the mind that sees the stone as a ti-ger and employ them, he will benefit in all respects.[61]

昔人有夜行者. 見寢石以爲伏虎也. 以石爲虎是有懼心也. 援弓而射之. 一發沒矢. 下視之乃石也. 却而復射則矢躍無跡. 夫射一也而中否異焉. 一中一否何哉? 以虎視石則其心有不免之懼. 以石視石則恬不知怪而以 戲處之者也. 故夫有功處事以石視石者也. 以過處事以虎視石者也. 人君 之用人也能得以虎視石之心者而用之, 亦何所不濟哉!

The parable of the stone was not of Chen Fuliang's devising. He para-phrased a story that Liu Xiang 劉向 (ca. 79–ca. 6 BCE) had selected for his collection of anecdotes a thousand years earlier. Chen Fuliang's rendition of the parable expanded on the original; he presented it with a supplementary explanation of its psychological and political implica-

---

61. *LXSC*, 5.102b–103a. The original version of the story appears in Liu Xiang, *Xin xu*, 4.12b.

tions. Chen's selection and lengthy rendition of this story from Liu Xiang's collection of anecdotes fit into his views on good government.

In the third and last part of the exposition, Chen used the analysis and diagnosis of the motivations of historical actors as evidence for the substantiation of principles of good government. Chen's explorations into the circumstances of a historical event and the psychology of the actors involved in the events were guided by his belief in general principles of government.

Chen Fuliang elaborated on Liu Xiang's psychological observations to explain a political principle that epitomized Yongjia political reasoning. Chen argued that a calculating ruler could benefit by capitalizing on the energy unleashed by the fear of an official in danger of losing his position. Chen made the same argument in "Employing Those with a Blemished Record."[62] The principle "One should not dismiss a scholar-official because of one imperfection," defended in *To the Point in All Cases (Bamian feng* 八面鋒), a Yongjia manual discussed in Chapter 4, also highlighted the benefits of retaining officials with a failure on their career records.[63] In sharp contrast, Zhu Xi dismissed this pragmatic approach to the recruitment of officials and argued that selection should serve to distinguish between the morally inferior and the morally superior (Chapters 6 and 7).

The selection of effective officials was, according to the political vision embedded in Chen Fuliang's examination essays, the core responsibility of the ruler. In turn, the official's major responsibility consisted of providing truthful advice in response to the needs of the times. In his historical reconstruction of the first exchange between Shengong and Emperor Wu, Chen highlighted the official's duty to provide advice based on an understanding of the current political climate. He read Shengong's pithy line of advice as a conscious refusal to discuss government without sufficient knowledge of current policy making. Shengong's response to the young emperor illustrated the more general political principle that it was inappropriate to advise a ruler in general terms without discussing the pros and cons of concrete policies.

---

62. *Shi xiansheng aolun zhu*, houji, 8.9a–10b.
63. *BMF*, 12.89–90; cf. 11.82–83.

Knowledge of current affairs was a primary criterion in Chen's evaluation of the effectiveness of an official's advice. A second was his ability to formulate proposals that addressed administrative questions based on the needs of the time rather than on the dictates of moral universals or ritual practice. Like his Yongjia associates, Chen Fuliang was known for his strong interest in administrative institutions. Surprisingly few of his examination expositions dealt with institutional matters, but this has to do more with the requirements of the genre than with Chen's interests. Policy questions and response essays were the more appropriate genres for discussing socioeconomic, military, and political institutions. The general principle that institutions and administrative measures should be designed and changed in accord with the trends of the time, however, was part of the political vision that Chen Fuliang elaborated in his examination expositions.

Chen advocated contemporary solutions to contemporary problems and criticized traditionalism in an examination exposition on the topic "What are the methods of the kings like?" The question came from a series of questions Emperor Cheng 成帝 (r. 33–7 BCE) asked his minister Du Qin 杜欽[64] in an examination policy question:[65]

What is valuable in the Way of Heaven and Earth? What are the methods of the kings like? What is of the greatest importance in the meaning of the Six Classics? What comes first in the behavior of human beings? On what should the methods for recruiting people be based? To what should we devote ourselves in contemporary government? Answer each of these on the basis of the Classics.

策曰: 天地之道何貴? 王者之法何如? 六經之義何上? 人之行何先? 取人之術何以? 當世之治何務? 各以經對.[66]

Chen's argument read:

In my opinion, Chengdi asked a question he should not have asked. As for Du Qin's answer, he also answered what he should not have answered.

吾意, 成帝問其所不當問. 杜欽之答又答其所不當答.

---

64. For Du Qin and the events recounted by Chen Fuliang, see Loewe, *A Biographical Dictionary*, 81–82.

65. *LXSC*, 4.50a–55a.

66. Ban Gu, *Han shu*, 60.2673.

Chen criticized Emperor Cheng and his adviser for aspiring to the models of Antiquity while neglecting the models of the more recent Han past.

It is not that Yao, Shun, and the Three Dynasties are not worth aspiring to. However, in case our penal regulations cannot be properly organized, what do we gain from "painting marks"?[67] [Under Yu, instead of real, physical punishments, marks were painted on clothes and caps.] In case our governmental affairs cannot be put in order, what do we gain from "dancing with shields"?[68] [Conflicts were settled through ritual.] Even if we do not have measures for taxation and no methods for education, how different are the tax, mutual aid and share systems and the educational institutions of the Xia, Yin, and Zhou! Therefore, in government, on the whole, imitate what is not ancient.

彼堯舜三代非不足慕也. 吾刑罰不能清, 何有於畫象? 吾政事不能修, 何有於舞干? 吾賦取無度敎養無術, 何貢助徹校庠序之異哉! 故凡治效不古.

In the examples he marshaled to further illustrate his point, Chen was not concerned with the intrinsic pros and cons of particular institutions or their accordance with a set of general standards.[69] Chen Fuliang's evaluation of Du Qin derived from the political principle that ministers were responsible for designing institutions that met the specific needs of their time.

Chen Fuliang saw impartiality as a third basic virtue for effective officials. Partiality hampered officials' ability to focus their administrative efforts on the needs of the times. In his exposition on Emperor Cheng and Du Qin, he explained that Du Qin's failure to comply with the principle of meeting the needs of the times was caused by factionalism. Du Qin supported the clique of Wang Feng 王鳳 (d. 22 BCE), an uncle of Wang Mang 王莽 (r. 9–23) who had recommended Mang for

---

67. The commentator refers to Ban Gu, *Han shu*, 23.1097.

68. Legge, *The Chinese Classics*, vol. 3, *The Shoo King*, 66.

69. This argument indicates a basic assumption operative in his thinking about institutions. It does, however, not imply that Chen never argued about the pros and cons of particular institutions. For a discussion of his policy questions and essays, see the previous section "*Awaiting Reception* and *Presented Scrolls*" and Chapter 4.

office.[70] Wang Mang staged a palace revolt and temporarily overthrew the Han in 9 CE. In Chen's reconstruction of the scene, Du Qin feared that encouraging the emperor to follow the models of his Han predecessors would lead to the downfall of his patron. Wang Mang advocated a return to an idealized model of the Zhou dynasty.

The responsibilities Chen Fuliang assigned to the ruler and the official reveal that he saw the relationship between ruler and official as asymmetrical. The ruler was in charge, but once the ruler selected his advisers, it was up to them to initiate sound and timely policy proposals. The management of affairs as well as the management of the emperor was in the hands of his chosen advisers. In an exposition entitled "Wei Xiang Satisfied the Emperor's Ambitions," he wrote that the historian Ban Gu 班固 (32–92) had had good reasons for commending Wei Xiang 魏相 (?–59 BCE) exclusively for "satisfying the emperor's ambitions."[71] Chen showed that Wei Xiang had become regarded as an exemplary minister because his strictness equaled the emperor's. In Chen's estimation, Wei Xiang was not on a par with other famous ministers because he had failed to balance out the emperor's strictness, and so failed to respond to the needs of his time. Chen Fuliang maintained that the proper ruler-minister relationship was characterized not by consent but by complementarity. This implied that the minister needed to find ways to make up for the ruler's deficiencies.

## RHETORICAL STRATEGIES

Chen had studied current conventions for examination writing carefully. His comments on the structure of the exposition preserved in *The Art of the Exposition* (*Lun jue* 論訣) bear witness to his adherence to conventional standards in examination writing. Chen taught his students the proper sequence of the subdivisions of the exposition, replicating the introduction–main discussion–citation–conclusion model common by the 1150s. At the same time, Chen Fuliang's focus on historical circum-

---

70. Twitchett and Loewe, eds., *The Cambridge History of China*, 1: 215, 226. Wang Feng laid the foundation for the succession of scions of the Wang family in the most powerful offices in the central government.

71. *LXSC*, 5.94a–98b; Ban Gu, *Han shu*, 74.3135.

stances and the mental states of his subjects translated into conscious modifications of current conventions.[72] He developed compositional techniques to create effective historical cases for use in the discussion of administrative principles. Chen expressed his historical reimagination of events in writing through dramatic presentation. These strategies broke with current writing standards with regard to structure, grammar, and word usage. Chen's subtle violations of current standards lent liveliness and drama to his expositions that critics and students alike admired.

His repertoire of compositional techniques violated the conventions exemplified in Huang Huai's regulated exposition in almost every respect. My analysis of Huang's winning 1154 examination exposition highlighted the standardized format of regulated expositions from the twelfth century and the common use of parallel argumentation. The standard exposition from the 1150s was written in accordance with the requirements for each subdivision and the proper sequence of the subdivisions. In contrast to the clear demarcations between subdivisions in Huang Huai's exposition, Chen's expositions were marked by continuous sequences that blurred the transitions between sections. Take, for example, the introductory part of the exposition on Shengong's meeting with Emperor Wu:

One day a scholar met his ruler unexpectedly. Since he has not yet seen his ways of accepting and rejecting, how then should he advise him? [*poti*] Just trusting and endorsing him is certainly impossible. Just doubting and restraining him is also impermissible, . . . [*jieti*]

   When Shengong first saw Emperor Wu, he said, "Government does not lie in many words; just care about how to carry things out energetically." Shengong's intention was to wait and speak then. [end of the *maozi*, the first part]

   Really! He can be called a man of maturity and foresight! How so?[73] [beginning of the *yuanti*]

論曰: 士有一旦卒然遇其君者. 未見其趨舍之方, 則亦何以告之? 徒信而許之固不可; 徒疑而禁之又不可. . . .

---

72. Zhu Shangshu (*Songdai keju yu wenxue kaolun*, 220–25, 430–46) focuses his analysis of the expositions of Chen Fuliang on their stylistic features, using Chen Fuliang's discussion of composition as well as contemporary criticism of Chen's work as his guide. Zhu reads Chen's expositions as examples of standardization rather than as conscious manipulations of contemporary standards.

73. *LXSC*, 7.55b–56a.

申公之始見武帝也, 曰: 爲治不在多言, 顧力行何如耳. 公之意蓋有待而後言也.

嗚呼! 可謂老成長慮者矣! 是何也?

Unlike the continuation (*jieti*) in Huang Huai's exposition, Chen's second line was not a paraphrase of the introductory sentence (*poti*). It was instead a direct continuation of the first line. The opening line depicts Shengong on his first meeting with the emperor. The second line presents the uncertainties that crept into Shengong's mind. Chen similarly smoothed over the requirements of the third section (*xiaojiang*) in the introductory part (*maozi*). The standard formulas called for the author to cite the source of the quote and express his intention to explain the introductory statement of his argument typically as a means of concluding the introductory part. The omission of the author's personal voice in the concluding sentence of the introduction once more enhanced the continuity with the preceding line and bridged the transition to the citation (*yuanti*).

The introductory statements of the argument in the initial sentence and the continuation and the standard formulas marking off subdivisions were the result of a process of standardization. This process was galvanized by the scanning practices for which civil examiners had become notorious by the twelfth century. Since examiners were required to read through thousands of examination papers, their decision to consider a paper further depended on a cursory reading of the first subdivisions. Chen Fuliang's cautious modifications of these markers contributed to the dramatic presentation of his political arguments. The continuity of narrative across subdivisions he achieved in lines such as those quoted above focused the reader's attention on the historical case without intrusion of the author's personal voice or standard phrases common to examination writing. At the same time, Chen's narrative integration of standard subdivisions produced a rhythm that kept his readers riveted. Chen's techniques for discussing evidence produced similar effects and brought him similar acclaim.

Convention required that examination candidates demonstrate their mastery of the historical and literary tradition by citing abundant evidence. Candidates typically filled their essays with allusions, laid out in parallel style. Huang Huai illustrated the use of this technique by combining six historical references in three parallel lines:

Su [Qin] [?–317 BCE] and Zhang [Yi] [?–309 BCE] tried their luck by sowing political discord; Sun [Wu] [ca. 6th c. BCE] and Wu [Qi] [440–381 BCE][74] sold their strategic insight by publicizing military strategies; Shen [Buhai] [?–337 BCE][75] and [Shang] Yang [ca. 385–338 BCE][76] furthermore indulged their selfish desires by devising cruel administrative techniques.[77]

蘇張以口舌投其機. 孫吳以爪距媒其權. 申鞅又以刻剝之術逞其慾.

Chen Fuliang equally impressed his critics with his knowledge of the historical record but gained even wider acclaim for the way in which he staged his evidence:

In the past when Fu Yue [ca. 1200 BCE] met Gaozong [r. ca. 1200 BCE], the ruler and the minister enjoyed each other's company. However, Yue never spoke a word that touched upon the present generation. He asked the emperor to accept criticisms and urged him again to follow up on remonstrations. These were broad instructions.

When he spoke on how to lead the many officials, at first he explained it only on the basis of one or two cases.

Then, when he discussed the defilement of the sacrifices, he would have used up to several tens of words, but Gaozong did not wait until the end to praise his exposition. He sighed in praise, "Excellent!"[78]

昔者傅說之遇高宗, 其君臣甚相歡也. 而說未始有一辭及當世者. 命之以納誨, 復之以從諫, 皆大略之説也.

　及進之以率百官則始一二而言之.

　蓋至於黷祀之論, 累數十言而高宗不俟其終篇, 輒勤其説而有旨哉之歎.

The critic cited in *Standards for the Study of the Exposition* exclaimed: "For a single event he writes three sections. This is lively usage." In his essay on Shengong, Chen Fuliang used Fu Yue's service to Emperor Gaozong as evidence for the argument that, when first meeting his sovereign, a minister needs to exercise caution. Fu Yue was living as a recluse when summoned by Wuding 武丁, enshrined as Gaozong, a sovereign of the

---

74. For these dates, see Loewe, *Early Chinese Texts*, 449.

75. For an alternative dating of Shen Buhai's life, see Loewe, *Early Chinese Texts*, 394.

76. Ibid., 368, 394.

77. *LXSC*, 5.92a.

78. Ibid., 7.56b–57a.

Shang house in the early thirteenth century BCE.[79] Chen went beyond the regular practice of suggesting the simile by quoting phrases from the relevant history or classic. He broke "the single affair" down into stages not evident in the source text, in this case *The Book of Documents*.[80] Chen analyzed and staged his evidence in the same way he analyzed and staged the circumstances of the event referred to in the topic, in this example the meeting of Shengong and Emperor Wu discussed above.

Chen Fuliang's emphasis on the reconstruction of historical evidence as the foundation for political theory thus carried with it structural changes in examination writing. Unlike Huang Huai, Chen's popularity did not derive from the careful use of parallelism in the layout of his expositions or from the strict adherence to the requirements of each subdivision and the transition between subdivisions. Rather, his popularity was due to the skillful application of a broad spectrum of rhetorical strategies. At the grammatical level of the text, Chen's expositions utilized patterns that mirrored his efforts to break the strong hold of parallel composition at a structural level. The hallmark of Chen's influential "writing technique" (*wenfa* 文法) was the use of "chain-sentences." The opening sentences of his exposition "Shun and Yu Held Aloof from the Empire Even When They Were in Possession of It" clarify the meaning of this technique:

The reason why they [the sage-kings of Antiquity] did not regard the empire as big is because they had a modest opinion of themselves. Since, beginning with the tiniest affair, everybody is striving to achieve things, often without success, how could it be that *the empire can be obtained without reason! The empire cannot be obtained without reason.* However, *if one obtains it,* this must mean that *one greatly surpasses others in some way.* Pay heed! Only with the personal belief that his person truly *surpasses others in some way* and that he is *able to obtain the empire* will someone, when he *obtains the empire,* cherish it as a personal fortune and secure it as a private possession.[81]

----

79. For the dating of Wu Ding's reign, see Loewe and Shaughnessy, eds., *The Cambridge History of Ancient China,* 25, 181.

80. The dealings between the king and Fu Yue are recorded in the chapter "Yue ming" 説命 ("The Charge to Yue") in the *Book of Documents* (Legge, *The Chinese Classics,* vol. 3, *The Shoo King,* 248–63). The passages Chen Fuliang refers to appear in the first and the second part of the chapter (Legge, 253, 254, 257–58).

81. *LXSC,* 8.83a–86b; Lau, *The Analects,* 94.

論曰: 不見天下之爲大者, 其自視小者也. 夫自一介而上皆人之所役役焉. 求之而弗遇者. 況夫天下而有無故之獲哉! 天下不可以無故得也. 而儻然得之, 則若必有以大過人者. 嗚呼! 吾視吾身誠有以大過於人而能得天下, 則夫得天下者始可挾之以爲喜, 固之以爲私.

In mainstream twelfth-century examination writing, authors set up opposing arguments in the introduction and then worked these arguments out in parallel fashion in the main body of the text. The analysis of the introduction of Huang Huai's winning 1154 exposition in Chapter 2 demonstrates the use of this rhetorical strategy: two lines of argument (responding to circumstances with insight versus interfering with the natural course of things through clever scheming) shaped the layout of the exposition, and each was analyzed in parallel fashion. Chen Fuliang took the reader to his final argument via a course with "twists and turns." In the passage quoted above, he first stated the main argument that the sage-kings of Antiquity were able to hold on to the empire because of their humility. He then seemed to contradict this argument with the observation that nothing in the world can be achieved without conscious effort and that rulership over the empire implies that one excels others. He gave a twist to this observation in the following line by arguing that rulers who believed they excelled others would exploit the empire for their personal gain—the implication is that such rulers would not be able to hold on to the empire. In the words of a contemporary critic, the kind of intricate reasoning performed in this passage produced the "rounded and lively" quality of Chen Fuliang's examination writing.[82]

Although literal repetition was ordinarily read as a lack of sophistication, Chen Fuliang drove the reader toward the climax of his argument on the rhythm of crescendoing repetitions (italicized in the translation). Chen used repetition to present his arguments to great theatrical effect. In the essay on Emperor Cheng and Du Qin discussed in the previous section, he imputed factionalist motivations to Du Qin's advice with the following exclamation: "If he [Du Qin] would have spoken, the emperor would have been enlightened, but [Wang] Feng would have been thrown out. Would Qin have been willing to say this? . . . If he

---

82. *LXSC*, 8.83b.

would have spoken, the emperor would have been enlightened, but Feng would have been thrown out. Would Qin have been willing to say this?"[83] Such rhetorical means gave Chen Fuliang's evaluation of Du Qin's proposals a sense of drama rarely seen in examination expositions.

With regard to word usage, Chen Fuliang mixed classical bookish words with terminology pulled from a colloquial linguistic register. He incorporated terminology considered inappropriate and arrogant by the protectors of the traditional literary standards of his time. Instead of the more humble "I, in my state of ignorance" (*yu* 愚), he frequently inserted colloquial phrases like "in my view" (*wujian* 吾見) and "in my opinion" (*wuyi* 吾意). Such improprieties and their impact on the larger student body were bemoaned by examiners in the 1180s.[84] As subsequent chapters will demonstrate, Chen's efforts to break out of the more conventional rhetorical molds to present his interpretations of history and his views on government aroused both great enthusiasm and great dismay.

---

83. Ibid., 4.54b–55a.

84. *SHY, XJ,* 5.10b. For an example of a memorial criticizing vernacular expressions in examination papers, see Chapter 5.

# 4

---

## *Preparing for the Examinations (ca. 1150–ca. 1200): The "Yongjia" Curriculum*

Commercially printed anthologies featuring the expositions and policy response essays of Chen Fuliang and Ye Shi shaped the perception among their contemporaries that the Yongjia teachers were setting the standard for examination writing. The reputation of the Yongjia teachers was the result of a deliberate effort on these teachers' part. Printers simply capitalized on the high profile the Yongjia teachers had in the examination field. They taught composition, institutional history, administrative reasoning, and the textual analysis of the Classics, philosophers, and the histories as part of a reform program aimed at restoring the authority of the Song court over all its former territories. The reform program called for a retrenchment in government activity domestically, a retrenchment informed by knowledge of the history of court-bureaucracy relations, taxation, financial administration, and military organization.

The Yongjia teachers were not the only educators to develop examination curricula. In nearby Wuzhou 婺州 prefecture, Lü Zuqian acquired comparable fame in examination teaching. Lü's curriculum overlapped to some extent with that of the Yongjia teachers, so much so that Zhu Xi, a contemporary critic, labeled his efforts "Yongjia" work. This chapter analyzes manuals attributed to the Yongjia teachers and Lü Zuqian and attempts to reconstruct the curricula for which hundreds of students enrolled at a time. It highlights the similarities in teaching practice among examination teachers in Liang Zhe East

Circuit, which comprised both Wenzhou (Yongjia) and Wuzhou. The curricula developed in Wenzhou and Wuzhou in the 1160s and 1170s played a major role in the definition of the standards for examination success in the late twelfth century.

This chapter also draws attention to the different emphases in examination teaching in Wenzhou and Wuzhou. Despite similarities, diverging visions of reform inspired different analytical approaches and a different assessment of the role of moral philosophy in history and government. Discussion of the similarities and differences in the curricular activities of the teachers of East Zhe illustrates two aspects of the twelfth-century examination field: the appearance of local teachers as authorities in the shaping of examination standards and their participation in examination preparation as an arena in which intellectual formations debated standards not only for the examinations but for elite status more generally.

## History and Administrative Reasoning

The eastern prefectures of Liang Zhe built a reputation for examination success in the twelfth century. As noted above, Wenzhou prefecture produced the second highest number of *jinshi* degree–holders in the Southern Song territories. Wuzhou enjoyed comparable success. Peter Bol estimates that nine or ten candidates from Wuzhou obtained the final degree during each triennial examination after 1150.[1] The competitiveness of both prefectures is evident in a decree issued in 1156 setting the ratio of graduates to candidates at 1/100 for all prefectures except the prefectures of Wenzhou, Wuzhou, and neighboring Taizhou, for which the quota was set at 1/200.[2] The examination success and the competitiveness of East Zhe were linked to a vibrant examination culture. Hundreds of examination candidates gathered at the schools of teachers like Lü Zuqian and Chen Fuliang. Many more bought or copied the manuals they and their pupils compiled.

An emphasis on institutional history and administrative thinking was the hallmark of examination teaching in Wenzhou and Wuzhou. Con-

---

1. Bol, "Neo-Confucianism and Local Society," 257.
2. Chaffee, *The Thorny Gates*, 125, 155.

temporaries associated this aspect of East Zhe teaching with utilitarian statecraft (*shigong* 事功 / *gongli* 功利). Yongjia scholars and those associated with them studied history, the history of governmental regulations and institutions in particular, to assess the applicability of policies and institutions to present conditions. Policy response manuals published in East Zhe attest to the interest in governmental affairs and the historical record, as well as the result-oriented approach to historical study. In the first subsection below, I discuss the interpretation of history and the administrative reasoning taught in East Zhe based on two extant policy response manuals, *The Yongjia Master's "To the Point in All Cases"* (*Yongjia xiansheng bamian feng* 永嘉先生八面鋒) and *Detailed Explanations of Institutions Throughout the Ages* (*Lidai zhidu xiangshuo* 歷代制度詳説).

The impact of these and similar manuals was felt outside the East Zhe region from the 1160s on. Zhu Xi, a contemporary critic of the utilitarian bent of "Yongjia" examination preparation, noted in 1173 that one of Lü Zuqian's manuals was available in print in Jianyang, a relatively marginal area in Fujian known for its printing activity.[3] He later commented that from the 1160s to the 1180s examination writing was dominated by the utilitarian discourse emanating from the teachers of East Zhe.[4] The following discussion of policy response manuals reconstructs the training of examination candidates following the "Yongjia" curriculum and demonstrates that the "Yongjia" teachers' professional involvement in examination preparation laid the foundation for their authority in the examination field (Chapter 3).

## Administrative Reasoning: *The Yongjia Master's "To the Point in All Cases"*

*The Yongjia Master's "To the Point in All Cases"* illustrates a central feature of "Yongjia" utilitarian discourse, namely, the computation of profit and damage (*ji libing* 計利病). The goal of an administrator facing any kind of administrative challenge was, according to the "Yongjia" model, to achieve maximum benefit for the state and for the people. The tactic for achieving maximum benefit was to analyze all factors that would

---

3. Ichiki, *Shu Ki monjin*, 306.
4. *ZZYL*, 109.2701.

impact the outcome of an administrative intervention, favorably or adversely. The manual immersed students in a model of reasoning based on the conviction that those in government constantly confront circumstances to which they must respond by intervening or not intervening. Administrators navigate a natural and administrative web. Natural conditions and social circumstances are subject to unpredictable change; administrators have to monitor such changes and assess their impact. They have to relate potential policy initiatives to the larger political context. The import of the matter at hand has to be weighed against other matters, and the consequences of a potential policy initiative must be considered meticulously. Political action is necessary but it is legitimate only if the benefits of the initiative outweigh the potential damages:

In the subcelestial realm there has never been such a thing as 100 percent profit. If you are waiting for full profit in affairs, then it will always be impossible. When the sage deals with affairs, he does not dare act if the ratio of benefit to harm is one to ten. If the ratio is ten to one, he will surely act. If profit and damage are equal, he will not dare act. Because, if the damage is the same, even though you get some profit, the damage will cancel it out. Then it is more profitable to be content with nonintervention.[5]

天下未嘗有百全之利也. 舉事而待其百全, 則亦無時而可矣. 聖人之舉事也, 利一而害十, 有所不忍爲. 利十而害一, 當有所必爲. 利害之相當, 有所不能爲. 以其害之相當, 雖得其利, 而其爲害亦足以償矣. 不若安於無事之爲愈也.

*To the Point in All Cases* taught students methods of presenting administrative proposals founded on a risk-benefit analysis. The calculation did not involve statistical models; rather, it was based on the application of principles of administrative reasoning. To help students navigate the web of factors to be taken into account in policy making, the "Yongjia" teachers developed administrative principles. At the core of *To the Point in All Cases*, and at the core of "Yongjia" examination teaching more generally, lay a set of guidelines.

The collection is organized on the basis of these principles. It contains policy response essays, for the most part extracts from such essays, organized by title. The titles are short (seven to nine characters long)

---

5. *BMF*, 1.2.

and convey administrative tenets. Each title functions as a rubric drawing the students' attention to the gist of the essays included under it. Taken as a whole, the ninety-three rubrics provided a comprehensive reservoir of administrative strategies that could be applied to policy questions in examinations and in the official world. Qing scholars noted that the organization of the text, under the main rubrics and in further subdivisions, attested to its use in examination preparation.[6] The structuring of the rubrics as sentences further suggests that they were composed to facilitate memorization. The titles are brief and conform in length; the cadence breaks them up into two, often parallel clauses (see the example quoted below).

As in the example quoted below, the title/administrative principle is followed by a list of subthemes in the majority of the entries. The subthemes marked the areas of administration to which the tenet could be applied:

"When bringing about major profit, don't count in minor damage."

興大利者, 不計小害.

*supernumerary officials; supernumerary troops; imperial sacrifices; receiving taxes; practicing archery; mobilizing troops; water conservation; local militia*[7]

冗官 冗兵 郊賞 入粟 習射 用兵 水利 民兵

In this example, the compiler suggested that the general admonition not to worry about trifles when planning major programs could be applied in answering questions about supernumerary officials, supernumerary troops, imperial sacrifices, etc. The essays and extracts presented under a rubric do not necessarily illustrate the principle's application to all the areas listed. In the above example, the essays applied the principle of discounting minor risks and damages to military mobilization and the use of financial resources only. The manual thus provided students with a set of general administrative principles and trained them in the application of these principles to all areas of government responsibility; the

---

6. The abstract in the front matter of the SKQS edition mentions that there are ninety-three categories, which accords with my count. The entry in *Siku zongmu tiyao* (*SKTY*, 26.2798–99), revised by Ji Yun, gives eighty-eight for the total number of categories.

7. *BMF*, 1.2.

goal was evidently not to supply ready-made essays for memorization and rote reproduction.

The political and intellectual message embedded in the principles and essays included in *To the Point in All Cases* echoed the reform package articulated in the policy essay collections of Chen Fuliang and Ye Shi discussed in the previous chapter. The "Yongjia" teacher whose course in administrative thinking and policy response writing resulted in the compilation of this manual envisioned an articulate class of scholars who advocated fiscal conservatism, retrenchment at the level of the central government, and the delegation of decision-making power to local authorities.

One matter of great import in *To the Point in All Cases* is the assumption that students should engage in political debate. The "Yongjia" scholars' position on the political participation of non-officeholders was rooted in their belief in shared governance and the emergence of a political subjectivity among the scholar-official class. According to Yü Yingshi, from the early eleventh century on, scholar-officials ceased to see themselves as mere recipients of imperial permission to assume the task of government. Governing the empire not only was a matter of personal responsibility for emblematic statesmen like Fan Zhongyan but also had become part of the collective identity of the scholar-official class. As a class, rather than as individuals, they regarded the governance of the empire as a responsibility they shared with the emperor. Implicit in this was the duty to set standards for participating in the administration of the realm.[8] Although I agree with Yü Yingshi that this kind of political subjectivity inspired the lives and careers of broad sections of the scholar-official class, the participation of non-officeholders in policy debates was controversial and merits further analysis.

Many among the office-holding elite perceived criticism of court policy in local government and private schools as an infringement on the prerogatives of the Censorate, the government agency that theoretically monopolized the evaluation of government policies and officials. The administrative philosophy articulated in *To the Point in All Cases*

---

8. Yü Yingshi, *Zhu Xi de lishi shijie*, vol. 1, esp. chap. 3.

outlined a governmental structure in which scholars, including those who did not hold official titles, were assigned specific political roles:

As for the state of the realm, it is not at times of disorder that we fear; it is during times of order that we fear. It is not at times of upheaval that we worry; it is during times of peace that we worry. It is the task of those who are scholars to engage in learning by discussing together techniques to maintain peace. It is the task of those who are high officials to prove their loyalty by proposing together techniques to maintain order. It is the task of the sovereign to govern by unceasingly searching out techniques to maintain order.[9]

天下非未治之可畏; 已治之可畏也. 非未安之可憂; 已安之可憂也. 方天下之未治未安, 為士者相與講治安之術而為學; 為公卿大夫者相與進治安之術而為忠; 為人主者則又日夜求治安之術而為政.

This passage was a comment on the principle "Bringing about order is not difficult; maintaining order is difficult." It interpreted the principle as justifying the participation of scholars in maintaining order at the local as well as at the central level. According to the governmental structure outlined in this passage and applied throughout the manual, the political hierarchy consisted of three functionally distinct levels in which the higher strata fed on information provided by the lower level(s). The emperor stood at the top of this hierarchy and, theoretically, made the final decisions. Policy response essays were, originally and in principle, directly addressed to him.[10] The sovereign's role was ambivalent, however, and carried strong symbolic overtones. In Yongjia political theory, the emperor maintained and represented unity; he "tied people's hearts." To this end, he had to avoid entanglement in the details of the administration of the empire.[11] On the other hand, the maintenance of unity depended on the emperor's judicious selection of top advisers and broad consultation of opinion. He selected the grand councilors, but it was up to them to lead the bureaucratic apparatus and

---

9. *BMF*, 7.56.

10. The earliest models for Song literati were Han intellectuals' long essays in response to their emperors' policy questions. Song policy questions typically contain a formulaic closing in which the examination officials exhorted students to come up with practical proposals with the promise that their essays would be forwarded to the emperor for ratification.

11. *BMF*, 4.29, 7.51, 13.99.

make policy proposals.[12] The centrality of the broad consultation of opinion in "Yongjia" political theory manifested itself in the description of ministerial power. The grand councilors and other high-ranking court officials were supposed to base their policy proposals on the discussions held among scholars and officials.

In "Yongjia" texts, the misgivings of contemporaries about the profusion of argumentative writing (*yilun* 議論) among local elites were dismissed with the argument that detailed discussions of "profit and damage" and "right and wrong" laid the foundations for decision making at higher levels.[13] Diversity of opinion was considered an asset and in line with the administrative practice of Antiquity. *To the Point in All Cases* suggested that the sage-rulers of Antiquity let people fully express their minds, for an intelligent examination of different ideas helped them decide on the appropriate course of action: "As more words are presented, the proper principle becomes clearer, and [the sovereign's] insight becomes more judicious."[14] The diversity of opinion voiced in the policy response essay and related genres and submitted by local officials and non-officeholding elites was thus presented as the raw material for policy making at higher levels. Yet, in a typical move against inadvertent miscalculations, the manual also warned scholars against overestimating their capacities: "Of all the afflictions in the realm, none is worse than not measuring the extent of the power of your learning and applying it carelessly."[15]

This undertone of caution surfaces elsewhere in the manual. *To the Point in All Cases* expressed the "Yongjia" scholars' effort to engage examination students in the discussion of policy, which they defined as the discussion of the benefits and risks. It explicitly warned against the development of a specific action program. Instead, the manual advanced the proposition that it is better not to announce new measures in order to avoid provoking reactions by others. To put forward a pro-

---

12. *BMF*, 12.90.

13. The author regarded politicians from the Qingli 慶曆 (1041–48) era as models for the open discussion of administrative measures. For a different point of view on this matter, see the thirteenth-century manual by Liu Dake, *Bishui*, 22.1a–10b.

14. *BMF*, 6.42.

15. *BMF*, 8.63.

gram was like "setting up a target for others to shoot at." The elaboration on the principle "Those who are good at bringing about profit (*li*) only eliminate bad side effects (*hai*)" offered an explicit argument against campaigning for new regulations.[16]

This principle of caution can be read both as an attempt to avoid partisanship in the bureaucracy and among the scholarly elite and as a shift away from uniform central solutions to local problems. The warning not to set up programs because they became targets for others matched the mediating role taken on by Yongjia scholars in the factional struggles concerning the Learning of the Way later in the 1180s and 1190s. The example used to illustrate the dangers of articulating new measures implemented empire-wide were the New Policies of Wang Anshi. Despite occasional support for some of Wang Anshi's innovations in Yongjia, his attempts to institute loans to farmers (the "Green Sprouts" program) and his educational reforms were dismissed because these projects caused new problems without solving existing ones.[17] They had also led to intense factional struggles, which shaped the political history of the late eleventh and early twelfth centuries. The calculation of benefits and damages oriented students' attention away from the development of comprehensive central policy programs to the formulation of local solutions to local problems based on an assessment of local conditions.

The manual's analysis of Southern Song security problems exemplifies the preference for local and customized solutions. As a result of the tensions between the Jin and the Song courts, local unrest and vagrancy had become persistent problems in the Liang Huai 兩淮 region, the area forming the eastern border between the Jin and Song territories in the twelfth and thirteenth centuries. Renewed fighting along the border in the early 1160s had led to death and migration among the local population and had given rise to local banditry. Illustrating the principle "To get something done, one does not need a name for it," one of the

---

16. *BMF*, 9.70.

17. Ibid. For a discussion of Chen Fuliang's critique of Wang Anshi's reforms as well as the reaction of other scholars in East Zhe, see Li Huarui, *Wang Anshi bianfa yanjiu shi*, 49–64.

sample essays argued that it was politically more advantageous to settle the problem locally.

The essay proposed that a combination of ad hoc measures would solve the problems of local unrest and vagrancy. The author suggested co-opting local strongmen to pacify the bandits, redistributing available land, and inviting poor people from other areas to repopulate the area.[18] He preferred such specific policies over grander efforts to develop integrated defense plans for all border regions. The Yongjia teachers generally opposed the stationing of more government troops along the borders and advocated less cost-intensive defense programs, such as the use of local militia and military colonies. The author of *To the Point in All Cases* likewise applauded the cost-effectiveness of local militia and self-sufficient military colonies but argued that in the current political climate explicit support for them would automatically generate opposition. Factoring in political sentiments as well as fiscal resources, the author concluded that a series of small measures was more likely to generate maximum benefit for the state as well as the local population. In his estimation, his program of enlisting local strongmen in policing and re-opening the land through land distribution among the remaining population and poor peasants from other areas would result in benefits similar to those of establishing militias and colonies, but because the sensitive issue of local militias would not be raised in court politics, the ad hoc measures had a better chance of success.

The priority assigned to calculating benefits and to the effectiveness of policy proposals makes it difficult to pin down to a specific "Yongjia" political agenda. As the above example illustrates, flexibility was central to "Yongjia" administrative reasoning. Even so, in the second half of the twelfth century the "Yongjia" teachers consistently advocated less state intervention. The proposed solution to the vagrancy problem in the Liang Huai region is indicative of the preference for a smaller army and the recruitment of local resources to solve local problems. The advocacy of fiscal responsibility and even of fiscal conservatism in all policy making was a second feature of the "Yongjia" political agenda, one that was also central to the policy essay collections of Chen Fuliang and Ye Shi. According to "Yongjia" administrative philosophy,

18. *BMF*, 10.75–76.

a cautious government should measure its own resources against the estimated cost of future ventures. In "When bringing about major profit, don't count in minor damage," the author warned against Emperor Xiaozong's 孝宗 (r. 1163–89) efforts to recapture the north because of fiscal reasons. He argued that when danger is too close at hand, as in the contemporary threat of the Jin armies, a government cannot afford to sacrifice its last resources.[19]

*To the Point in All Cases* not only shared the political stance and the interest in principles of government of Chen Fuliang and Ye Shi, the two principal representatives of Yongjia examination success in the 1170s, but also taught the methods of argumentation applied in their examination writing. As shown in the analysis of Chen Fuliang's expositions in Chapter 3, the dramatization of historical evidence and the revitalized Ancient Prose style exercised a great appeal. The essays anthologized in *To the Point in All Cases* applied the same methods to the elaboration of principles of government. I discuss the use of historical evidence below and the teaching of Ancient Prose in the last section in this chapter.

The author made ample use of historical evidence to illustrate general principles of administration, positively or negatively, drawing most of his evidence from the Warring States period and the Han, Tang, and Northern Song dynasties. The Zhou dynasty figured prominently in "Yongjia" texts on institutional history, but the idealized picture of Zhou government in classical sources made Zhou rulers less suitable for discussions of historical contingency and the algebra of good government than the richer and far more cautionary exempla of Warring States, Han, Tang, and Northern Song rulers and ministers.[20] The manual taught students to contextualize history by speculating about the motivations of historical actors and the consequences of their acts.

The contextualization of historical evidence was the first step in the computation of the benefits and risks of governmental interventions. In "When bringing about major profit, don't count in minor damage," the

---

19. *BMF*, 1.3.

20. On the grounds for Chen Liang's interest in the Han and Tang dynasties, see Tillman, *Utilitarian Confucianism*, 134–43, and *passim*; and idem, *Confucian Discourse*, 168–86.

author analyzed an early third-century BCE deal between Liu Bang 劉邦 (ca. 256–195 BCE) and Chen Ping 陳平 (?–178 BCE). Liu Bang provided Chen Ping with 40,000 pounds of gold to be used for bribing advisers to Xiang Yu 項羽 (232–202 BCE), Liu Bang's main rival. The amount seemed inordinate. However, this transaction paved the way for Liu Bang's ultimate success.[21] In the author's reading, the bribe effected the alienation between Xiang Yu and his former ally Fan Zeng 范增 (277–204 BCE). The author's reinterpretation illustrated the principle that a high investment is justified if it results in even greater benefits. "Yongjia" administrative thinking demonstrated little interest in the moral evaluation of historical events and actors; past administrative behavior was judged on the basis of its net effects on the body politic and the welfare of the people. Liu Bang's bribe, in the long run, led to the end of civil war and the establishment of the Han dynasty.

In contrast, the author read Han Emperor Wu's military expenditures as a case in which the benefits did not justify the action. Emperor Wu captured many prisoners of war and killed many enemies in his northern and western expeditions, but these victories were canceled out by the huge losses in men and resources. The Han forces expanded Han territory in Central Asia, but these gains were temporary. The author was not averse to military action per se; he considered Emperor Wu's military campaigns ill advised because they were of little benefit to the state and the people.

Because of the similarities between *To the Point in All Cases* and the work of Chen Fuliang and Ye Shi in terms of political stance, scholarship, and argumentation, it has been attributed to both authors. Apart from their intellectual and political endeavors, Chen and Ye allegedly shared the epitaph "Master of Yongjia." The manual was most commonly attributed to Chen Fuliang and is still listed as his work in recent scholarship. According to Zhang Yi 張益 (1395–1449), a fifteenth-century editor, the copy he used for his edition had at one point contained a preface by Chen Fuliang. He interpreted the existence of this preface as evidence that Ye Shi must have been the author, since authors only wrote "introductions by the author" (*zi xu* 自序) and not

---

21. Sima Qian, *Shi ji*, 8.373; Loewe, *A Biographical Dictionary*, 35.

prefaces (*xu* 序).[22] In 1778 one of the editors on the Siku quanshu project wrote the most extensive bibliographic review of *The Yongjia Master's "To the Point in All Cases"* to date. Even though Chen Fuliang is listed as the author in the series, the editor expressed doubt about the attribution. I agree with him that there is no evidence that links Chen Fuliang, or Ye Shi, directly to this manual. From internal evidence, it is clear that the manual dates to the Southern Song period. However, the reference to the Master of Yongjia does not mean that the work can be attributed to the most famous representatives of Yongjia examination teaching. The reference to the Master of Yongjia was in this case probably a sales gimmick. It indicates that "Yongjia" stood for more than the teaching and work of one or two particular teachers. "Yongjia" connoted a style of scholarship that included historical analysis and administrative reasoning, Ancient Prose composition, a political stance and a style of scholarship that were shared by Chen Fuliang and Ye Shi but not limited to them. This manual was part of the larger output of examination materials in Yongjia and East Zhe and attests to the authority of "Yongjia" scholarship in the examination field.

## INSTITUTIONAL HISTORY: *DETAILED EXPLANATIONS OF INSTITUTIONS THROUGHOUT THE AGES*

Yongjia was well known for its scholarship on institutional history. Twelfth-century teachers from the area wrote commentaries or studies on *The Rites of Zhou*, a classical text outlining the institutions that supposedly governed the Chinese territories in early Zhou times. Some wrote monographs on the history of institutions. Xue Jixuan, for example, published a study of the military in the Han dynasty. Chen Fuliang, his disciple, completed an overview of military institutions from the Zhou through the Song dynasties. Institutional history was a major

---

22. This preface, as well as Du Mu's cited in the Appendix, can be found in an 1844 Japanese edition preserved at Shanghai Library. It also includes a postface by Chen Chun 陳春 written on the occasion of a reprint in 1819. For a more extensive discussion of dating and editions, see the Appendix. Chen Zhenbo ("*Yongjia xiansheng Bamian feng xin tanxi*") makes a case that *To the Point in All Cases* was written by Chen Fuliang after he obtained the *jinshi* degree in 1172 and returned home to teach. The evidence presented is inconclusive.

focus in examination teaching not only in Yongjia but also in other counties and prefectures in East Zhe. In the last decades of the twelfth century and the first half of the thirteenth century, teachers and students from the East Zhe region produced a host of encyclopedias collecting vast amounts of information on institutions.[23]

Institutional history was a particularly relevant subject for students preparing for the policy response session. Policy questions often included long lists of subquestions asking examinees to explain the rationale behind the establishment of institutions and regulations and to evaluate their use in different periods. *Detailed Explanations of Institutions Throughout the Ages* offers a rare glimpse into the world of institutional history instruction in East Zhe.

The manual has traditionally been attributed to Lü Zuqian. Lü hailed from Jinhua 金華 (Wuzhou prefecture) and became a supporter of Chen Fuliang's work when Chen entered the Imperial College in 1170. Many have questioned the attribution because the title does not appear on established lists of Lü's publications. The fact that publishing houses utilized Lü Zuqian's name for examination compilations because of his popularity as a teacher is another factor that makes the attribution suspicious. However, excerpts from *Detailed Explanations of Institutions Throughout the Ages* were attributed to Lü Zuqian in at least two other early thirteenth-century encyclopedias.[24] This evidence, combined with the compatibility between this manual and Lü's other work, suggests to me that *Detailed Explanations of Institutions* was based on lectures given by Lü, which were collected and printed after his death in 1181.[25] Given the complex origins of the text, I will refer to the speaker in the text as "the lecturer" rather than "the author" or "Lü Zuqian." Even though the attributions are questionable in several respects, I also refer to these and other manuals attributed to Lü Zuqian as "Lü Zuqian's" examination

---

23. Bol, "Zhang Ruyu, the *Qunshu kaosuo*, and Diversity in Intellectual Culture"; Zhou Mengjiang, *Ye Shi yu Yongjia xuepai*, 304–5; Song Jaeyoon, "Tensions and Balance."

24. For instance, *Yuanliu zhilun*, xuji, 2.9b, citing a passage found in *Xiangshuo*, 2.9b; 4.1a–5b, citing 6.2a–3a; and 4.16, citing 5.5a–6b; *Bishui*, 79.4, citing 4.8a–b. For more information on both of these encyclopedias, see Chapter 6 and the Appendix.

25. See the Appendix for a more detailed discussion of editions, authorship, and dating.

teaching because contemporaries associated them with Lü Zuqian's pedagogy and because they share characteristics with practices credited to him (Appendix A).[26]

The fifteen-chapter manual covers governmental institutions and regulations in a wide range of areas: the civil service examinations, schools, taxation, grain transport, the salt monopoly, the monopoly on liquor, currency, famine relief, land distribution, military colonies, the military, the horse administration, personnel evaluation, the dynastic house, and sacrifices. The pedagogical goals of this private encyclopedia are also reflected in the organization of individual entries.[27] Each chapter/entry is divided into two parts: quotations from primary sources and explanations. The first section quotes primary source materials on the topic, usually arranged by subheading and in chronological order. The explanatory part reviews trends in the history of the institution concerned. This part presents an overview of the history of the institution and raises problems related to the regulations developed to organize the institution, past and present.

The organization of *Explanations of Institutions* may reflect the structure of classes on institutional history in East Zhe schools. The section on primary sources relates to the secondary text much as does a handout to a lecture. In the explanatory part, the author/lecturer explicitly refers to passages in the primary-source texts. Those references are typically rendered in classical Chinese that incorporates colloquial expressions. For example, passages such as "When we investigate in detail the paragraph mentioned above . . ." or "When you look just at the character *bin* (guest) . . ." evoke the lecturer inviting students to reflect on a passage just read.[28] The lecturer also occasionally provides paraphrases of expressions in the primary sources that may have been hard to comprehend upon a first reading.[29] This suggests that students were presented with the texts of primary source materials forming the basis

---

26. See also De Weerdt, "Content and Composition."

27. I contrast the features of *Detailed Explanations* as a private encyclopedia with those of imperial encyclopedic projects in "The Encyclopedia as Textbook."

28. *Xiangshuo*, 1.5a. I relate the separation of primary sources from interpretive essays to the emergence of an archival mentality in "The Encyclopedia as Textbook."

29. *Xiangshuo*, 1.6b.

for the lecture. Class notes, including the primary source materials and the associated lectures, were marketable items in the twelfth century, especially if they recorded the teaching of high-profile teachers like Lü Zuqian.

The classes and the resulting manuals introduced students to the main topics, sources, and methodologies in institutional history. The chapter titles in *Detailed Explanations*, as in other manuals discussed in Chapter 7, provided an index to the main topics discussed in class; these topics were also the main issues brought up in policy questions. In *Detailed Explanations* the lecturer exposed students to the main primary sources for studying the history of educational, economic, military, and political institutions. Under each section, the student found a chronological list of quotations drawn occasionally from the Classics and the philosophers but mostly from the dynastic histories, Du You's 杜佑 (735–812) *The Comprehensive Statutes* (*Tongdian* 通典; 801), Sima Guang's *Comprehensive Mirror for Aid in Government* (*Zizhi tongjian* 資治通鑒; 1084), and, for more recent developments, Song official sources such as the *Collected Essentials* (*Huiyao* 會要).

In the lectures, the teacher taught the students the methodology of institutional history. For the lecturer in *Detailed Explanations*, a comprehensive and chronological survey of institutions in each of the areas covered was the basis for policy writing. He brought materials from different sources and different periods together to sketch the broad lines of historical development. In the context of the chronological development of institutions, students then learned to compare the advantages and disadvantages of various institutions and policies in relation to contemporary problems. The selection of the primary sources, arranged in chronological order and covering all of Chinese dynastic history up until 1180 or so, conveyed the importance the teacher attached to chronology. This approach was also reflected in the lectures accompanying the primary source selections. The lectures usually began with a chronological description of the institutions relevant to the main subject area, covering all periods of Chinese history.

The priority assigned the comprehensive knowledge of the whole past in *Detailed Explanations* fits in with Lü Zuqian's philosophy of teaching history and his own historiography. "As for history, you have to read from Master Zuo [*Zuo's Commentary*, *Zuozhuan* 左傳; ca. 4th

c. BCE] to the history of the Five Dynasties. If you read them in sequence, you will gain a profound understanding of [the connections between] what came before and what happened after, between the beginning and the end."[30] His *Chronicle of Major Events (Dashi ji* 大事記) is a chronological survey of the period from 481 to 90 BCE. He aimed to extend the work to the period of the Five Dynasties and the founding of the Song, but illness prevented him from producing his own *Comprehensive Mirror for Aid in Government,* the historical survey of Chinese history composed by Sima Guang in the eleventh century.[31] In other historical publications, Lü Zuqian also emphasized the study of primary source materials from the past through the present. To this end, he compiled two anthologies of memorials, one covering the period up to the Song, *Memorials Throughout the Ages (Lidai zouyi* 歷代奏議) and one covering the Song dynasty, *Memorials of Famous Officials of Our Reigning Dynasty (Guochao mingchen zouyi* 國朝名臣奏議).

In *Detailed Explanations,* the primary source materials and the chronological survey of the major developments in the history of particular institutions prepared the stage for a discussion of current problems. History was used to evaluate current policy and to formulate alternative solutions. The lecturer aimed to demonstrate that an understanding of long-term developments was required to solve current policy questions. In the lecture on the history of the organization of the military, for example, he argued that current military and financial problems derived from trends that began under the late Tang and Five Dynasties. In his view, the abolition of the state militia system in 749 and the rise of military governors after the An Lushan rebellion marked a nadir in the history of military institutions, a low point from which subsequent ruling houses had never fully recovered. The military governors maintained large professional armies over an extended period of time, a fact unprecedented in Chinese military history. The governors' scramble to collect taxes to maintain their armies led to the demise of the Tang. The Five Dynasties rose and fell for the same reason.

---

30. Cited in Pan Fuen and Xu Yuqing, *Lü Zuqian pingzhuan,* 435. For the dating of *Zuo's Commentary,* see Schaberg, *A Patterned Past,* appendix, 315–24; and Loewe, ed., *Early Chinese Texts,* 67–72. I discuss this work below.

31. Hervouet, *A Sung Bibliography,* 76; Tillman, *Confucian Discourse,* 99.

During the first decades of Song reign, the lecturer continued, the trend toward increased militarization had temporarily halted. Emperor Taizu 太祖 (r. 960–76) strictly regulated the chain of military command and established peace over a vast empire with an army, the lecturer estimated, numbering only 150,000 men. However, since the reign of Taizu's successor, officials had been using the excuse of foreign threat and internal rebellion to recruit more and more soldiers. At the time, around 1180, the lecturer estimated the total size of the army at 800,000.[32] The drastic increase was not warranted in his view, since there had been no military confrontations in recent years. The expense of maintaining such a vast army lay at the heart of the financial and bureaucratic malaise of the Song government.

The trend toward increased militarization needed to be countered, the lecturer argued, rather than reinforced. His account of the reasons for the military and financial problems of the Tang and later dynasties argued against contemporary proposals. Like their Tang and Five Dynasties predecessors, Song officials, blinded by the cumulative effect of militarization, continued to extort ever more resources from the population to fund the military. The lecturer argued against proposals to mobilize more troops. Instead, he recommended the incremental realization of three goals: no more recruiting for the next few years, better training of existing forces, and the restoration of militias. These three measures coincided with the stance of the Yongjia teachers on the state budget. As shown in Chapter 3 and the earlier part of this chapter, they, too, advocated a halt to increases in the size of the Song armies and the implementation of less cost-intensive solutions such as local militias.

The interest in institutional history and the advocacy of a smaller army and state budget are indicative of the broad similarities in political stance and scholarship between the Yongjia teachers and other teachers from East Zhe like Lü Zuqian. Zhu Xi, who monitored Lü Zuqian's activity in examination teaching, reprimanded him for contributing to the authority of "Yongjia" scholarship: "Your explanations of Master

---

32. Lü Zuqian's numbers are on the lower end. The numbers cited in the "Treatise on Military Affairs" in *The Dynastic History of the Song* (*SS*, 187.4576) are slightly higher, as are Chen Fuliang's estimates in *Lidai bingzhi*; cf. Chapter 3.

Zuo[33] and the memorials are all about the benefit and risk in it for current affairs; these are not urgent matters that speak to the scholars' personal morality."[34] In the minds of his contemporaries, Lü Zuqian's historical studies and guides were closely associated with the risk-benefit analysis central to "Yongjia" teaching.

There were also marked differences between the manuals originating in Yongjia and those from other East Zhe prefectures. The differences discussed below underscore the lack of uniform standards in examination curricula and the centrality of local teachers in shaping examination preparation. Lü Zuqian, in his examination teaching, integrated the moral philosophy of the Cheng brothers with the result-oriented administrative thinking characteristic of the "Yongjia" teachers. He believed in the universal value of the moral standards attributed to the sage-rulers of Antiquity. The chronological surveys in *Detailed Explanations* started with and always referred back to the ideal institutions and practices of the period of the Three Dynasties. The moral philosophy of the Cheng brothers offered the explanation for the question why later institutions and regulations failed to match the ideal. After the reign of the sage-rulers had ended, the pursuit of selfish desires replaced benevolent government.

The chronological surveys in *Detailed Explanations* were narratives of decline. The ideal systems were located in the period of the Three Dynasties. The lecturer's account of the development of military organization, for example, traced a path of gradual decline after Antiquity and marked off climaxes in degeneration and moments of restoration. Military organization was originally part of the process of civilization, and it involved all members of society. The sages instituted civilization among an uncivilized people. They taught the people to till the land to feed themselves and to build houses. When the need for self-defense arose, they taught the people to be soldiers. The lecturer pointed out that it was not the case, as some had claimed, that the rulers of the Three Dynasties "lodged soldiers among the people," that is, that they estab-

---

33. Referring to *Zuoshi bo yi* 左氏博議, a manual published by Lü in 1168. This manual is discussed below.

34. Zhu Xi, *Zhu Xi ji*, 35.1535; cited in Lin Sufen, "Lü Zuqian de cizhang zhi xue," 148, 154.

lished a professional army. In his interpretation of the historical record, in Antiquity everyone became a soldier when the need for self-defense arose.

The first step in the decline of the military occurred following the rule of sage-kings. The lecturer attributed the decline and transformation of the armies to the selfish motivations of political leaders from the late Zhou period on. The Zhou feudal lords mobilized their people not to protect them but to pursue selfish interests. The rulers of the Qin, Han, Jin, and the Northern and Southern Dynasties constantly engaged in warfare for the same reason, robbing their people's possessions and sending them into death. The lecturer noted a brief period of respite during the Tang dynasty. Under the state militia system established during the Tang period, a fixed number of males in each administrative unit were conscripted but exempted from taxes while registered as conscripts. Although this system did not come near the ideal of universal mobilization for self-protection in Antiquity, it was, according to the lecturer, a model of a kind. With no sage-rulers on the throne, people were compelled to fight for reasons other than legitimate self-defense. The state militia system had the double advantage of producing a regular force without placing too heavy a burden on the population. After the An Lushan rebellion, the history of military organization progressed along its course of decline, with a momentum that had only increased in the twelfth century. As recounted earlier, the lecturer read the increasing size of the Song standing army as an unmistakable sign of deterioration.

Moral philosophy plays a secondary role in *Detailed Explanations*. It explains the gap between the ideal government of Antiquity and that of later ages, but the lecturer did not use it to solve administrative questions. The lectures focused on the transformations of institutions after the Three Dynasties and the contemporary problems that resulted from the long-term effects of institutional changes. The lecturer taught that a comprehensive understanding of historical development, rather than moral philosophy per se, was the best way to gradually move back toward ancient ideals. The measures proposed to curb the trend toward increased militarization in the twelfth century (manpower cuts, improved training, and the establishment of militia) illustrate the conviction that an understanding of institutional history and policies suited to the needs of the time formed the basis for rehabilitating the government.

Moral philosophy played a central role in other examination manuals attributed to Lü Zuqian. *Extensive Deliberations on Master Zuo* (*Zuoshi bo yi* 左氏博議; 1168) is a collection of essays on *Zuo's Commentary*, a collection of historical narratives and commentary related to *The Spring and Autumn Annals* and covering the period from 722 to 479 BCE.[35] The collection contains model essays used in school examinations. In these essays, Lü Zuqian explained and evaluated historical events related in *Zuo's Commentary* on the basis of ethical principles deriving from the philosophy of the mind developed by the Cheng brothers in the eleventh century.[36] He explicitly rejected purely situational explanations of the actions of rulers, ministers, and their associates and tried to uncover and evaluate the workings of the moral and the immoral mind in events.

Lü's efforts to integrate the moral philosophy of the Cheng brothers in his examination teaching were not appreciated by one of the main interpreters of their work. Zhu Xi was not pleased with Lü Zuqian's work on *Zuo's Commentary*. He read Lü's attempt to read moral principles back into the stories as a misguided *tour de force*. To Zhu, Master Zuo was an opportunist with no inkling of moral principle who was always mistaken about the obligations of human relationships.[37] He regarded Lü's explanations of the stories related in *Zuo's Commentary* as too intricate.[38] Moral principles should be readily intelligible. Zhu Xi admonished Lü to focus on the Classics, *Mencius*, and *The Analects* in his teaching. These books would provide students with the moral knowledge necessary to tackle historical questions. In Lü Zuqian's mind, in contrast, coming to grips with the successes and the failures in history was a more attractive path to self-awareness and sociopolitical rehabilitation:

---

35. For an excellent interpretation of the historiographical practice of the fourth century BCE as reflected in *Zuo's Commentary*, see Schaberg, *A Patterned Past*. See also Ichiki, *Shu Ki monjin*, 298.

36. For an analysis of some of the essays showing how Lü's inward-oriented ethics was complemented by the strengthening of hierarchy in human relationships, in particular, the ruler-minister relationship, see Ji Xiaobin, "Inward-Oriented Ethical Tension," chap. 3. Cf. Xu Ruzong, *Wuxue zhi zong*, 85–96.

37. *ZZYL*, 123.2959.

38. Ibid., 2160.

When, in the end, it was made clear to the world once more that the hierarchical difference between the ruler and the minister is as vast as the difference in height between Heaven and Earth, whose merit was it after all? Take note! That the central states did not perish in the centuries between the depletion of the blessings of King Wen, King Wu, and the Duke of Zhou, and the emergence of Confucius, the sage, is entirely due to the perseverance of the official historians![39]

終使君臣之分, 天高地下, 再明於下, 是果誰之功哉? 嗚呼! 文武周公之澤既竭, 仲尼之聖未生, 是數百年間, 中國所以不淪於夷狄者, 皆史官扶持之力也!

Lü's attempt to apply the moral philosophy of the Cheng brothers to historical study and the exchange between Lü and Zhu concerning moral philosophy's significance in examination learning signal the relevance of a second major scholarly tradition in the twelfth-century examination field. I will return to the impact of the moral philosophy of the Cheng brothers and Zhu Xi's Learning of the Way in Chapters 6 and 7. In contrast to Zhu Xi and his followers, Lü and the Yongjia teachers, who were equally familiar with the work of the Cheng brothers, did not make moral philosophy the foundation of their curricula. In a policy response written for a special examination to select academicians (*guanzhi ce* / *guan'ge ce* 館職策/館閣策), Lü showed his unwillingness to expound on moral doctrine. He charged Jia Yi 賈誼 (201–169 BCE) and Yao Chong 姚崇 (651–721), both imperial advisers, for not instructing the emperors they served on "the great root of government," that is, the cultivation of the emperor's mind, but he concluded: "There must be people who can be summoned to explain where the great origin resides; I dare not presume to make proposals on this."[40] More antagonistic followers of Zhu Xi were ready to step in and provide a systematic account of the essentials of the Learning of the Way in their manuals.

---

39. Lü Zuqian, *Zuoshi bo yi*, 8.11b; idem, "Cao Gui jian guan she," in *Donglai bo yi*, 3.217; cited in Pan Fuen and Xu Yuqing, *Lü Zuqian pingzhuan*, 433; and Ji Xiaobin, "Inward-Oriented Ethical Tension," 48.

40. Lü Zuqian, *Donglai ji*, 5.3a–12b. For Zhu Xi's evaluation of this piece, see *ZZYL*, 122.2953, 2954.

### Reading and Writing

Besides administrative reasoning and institutional history, composition—the acquisition of rhetorical strategies—constituted a third major component in "Yongjia" teachers' examination curricula. The teaching of composition pervaded all other subjects. Teachers included training in rhetorical techniques in the courses and manuals on administrative reasoning and institutional history. Composition also developed into an independent area of pedagogical interest in the twelfth century. "Yongjia" teachers taught students how to read and analyze model texts and trained them in the application of rhetorical strategies in their own writing. Through their courses and manuals, they made significant contributions to the formation and the institutionalization of the "Ancient Prose" (*guwen* 古文) canon.

Ancient Prose first developed in the late eighth and early ninth centuries. It was then defined as a way of writing intentionally modeled on classical texts and was closely associated with the restoration of "the Way of Antiquity." Antiquity was defined by its imputed moral and cultural values. The classical models of the early Ancient Prose authors broadly encompassed the literature produced in the centuries leading up to and including the early Chinese empire, up to the second century CE. Han Yu was the first major advocate of Ancient Prose. The main goal of Han Yu's literary reform was to persuade literati to express themselves in a more straightforward manner, unrestricted by the demands of genteel parallel prose. Han Yu compiled a list of classical authors who, in his mind, embodied the virtues of clear and free expression.[41] By the eleventh century Han Yu's list of ancient writers had expanded to include authors born after the classical period, such as himself and subsequent authors writing in a similar style based on classical models. At the same time divergent and competing interpretations of Ancient Prose had emerged in elite circles; the different interpretations of Ancient Prose impacted the eleventh-century examination field.

Early Northern Song proponents of Ancient Prose advocated a return to the Way of Antiquity. The Way of the sages symbolized the

41. For Han Yu's interpretation of "Ancient Prose," see Hartman, *Han Yü and the T'ang Search for Unity*, esp. 220–25; and Bol, *"This Culture of Ours,"* esp. 131–36.

yearning for an ethic that reunited the several domains of scholarly life (personal morality, literary achievement, and sociopolitical responsibility), which to eleventh-century elites, seemed to have become separate endeavors. Early Ancient Prose advocates blamed the focus on contemporary literary fashions for the disassociation among the various domains constituting elite status. They argued that the holistic moral values, which had structured classical civilization, could be accessed through the effort to think through the work of a select group of transmitters of the Way of Antiquity. The genealogies defined by Liu Kai 柳開, Sun Fu 孫復, Shi Jie 石介, and Zu Wuze 祖無擇 (1006–85) included Confucius, Mencius, Xunzi, Jia Yi, Dong Zhongshu, Yang Xiong, Wang Tong, and Han Yu. Despite intellectual differences, all these men had allegedly modeled their work on the Classics of Antiquity in intent as well as in expression. Early Song Ancient Prose advocates similarly aspired to break with contemporary literary conventions by modeling their writing on the "simple" language of the classics.[42]

Their writing provoked the criticism and ridicule of contemporaries because their imitation of the earliest classical writing came across as both odd and impenetrable. Despite its antiquarian overtones, however, the Ancient Prose of these early Song masters found adherents at the central institutions of learning in the 1040s and 1050s. While holding positions as auxiliary lecturers at the Directorate of Education (Guozi jian 國子監), Sun Fu (1042–45, 1055–57) and Shi Jie (1042–44) introduced Imperial College students to Ancient Prose, and through their efforts Ancient Prose entered the examination preparation curriculum for the first time.[43]

Ouyang Xiu initiated a major change in the definition of Ancient Prose as well as in its role in examination preparation and examination writing. Starting with Ouyang Xiu, Ancient Prose advocates propagated

---

42. Bol, *"This Culture of Ours,"* 165.

43. Azuma, "'Taigakutai' kō," 100; Wang Shuizhao, *Songdai wenxue tonglun*, 205; Takatsu, *Keju yu shiyi*, "Bei Song wenxue zhi fazhan yu Taixue ti"; Zhu Shangshu, *Songdai keju yu wenxue kaolun*, 383–92. My thanks to Michael Fuller for suggesting that I do more research on the meaning of "the Imperial College style" (Taixue *ti* 太學體) in the mid-eleventh century.

genealogies of written texts that included authors not listed in Ancient Prose genealogies of the Way of Antiquity. In the estimation of Ouyang Xiu and Su Xun 蘇洵 (1009–66), whose work Ouyang upheld as the new model for Ancient Prose, the list of Ancient Prose authors should include writers such as the historians Sima Qian 司馬遷 (145–90 BCE) and Ban Gu for their "potent and firm" writing style or the Warring States Period military strategists Sun Wu 孫吳 (6th c. BCE) and Wu Qi 吳起 (d. 381 BCE) for the "concise and cutting" qualities of their work.[44] The adoption of these models reflected the wider intellectual and political interests of Ouyang Xiu and Su Xun; it also signified a transformation in the Song Ancient Prose movement. After Ouyang and Su Xun, Ancient Prose composition became the object of concentrated analysis. Twelfth-century literati increasingly conceived of Ancient Prose as a set of techniques and qualities of good writing best illustrated in selections of texts from Tang and Song authors. The literary models of Ouyang Xiu and Su Xun prefigured the formation of the Ancient Prose literary canon in the twelfth century.

Ouyang Xiu campaigned against the more archaic form of Ancient Prose promoted by Sun Fu and Shi Jie. In the departmental examination of 1057, he penalized those who wrote in anachronistic and abstruse classicist prose and selected papers that adopted the discursive strengths of classical historical and philosophical writing but modified their language in accord with contemporary usage.[45] Ouyang's intervention marked the beginning of a shift toward more contemporary literary models in the Ancient Prose tradition. He promoted Han Yu's prose as a means to restore the commonality and ease that was, in his mind, a core characteristic of classical writing.[46] Ouyang Xiu's partiality to examination papers written in this new version of Ancient Prose provoked loud protest in 1057. However, his intervention contributed to the growing popularity of the work of Han Yu and later Su Shi among literati and students in the late eleventh century.[47]

---

44. He Jipeng, *Tang Song guwen xin tan*, 273, 276.

45. Azuma, " 'Taigakutai' kō," 104–5.

46. On Ouyang's appreciation of Han Yu's prose and his mixed feelings about Song Ancient Prose proponents, see James Liu, *Ou-yang Hsiu*, 143–45.

47. Ibid., 151–52; Bol, *"This Culture of Ours,"* 192–93.

The major breakthrough in the formation and the spread of the Ancient Prose canon occurred in the twelfth century. The best evidence of this breakthrough is the number of anthologies produced. Virtually none are extant for the eleventh century; more than a dozen anthologies published between the 1130s and the 1270s survive.[48] In the collections published between the mid-twelfth and mid-thirteenth centuries, an informal canon of authors and prose pieces from these authors took shape. Twelfth-century teachers and commercial printers taught the language of Ancient Prose through the development of courses and the publication of manuals analyzing the argumentative writing of a select group of Tang and Song authors. These anthologies served as texts for the instruction of "classical Chinese" throughout the imperial period and continue to do so today. Three anthologies published during this period remain bestsellers. They have been republished in countless editions in East Asia and have become authoritative texts for introducing students to classical Chinese and literary language.[49] This section examines the formation of the Ancient Prose canon and especially the role of East Zhe examination preparation in this process of canon formation. It focuses on the first stage in the process of canon formation; the transformation of the Ancient Prose canon in the thirteenth century is discussed in Chapter 6.

## THE CANONIZATION
## OF ANCIENT PROSE

The main characteristics of the Ancient Prose anthologies published in the twelfth and thirteenth centuries illustrate how the Ancient Prose canon that emerged from these anthologies was shaped by the pedagogical interests and the pedagogical claims of examination teachers and commercial printers. The first characteristic shared by these

---

48. For an overview of anthologies from the Song period, see Wang Ruisheng, "Jin cun Songdai zongji kao"; Poon, "Books and Printing in Sung China," 394–402; and Chia, *Printing for Profit*, 140. The increase in the number of anthologies corresponds to Chia's finding that "commercial printing began in the Northern Song in the eleventh century but accelerated dramatically in the Southern Song" (ibid., 66).

49. These are *Guwen guanjian*, *Wenzhang zhengzong*, and *Wenzhang guifan*. The first is discussed in this chapter; the last two are discussed in Chapter 6.

anthologies is the focus on writings in the argumentative mode (*yilun* 議論). Writings in diverse genres could be cast in the argumentative mode. Ancient Prose anthologies typically included expositions, prefaces, occasional pieces, and letters. The prominence given to pieces in the argumentative mode coincided with a shift in emphasis from poetry to prose in the examinations. In the twelfth century, the ability to write argumentative prose was key to success in the examinations—essay writing was required at all levels of the examinations, and the majority of the papers were prose pieces.

The collection of writings in the argumentative mode marked a departure from traditional patterns in literary anthologizing. Most earlier Song anthologies had been modeled on standard collections such as the sixth-century *Selection of Texts* (*Wenxuan* 文選) and included texts in a wider variety of literary genres. For example, the imperially commissioned *Mirror of Literature* ([*Huangchao*] *wen jian* 皇朝文鑑; 1179), for which Lü Zuqian served as the head editor, was set up as an index to the best Song works in all current genres. Similarly, the anthology compiled by Lü Zuqian's teacher Lin Zhiqi 林之奇 (1112–76), *Observing Foamy Waves* (*Guan lan wenji* 觀瀾文集), provided an overview of the literary history since the Warring States period by presenting texts through the Song period in a matrix of eighteen poetic and prose genres.

The growing preference for texts in the argumentative mode resulted from the emergence of a professional interest in the instruction of rhetoric. The structural and stylistic analysis of selected Ancient Prose pieces in the argumentative mode taught students how to express ideas about social, political, and cultural issues. Lü Zuqian's work as an examination teacher, rather than as a compiler for the court, provides one of the earliest examples of the emphasis on argumentative writing in the teaching of composition. Anthologies attributed to Lü Zuqian, such as *The Key to Ancient Prose* (*Guwen guanjian* 古文關鍵; published between 1160 and 1180) and *The Collected Writings of Su Xun—With Notes by Lü Zuqian* (*Donglai biaozhu Laoquan xiansheng wenji* 東萊標註老泉先生文集),[50] featured argumentative pieces. Lü Zuqian defined "argumentative

---

50. I used a Song edition, printed under the supervision of Wu Yan 吳炎, professor of the district school of Guiyang, around 1193. This edition is preserved in the Beijing

writing" as "useful writing" (*youyong wenzi* 有用文字).[51] He believed that the structural and stylistic analysis of expositions, prefaces, descriptions, and letters by Ancient Prose masters was the best way to teach students how to convey their thoughts on issues concerning scholarship and government.

Second, the publication of Ancient Prose anthologies was accompanied by the development of a critical apparatus. Ancient Prose texts became the subject of detailed analysis. Examination teachers in East Zhe developed a repertoire of commentarial techniques and a critical apparatus to analyze the structure of prose texts, syntactic patterns, and effective word usage. These techniques were intended to steer student reading and writing—the formation of the Ancient Prose canon was not just about the selection of texts; it was also about delineating appropriate ways of reading and applying these texts.[52]

*The Key to Ancient Prose* illustrates the concepts and methods "Yongjia" teachers used in the instruction of composition. Lü Zuqian wanted his students to read the Ancient Prose texts in his selection in a new way and prefaced the texts with instructions on how to read Ancient Prose. These instructions encouraged silent, structured reading as opposed to memorization. Lü Zuqian taught students to "look at" (*kan* 看) the texts and analyze the semantic and stylistic organization of the whole piece; he explicitly cautioned against mere memorization (*song* 誦):

> First, look at the overall argument.
> Second, look at the organization behind the movement of the text.
> Third, look at the crucial junctures in the scheme of the text.

How do the introduction and the conclusion correspond to each other with reference to the main point? How is the exposition in the main body of the text organized? How about the structural alternations between "opposing and lifting" and "opening up and mixing together"?[53]

---

Library. For a bibliographic description, see Zhu Shangshu, *Song ren bieji xu lu*, 1: 222–24. For a discussion of Lü's anthologies, see Du Haijun, *Lü Zuqian wenxue yanjiu*.

51. Lin Sufen, "Lü Zuqian de cizhang zhi xue," 152.

52. For the history of the use of critical marks in Song and Yuan works, see Takatsu, "Sō Gen hyōten kō." See also Zhu Shangshu, *Songdai keju yu wenxue kaolun*, 284–301.

53. *Kaihe* 開合 means raising a new line of argument and then linking it back to the main argument. *Yiyang* 抑揚 refers to the frequent technique of first praising something to criticize it later, or, conversely, criticizing something first to praise it later; see Chen

Four, look at the phrasing of striking passages.[54]

What are the striking passages of the text? How about those passages whose phrasing or word selection is very powerful? Which passages have shifts in the argument that are particularly beautiful? What does a powerful conclusion look like? In which passages are windiness and conciseness fused together effectively? In what passages does the author show real understanding of the topic?[55]

第一看大概主張
第二看文勢規模
第三看綱目關鍵
如何是主意首尾相應? 如何是一篇鋪叙次第? 如何是抑揚開合處?
第四看警策句法
如何是一篇警策? 如何是下句下字有力處? 如何是起頭換頭佳處? 如何是繳結有力處? 如何是融化屈折剪截有力? 如何是實體貼題目處?

Lü's anthologies encouraged a kind of reading different from that students had theretofore confronted in the work of the Ancient Prose masters. Philological rather than rhetorical annotations had been customary in previous editions of the prose of the masters. In the 1160s, printers from Masha 麻沙 (Jianning 建寧 prefecture, Fujian), famous for its cheap commercial editions, tried to attract customers for a small-character edition of the collected works of Liu Zongyuan, the second major representative of the Ancient Prose movement in the ninth century, by highlighting the phonetic and philological annotations the edition provided. *The Collected Writings of Master Liu from the Tang Dynasty—With More Notes and Phonetic Clarifications (Zengguang zhushi yinbian Tang Liu xiansheng ji* 增廣註釋音辯唐柳先生集) featured Song dynasty exegetical work on Liu's collected writings.[56] The author of the preface explained the rationale behind the publication as follows: "When I read the work of Han [Yu] and Liu [Zongyuan], I often have the feeling that

---

Yizeng, *Wenzhang ouye*, guwen pu, 3.10a–b, 11a. On *kaihe*, see Zhou Zhenfu, *Wenzhang lihua*, 143–49.

54. Lü Zuqian, *Guwen guanjian*, pt. I, 1a–b. The edition I used is a reprint of a Jiangsu shuju re-edition, ca. 1898, of Xu Shuping's collated edition from the Kangxi period (1662–1722).

55. Ibid.

56. I consulted the original Song edition at the library of the Palace Museum in Taibei.

the extraordinary characters of the ancients confuse my eyes and block my mouth. Since, as soon as I start reading, I have to look up the pronunciations for all sounds, I cannot get through a single page in one day."[57] The editors modeled their work on similar editions of Han Yu's collected writings, in which "the pronunciation was indicated by the *fanqie* system,[58] and the difficult characters were explained."[59] The comments on one of Liu's more celebrated pieces, "On Feudalism," were of four kinds: the pronunciation of difficult characters; variant readings; citations for the sources of some of Liu's arguments and historical examples; and definitions of terms and historical background information on events.[60] Such editions facilitated the recitation and memorization of the work of the Ancient Prose masters.

Philology remained an important part of the anthologies associated with Lü Zuqian. *The Key* incorporated the philological comments collected in the Masha edition and cited additional primary sources in order to contextualize Liu's arguments. Lü was credited with the annotation of a collection of classical texts compiled by his teacher Lin Zhiqi. The notes in *Observing Foamy Waves; A Collection of Texts—Organized by Category and with Notes Assembled by Lü Zuqian* (*Donglai jizhu leibian Guan lan wenji* 東萊集註類編觀瀾文集)[61] were of a purely philological nature. The comments on "On Feudalism" in this collection were identical to the philological footnotes to this text in *The Key*,[62] but the structural and stylistic comments in the latter collection were absent. In Lü's

---

57. Liu Zongyuan, *Zengguang zhushi yinbian Tang Liu xiansheng ji*, 1168 preface by Lu Zhiyuan 陸之淵.

58. In this system, the pronunciation of a character was indicated by two other characters, the first of which represented the opening sound of the syllable and the second of which represented the ending sound.

59. Liu Zongyuan, *Zengguang zhushi yinbian Tang Liu xiansheng ji*, preface.

60. Ibid., 3.1a–4b.

61. I looked at the Song edition preserved at Beijing Library. In naming his anthology *Observing Foamy Waves*, Lin Zhiqi was probably referring to *Mengzi* VIIA:24, cf. Legge, *The Chinese Classics*, vol. 2, *The Works of Mencius*, 463. Mencius argued there that things that seem big can look small if one adopts a higher perspective. A person who has seen the ocean will not be impressed with small streams. By adopting this title, Lin insinuated that he had collected the very best pieces of writing. For Lin Zhiqi, see *SB*, 606–7.

62. Lin Zhiqi, *Guan lan wenji*, 7.1a–6a; Lü Zuqian, *Guwen guanjian*, 1.27b–33a.

annotated edition of Su Xun's collected works from 1193 mentioned above, comments provided historical background information or explained the connections between different parts of the text to help the reader follow the argument.

Despite his continuation of the tradition of philological commentary, Lü was better known for the rhetorical analysis he applied to Ancient Prose texts. The general instructions on reading translated above were echoed in the comments on each of the anthologized texts. Lü supplied interlinear comments, comments in the margins, and evaluations of the piece either preceding or following it. The overall comment on Liu Zongyuan's "On Feudalism" reads: "This demonstrates the method of staging the exposition." "Staging" was just one entry in the expanding vocabulary of technical terms used to analyze prose texts. It referred to the gradual development of an argument, the succession of examples or subarguments building to a broader conclusion. Comments in small characters between the columns of the main text indicated the steps of the argument. Lü marked the first step in Liu's overview of the history of feudalism by placing the characters for "staging" after the text's discussion of Zhou feudalism. "Staging" reappeared after the discussion of Han feudalism.

The textual commentary reinforced Lü's reading principles. After having scanned the text for the overall argument and the general organization of the text, the reader was directed to the lower-level structural and stylistic features of the text: the junctures and the striking passages. Within each of the steps demarcated by "staging," Lü drew attention to the function of individual sentences. He highlighted those passages in which the author "concludes one line of thought," "gives a definition of feudalism," "links the Zhou house to the discussion," "raises a problem in his argument," "responds to it," "raises more objections to his arguments," and "finishes by tracing the cause back to feudalism."[63] He drew attention to "the up- and downward movement" of the argument, a reference to the author's presentation of opposite interpretations of historical events. In one passage, Liu Zongyuan

---

63. Lü Zuqian, *Guwen guanjian*, 1.27b–33a. For a partial translation of Liu's essay, see de Bary et al., eds., *Sources of Chinese Tradition*, 559–64; for a brief discussion, see Chen Jo-shui, *Liu Tsung-Yüan and Intellectual Change*, 96.

entertained an objection to his overall argument that the history of feudalism is dependent on impersonal historical forces and that its appearance or disappearance does not depend on the existence and the intentions of sage-kings. He suggested that some might argue for the
enduring validity of the feudal order based on the contention that the
sage-kings of the Shang and Zhou dynasties purposefully continued
the system they had inherited. Liu then reversed this opinion and contended that the sage-kings had been forced to continue the feudal system by circumstance, because of pre-existing conditions and not because of their own belief in the enduring value of the system. Lü's
comments further highlighted skillful transitions between paragraphs
and memorable expressions (see Fig. 2).

In Song editions, these comments were reinforced with critical
marks next to the characters. Lü Zuqian's marks were either omitted in
later editions or mixed with other critics' evaluations in anthologies
combining the commentarial work of several Song critics.[64] Xu Shuping
徐樹屏 (*js.* 1712), for example, noted in his collated edition of *The Key*
that the two Song editions he had used were so marked.[65] The marks
served mainly to indicate the main lines of the argument and to divide
paragraphs. Very occasionally, dots were used to highlight word usage.
The stress marks thus functioned as a visual grammar superimposed on
the printed text.[66]

*The Collected Writings of Su Xun—With Notes by Lü Zuqian* provides direct evidence of the use of lines and jots in twelfth-century Ancient
Prose composition manuals. Unbroken lines outlined the skeleton of
the text. They marked either conjunctions or short clauses, two to three
characters long, to highlight transitions in the argument, or, in longer
sentences, to link the main points of the argument together. The editor

---

64. This was the case in *The Legitimate Seal of Compositions from the Past and Present—A
New Collection with Several Scholars' Criticism*, discussed in Chapter 6.

65. Lü Zuqian, *Guwen guanjian*, editorial principles, 1a–2a. Cf. the postface by Yu Yue
俞樾 (1821–1907) from 1898. Chen Zhensun (*Zhizhai shulu jieti*, 15.451) mentions that Lü
Zuqian provided marks and comments.

66. For the notion of punctuation marks as a visual grammar, see Parkes, "The Influence of the Concepts of *Ordinatio* and *Compilatio*"; cited in Carruthers, *The Book of
Memory*, 224.

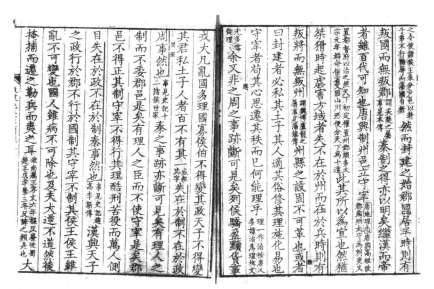

Fig. 2  Intralinear and interlinear annotations on Liu Zongyuan's "On Feudalism" attributed to Lü Zuqian in *The Key to Ancient Prose* (late 12th c.) (Qing edition [19th c.] held at the Harvard-Yenching Library).

also drew attention to these passages by registering the main points in the top margin. These comments consisted of indices of the structure of the piece (e.g., "the main point") or abstracts of the argument of the relevant passage. Jots, short tick marks, which signaled great lines and eye-catching phrases, were used sparingly, as in the Song editions of *The Key*.

Lü Zuqian was neither the first nor the only person to use marks in the twelfth century. The use of punctuation marks in Chinese literature had a long history. However, Lü's use of these marks as instructional tools for composition in printed editions of his Ancient Prose anthologies inaugurated a technique that was soon applied to prose and poetry anthologies and, in subsequent dynasties, to other literary genres as well.[67] The addition of comments and marks proved to be a major

67. For a general history of the use of punctuation marks for reading and critical purposes, see Guan Xihua, *Zhongguo gudai biaodian fuhao fazhan shi*. For an interpretation of the factors behind the emergence and expansion of their use, see Zhang Bowei, "Pingdian silun." For their use in Song times, see Ren Yuan, "Songdai jingdu zhi chuxin yu biduan," 59; and Takatsu, "Sō Gen hyōten kō." For later examples, see Chia, "Printing for Profit," 96–97. See also Guo Shaoyu, *Zhongguo wenxue piping shi*, 458; Luo

selling point for commercial editions of prose anthologies and encyclopedias. Lucille Chia, in her work on the commercial printers of Jianyang (Jianning prefecture, Fujian), demonstrated that competition in the market for examination manuals led to the expansion of marketing techniques. Publishers routinely pointed to the addition of new commentaries to canonical texts, the distribution of commentary across the page (in the margins and between columns), and the use of stress marks as selling points for their publications.[68]

Lü Zuqian's case demonstrates that the "Yongjia" teachers contributed to the incorporation of commentarial tools in printed examination manuals. Their own work also became the object of annotated commercial editions. One blurb from a thirteenth-century commercial printer highlighted the use of punctuation marks as a way to help students acquire the language and grammar of Ancient Prose composition (see Fig. 3).[69] This blurb advertised the 1212 edition in two installments of selected prose by Ye Shi and Chen Liang entitled *The Best of Chen Liang and Ye Shi—With Stress Marks* (*Quandian Longchuan Shuixin xiansheng wencui* 圈點龍川水心先生文粹).[70] Ye Shi, as noted in Chapter 3, rivaled Chen Fuliang as the most famous representative of Yongjia examination learning and success in the last three decades of the twelfth century. Chen Liang, who, like Lü Zuqian, was a native of Wuzhou prefecture, shared the intellectual interests of the Yongjia masters in institutional history and Ancient Prose. Chen Fuliang's prose received similar treatment. The earliest extant collection of his regulated expositions

---

Genze, *Zhongguo wenxue piping shi*, 875–80; and Cherniack, "Book Culture and Textual Transmission in Sung China."

68. Chia, *Printing for Profit*, 46–50. Chow Kai-wing (*Publishing, Culture, and Power in Early Modern China*, 12–14, 109–23, and chap. 4) identifies these and other types of additions to the main text as "paratext" and interprets them as a textual space in which reading publics are formed and through which political power is exercised. His analysis illuminates how paratexts generate publicity and power in the literary field; it is less successful in linking them to political practice.

69. *Quandian Longchuan Shuixin xiansheng wen cui* (*qianji, houji*), mulu 1a.

70. Preserved at the National Library in Taibei. Cf. Tillman, *Utilitarian Confucianism*, 19–22; and idem, "Ch'en Liang on Statecraft." Note, however, that Rao Hui wrote the preface to the edition and was not the editor. Cf. Deng Guangming, "Chen Longchuan wenji banben kao," in Chen Liang, *Chen Liang ji* (*zengding ben*), 1–27.

Fig. 3 Blurb advertising the 1212 edition in two installments of selected prose by Ye Shi and Chen Liang, titled *The Best of Chen Liang and Ye Shi—With Stress Marks* (original preserved at the National Library in Taibei; microfilm of this edition at the Harvard-Yenching Library).

is Fang Fengchen's annotated edition from the 1260s, *The Founding Father of the Exposition, Chen Fuliang—A New Print with Comments by Fang Fengchen* (*Xinkan Jiaofeng pidian Zhizhai lunzu* 新刊蛟峰批點止齋論祖).[71]

The commercial anthologies of the works of twelfth-century "Yong-jia" teachers exemplify a third characteristic of Ancient Prose anthologies in the twelfth and early thirteenth centuries, namely, the emphasis on recent rather than classical literary models. The formation of the Ancient Prose canon during the Song dynasty was intimately connected with instruction in reading and writing skills. The question of which texts and authors were most useful in the classroom was answered differently in the eleventh and twelfth centuries. In contrast to their eleventh-century predecessors, teachers of Ancient Prose in the twelfth century selected pieces from a core group of Tang and Song authors as the central texts of Ancient Prose. The reproduction of these texts and authors in commercial anthologies formed the core of the Ancient Prose canon. At the same time, the emerging Ancient Prose canon was open. Famous contemporary authors and teachers were continuously added to the list of canonical authors.

Eleventh-century teachers of Ancient Prose composition identified as their main models for composition texts from the Warring States, Qin, Han, and Tang periods. Su Shi had his students read *Stratagems from the Warring States* (*Zhanguo ce* 戰國策; comp. 1st c. BCE) to learn how to discuss "benefit and harm." He had them read the memorials of the Han officials Jia Yi, Chao Cuo 晁錯 (200–154 BCE), and Zhao Chongguo 趙充國 (137–52 BCE) to learn how to discuss affairs and the *Zhuangzi* 莊子 to learn how to explain the nature of things. And he assigned them the work of Han Yu and Liu Zongyuan because their prose embodied all aspects of composition.[72] The multifaceted display of

---

71. According to the 1268 preface, this edition was based on earlier printed editions. Fang Fengchen's critical notes were added to this edition by Shao Qingsou. (Qingsou was a common sobriquet; I have not been able to find further information about Mr. Shao.) I read the preface attached to the (Ming?) edition preserved at the Naikaku bunko in Tokyo. The earliest edition I have seen a reference to is a Yuan edition listed in Qu Yong, *Tieqin tongjian lou cangshu mulu*, 21.17b.

72. Wang Gou, *Xiuci jianheng*, 35, citing Li Zhi 李廌 (1059–1109), an acquaintance of Su Shi.

rhetorical technique Su Shi admired in the work of the Tang authors partially explains why their work was singled out for special attention in classrooms from the twelfth century on.

In contrast to the eleventh-century teachers of Ancient Prose, Southern Song teachers and editors focused on selected essays of Tang and, especially, Northern and Southern Song Ancient Prose celebrities.[73] The Tang and Northern Song exemplars in Southern Song Ancient Prose anthologies (especially Han Yu, Liu Zongyuan, Ouyang Xiu, and Su Shi) had distilled a new grammar from the language of the classical sources to which they aspired. The features of classical writing were, from the perspective of Southern Song teachers, more accessible to students in the more recent writing of these Tang and Song authors than in the original classical texts.

The works of Han Yu, Liu Zongyuan, and the Su family were already well known by the mid-twelfth century. The publication history of the Ancient Prose masters' collected works shows that their writings had become part of examination curricula. Commercial prints of the works of Han Yu and Liu Zongyuan circulated in several editions.[74] Various editions appeared of the collected works of Ouyang Xiu, Su Xun, and Su Shi.[75] Besides editions of their complete works, commercial printers put out a variety of compilations of extracts from their oeuvre, such as *The Quintessence of Ouyang's Writing* (*Ouyang wen cui* 歐陽文粹; 1173) and *The Quintessence of the Three Sus' Writing* (*San Su wen cui* 三蘇文粹; the earliest extant edition is an expanded edition from the early 1190s).[76]

Lü Zuqian played a major role in the canonization of a select group of Tang and Song Ancient Prose authors. In his teaching and his anthologies, he selected from the works of those who were to become known in subsequent dynasties as the "great masters of Tang and Song prose," namely, Han Yu, Liu Zongyuan, Ouyang Xiu, Su Xun, Su Shi,

---

73. Lin Sufen, "Lü Zuqian de cizhang zhi xue," 152.

74. Liu Zhenlun, *Han Yu ji Song Yuan chuanben yanjiu.*

75. Chia, *Printing for Profit*, 139–40; cf. Bol, "Reading Su Shi in Southern Song Wuzhou."

76. For a fuller analysis of these and other anthologies, see Bol, "Reading Su Shi in Southern Song Wuzhou."

Su Che 蘇轍 (1039–1112), and some of Su Shi's disciples.[77] The work of Han Yu, Liu Zongyuan, Ouyang Xiu, and Su Shi formed the core of Lü's expository writing curriculum:

When learning how to write, one should familiarize oneself with Han [Yu], Liu [Zongyuan], Ou[yang Xiu] and Su [Shi]. First look at the structure of the text, then investigate throughout the text how the ancients conscientiously put down their phrases.[78]

學文, 須熟看韓柳歐蘇. 先見文字體式, 然後遍攷古人用意下句處.

Lü's anthologies, *The Key to Ancient Prose* and *The Collected Writings of Su Xun*, underscored the pedagogical claim that the work of this select group of Tang and Song Ancient Prose masters and the particular reading methods taught in East Zhe provided students "the key" to learning composition. In addition to his own compilations, Lü also promoted similar Ancient Prose anthologies to his student audience. A manual printed in Wuzhou, *The Skilled Cavalryman* (*Jingqi* 精騎), was, if wrongly attributed to Lü, reputedly recommended reading for his students.[79] It featured selections from the work of the same select group of Tang and Song Ancient Prose authors. Lü admonished his students not to plagiarize the work of the three Sus,[80] but *The Key* was designed to teach students to learn from their approach to the topic. Lü's critical comments showed students how his model authors crafted original arguments in response to particular questions.

Students of Lü Zuqian continued his approach to the instruction of composition. Lou Fang 樓昉 (*js.* 1193) taught in Jinhua and at the Imperial College in the capital. According to the thirteenth-century bibliographer Chen Zhensun 陳振孫 (ca. 1190–after 1249), students found his anthology of Ancient Prose texts "convenient."[81] Lou's anthology

---

77. For an alternative version of the history of the canonization of the Tang and Song Ancient Prose masters, see Takatsu, "The Selection of the 'Eight Great Prose Masters of the T'ang and Sung.'"

78. Lü Zuqian, *Guwen guanjian*, 1.1a.

79. I looked at the Song edition preserved at the National Library in Taibei. For a further description of this work and its context, see Bol, "Reading Su Shi in Southern Song Wuzhou," 75–77, 89–93.

80. Lü Zuqian, *Guwen guanjian*, 1.1a.

81. Chen Zhensun, *Zhizhai shulu jieti*, 15.452.

circulated under multiple titles. The work was listed in Chen's catalogue as *Lou Fang's Notes to Classical Texts* (*Yuzhai guwen biaozhu* 迂齋古文標註) and was also known under this title to Liu Kezhuang 劉克莊 (1187–1269), a well-known poet, literary critic, and politician.[82] In the early thirteenth century, it circulated under different titles. The titles of two extant Song editions are *Lou Fang's Comments on Famous Writers' Literary Works* (*Yuzhai biaozhu zhujia wenji* 迂齋標註諸家文集; preface from 1226) and *The Key to Upholding Ancient Prose—With Notes by Master Lou Fang* (*Yuzhai xiansheng biaozhu chong gu wen jue* 迂齋先生標註崇古文訣; preface from 1227, allegedly the first printed edition).[83] These editions attest to the popularity of this anthology in the thirteenth century. The publication history of the anthology also highlights the continued importance of the distribution of texts in manuscript form alongside printed editions.[84]

Chen Zhensun explained that Lou had gotten his inspiration from Lü Zuqian's *Key*, but that he had added more pieces. Even though Lou Fang's anthology contained more texts and introduced more authors, Ouyang Xiu, Han Yu, Su Shi, and Liu Zongyuan topped the list of the most frequently cited authors with twenty-six, twenty-five, twenty-two, and fourteen entries, respectively. More than half of the 168 texts, picked from forty-nine authors from all periods of Chinese history up to the Song, were by these four authors.

The increasing emphasis on more recent, late twelfth-century models is evident both in collections of the works of specific authors and in general anthologies, as twelfth-century teachers of Ancient Prose joined the ranks of the Ancient Prose masters. Compilations of the works of twelfth-century Ancient Prose writers appeared soon after the first commercial editions of excerpts from the eleventh-century masters. Passages from the work of Chen Liang and Ye Shi were published with annotations in Jian'an around 1212 as *The Best of Chen Liang and Ye Shi—With Stress Marks* (see Fig. 3 above). Two installments of annotated excerpts from the prose of Yang Wanli were published around 1259, probably in the same printing center, as *Excerpts from Yang Wanli,*

---

82. Takatsu, "Sō Gen hyōten kō," 133.
83. Both editions are preserved at Beijing Library.
84. Inoue, "Zōsho to dokusho."

*Annotated and Classified* (*Pidian fenlei Chengzhai xiansheng wenkuai* 批點分類誠齋先生文膾).[85] Fang Fengchen's annotated edition of Chen Fuliang's expositions from the 1260s provides another example of this trend. Model expositions of Lü Zuqian, Chen Fuliang, Ye Shi, Yang Wanli, and lesser known writers, many of them hailing from Yongjia and the surrounding districts of Wenzhou, such as Zheng Boxiong 鄭伯熊, Dai Xi 戴溪 (1141–1215), Chen Wu 陳武 (*js.* 1178), and Chen Qian 陳謙 (1144–1216), also circulated widely. Some of these teachers' series of expositions on particular themes seem to have circulated independently, in schools and among their friends, before they were collected into larger compilations such as *Awesome Expositions from Ten Masters, with Notes* (*Shi xiansheng aolun zhu* 十先生奧論註).[86]

In the twelfth and early thirteenth centuries, the Ancient Prose canon was open. The position of a select group of Tang and Northern Song masters was fixed, but the boundaries of the canon were permeable. A greater degree of openness in the canon translated into the inclusion of contemporary authors and the greater percentage of contemporary texts in the selection. Tellingly, the bulk of the prose writings in most Southern Song Ancient Prose anthologies were by Song writers. Approximately 80 percent of the 522 pieces in *A Complete Collection of Ancient Prose* (*Guwen jicheng* 古文集成) were Song texts.[87] Fang Yisun 方頤孫, in his 1242 guide to Ancient Prose, *One Hundred Pieces of Brocade of Exquisite Pattern* (*Fuzao wenzhang baiduan jin* 黼藻文章百段錦), included seventy examples from Northern Song writers (half of which, thirty-four, were from Su Shi's prose) and forty-two passages from Southern Song texts (most of which, twenty-nine, were by Lü Zuqian); there were only two examples of pre-Tang texts and four from Han Yu and Liu Zongyuan.[88]

By the late twelfth century, the Ancient Prose canon was characterized by the selection of a group of Tang and Song authors and texts as

---

85. A Yuan edition preserved at the Library of the Palace Museum, Taibei, contains a preface dated 1259 by Fang Fengchen. The compilation is attributed to Li Chengfu 李誠父 from Jian'an.

86. For a fuller description of this title, see the Appendix.

87. *SKTY*, 38.4163. I discuss this source in more detail in Chapter 6 and in the Appendix.

88. For further discussion of Fang Yisun's *One Hundred Pieces of Brocade*, see Chapter 6.

the core of Ancient Prose and by their use as a tool to teach language, reading, and writing. The teachers of East Zhe played a key role in the consolidation of this core group of texts and authors. The publication of Lü Zuqian's annotated anthologies and their derivatives solidified the genealogy of the Ancient Prose masters and, equally important, defined the informal Ancient Prose canon as a tool for the instruction of composition. Works of the teachers of East Zhe were added to the canon for similar reasons. The annotated expositions of Lü Zuqian, Chen Fuliang, and Ye Shi demonstrated to students preparing for the examination how Ancient Prose rhetorical strategies could be used in examination genres. In thirteenth-century anthologies, the goals, the selection, and the methods of teaching Ancient Prose were modified, reflecting both the growing impact of the Learning of the Way in the field of examination preparation and the continued commercial interest in anthologies. These developments are taken up in Chapter 6.

# Part III

---

## *The Court in the Examination Field*

# 5

---

# *Court Politics*
# *and Examination Standards*
# *(1127–1274)*

Upon taking the throne in 960, the first emperor of the Song dynasty, Taizu, re-established the civil service examinations as a means of recruitment. While transforming himself from a general in the army of the Later Zhou 周 kingdom (951–60) to the emperor of the Chinese territories, Taizu recognized the value of the civil service examinations in cultivating the loyalty of civil servants and literati and in creating a dynasty that would endure. His lasting contribution, the addition in 973 of the palace examination during which graduates answered a question posed by the emperor or his closest advisers, bears witness to the examinations' importance to imperial rule.

The examinations were ideologically intertwined with dynastic rule, but the center did not necessarily dominate the examination field. As we have seen, in the latter half of the twelfth century, the "Yongjia" teachers developed distinct curricula and became authorities in the definition of examination standards through the commercial publication of examination manuals. Teachers and printer/publishers thus participated in the definition of standards. This chapter investigates the role of the court and the central bureaucracy (emperors, chancellors, and examination officials) in the definition of examination standards during the Southern Song period. It analyzes the role of the central government and how it defined its position in the examination field. It also focuses on the ways in which the government modified its policies in accord with current conventions in the examination field and the shifting positions of other agents within it. The history of government regulation of the examinations between the 1130s and the 1270s traced in this chapter suggests that the government shifted from a policy of curricular ecumenism in the twelfth century to a policy of curricular standardization in the last half-century of Song dynastic rule. Viewed within the broader context of the Song civil service examinations, both policies underscore the Southern Song government's secondary role in shaping examination standards.

## The Policy of Great Impartiality

Throughout the second half of the twelfth century, the Southern Song government expressed its commitment to Great Impartiality (*da gong* 大公). This slogan embodied the court's intention of avoiding factional politics, which had characterized the five decades between 1070 and 1126 and were widely blamed for the Song court's inability to maintain control over the north against the Jurchen invasions. In order to reconstruct Song authority in the south, Gaozong 高宗, the first emperor reigning from the southern capital of Lin'an, refused to side with advocates of the old factions, either the supporters of Wang Anshi's New Policies (the faction in charge of the government in 1076–86 and 1094–1124) or the supporters of his principal critics, Cheng Yi, Sima Guang, and Su Shi (in charge, 1086–93). In the realm of education, the commitment to impartiality translated into a refusal to endorse partisan curricula. Until 1241, Gaozong and his successors continued to abide by the policy that a single political faction or intellectual formation should not determine examination curricula.

The Southern Song court's espousal of impartiality in examination curricula stood in sharp contrast to examination policy under Wang Anshi and his successors. Starting in 1071, Wang Anshi and his partisans launched a series of institutional and curricular reforms. These reforms were motivated by the overall goal of creating a unified corps of bureaucrats trained in classical scholarship and receptive to the relevance of selected classical texts to the bureaucrat's responsibilities in maintaining state control over society. The elimination of the variety of examination routes was the first step in this direction. Wang canceled the "various fields" examinations. Henceforth candidates in the regular examinations competed for the *jinshi* degree only.

Candidates for the *jinshi* degree had to master a curriculum that emphasized classical scholarship and administrative ability over literary skill. The first session no longer tested students on their ability to write poetry; instead, students had to write ten essays on the significance of passages quoted from the new government's selection of classical texts: *The Book of Songs, The Book of Documents, The Changes*, and the classics on ritual (*Rites of Zhou* and *The Book of Rites*). They were also asked to write ten essays on questions drawn from *Mencius* and *The Analects*.

After the first wave of opposition to his centralizing policy initiatives, which led to his resignation as chancellor in 1073, Wang Anshi responded to his critics with further curricular reforms. He oversaw the compilation of a series of standardized commentaries entitled *The New Meaning of the Three Classics (Sanjing xin yi* 三經新義). Wang wrote the commentary on *Rites of Zhou* himself and commissioned the commentaries on *The Book of Songs* and *The Book of Documents*. The commentaries incorporated a rationale for Wang's centralizing policies[1] and were intended to replace commentaries circulating among examination candidates. The court put the new commentaries on the curriculum of all government schools in 1075. This policy was particularly significant since Wang Anshi intended to integrate the empire-wide network of government schools with the examination system. Examination candidates were to be trained in government schools over an extended period of time, and Wang ultimately wanted the schools to take over the selection of bureaucrats.

In 1082, the court added Wang's dictionary, *Explanations of Characters (Zi shuo* 字説), to the curriculum of the government schools.[2] As Peter Bol's work on Wang Anshi suggests, through the emphasis on the analysis of classical texts in the examinations and the normative interpretations of them in *The New Meaning of the Three Classics* and *Explanations of Characters*, Wang "was developing a way of learning that justified a way of governing he believed to be true to [the Way of the sage-rulers of Antiquity]."[3]

---

1. Hervouet, *A Sung Bibliography*, 29.
2. Araki Toshikazu, *Sōdai kakyo seido kenkyū*, 383, 390, 399.
3. Bol, *"This Culture of Ours,"* 233.

The impact of Wang's curricular policies was immediate. As *The New Meaning of the Three Classics* and *Explanations of Characters* became the core curriculum in government schools and determined success in the examinations, candidates focused their efforts on mastering the new curriculum. Government schools were not the sole institutions furthering the adoption of the state-imposed curriculum; private tutors and commercial publishers also promoted it. The government school system in pre-twentieth-century China never fulfilled the role that Wang Anshi had envisioned for it. In the eleventh and twelfth centuries, the schools were unable to absorb the growing numbers of students preparing for the examinations. Private tutors and commercial printers provided training to those not enrolled in government schools. The number of commercial printers kept pace with the growing numbers of examination candidates.[4]

During the period when the New Policies were in effect, commercial publishers marketed the materials listed on the state-imposed curricula. A memorial from 1112 complained about the commercial publication of pocket editions of *The New Meaning of the Three Classics*; candidates were smuggling them into the examination halls and using them as cribs.[5] This evidence suggests that the expansion of commercial printing in the early twelfth century coincided with the creation neither of a pluralistic intellectual milieu nor of a pluralistic examination curriculum. The following discussion of curricular changes in the ensuing decades also demonstrates that commercial printing reflected rather than generated the competition over the definition of curricular standards among teachers and officials, and the intellectual and political formations with which they were associated.[6]

---

4. Chia, *Printing for Profit*, 66; Ozaki, "Songdai diaoban yinshua de fazhan"; Chaffee, *The Thorny Gates*, 35.

5. Chia, *Printing for Profit*, 121.

6. Chow Kai-wing (*Publishing, Culture, and Power in Early Modern China*, introduction and chap. 4, esp. 149–51) argues, by contrast, that the expansion of commercial printing contributed to the creation of an intellectual milieu that encouraged open and pluralistic interpretations of the Confucian canon in the examinations. In his view, commercial printing had this effect because it stimulated the growth of a class of literary professionals and because it provided publicity, and thus political power, for this class. His argument covers only the late Ming dynasty and does not address the earlier expansion of commercial printing in the twelfth and thirteenth centuries.

In contrast to the eventful history of institutional and curricular reform and counterreform in the eleventh and early twelfth centuries, the Southern Song government adopted a low profile in the regulation of the civil service examinations. In 1145, the court established an institutional framework for the examinations that remained firmly in place until the end of the dynasty. The new procedures of 1145 attempted to strike a compromise between the reform-minded followers of Wang Anshi and the conservative opposition.

As far as the content of the examinations was concerned, the court aimed to alleviate the curricular traumas of the decades of reform by underlining the value of "broad scholarship"[7] and by allowing competing interpretations. Because of this decision to avoid partisan curricula, the Song government's position in the examination field shifted. Its prescriptive stance gave way to occasional prohibitive interventions. It monitored rather than directed the book market and the examination halls. The rest of this section reviews Southern Song government regulations concerning the examinations and discusses the government's participation in the publication of examination materials. Although the excursion into government regulations and the government's publishing record below suggests a retreat of government activity in the examination field, the following section explains how the government repositioned itself in reaction to other agents' claims on some of the prerogatives that the Song court exercised during Wang Anshi's time.

Throughout the Song dynasty, the government issued examination regulations (*gongju geshi* 貢舉格式, *tiaoshi* 條式, *gongju shi* 貢舉式, *gongju tiaozhi* 貢舉條制). Some covered the examinations as a whole; others were for particular tracks or levels, such as Zhenzong's 眞宗 (r. 998–1022) "Regulations for *Jinshi* Palace Examinations" ("Qinshi jinshi tiaozhi" 親試進士條制) or the many versions of the regulations for the prefectural examinations (*fajie tiaozhi* 發解條制).[8] Most regulations

---

7. *SHY, XJ*, 4.40b–41a (1171), 5.6a (1183), 7b–8a (1185), 10a–11a (1187; this is the memorial by Hong Mai et al. quoted in translation on p. 197).

8. Araki Toshikazu, *Sōdai kakyo seido kenkyū*, 44, 50, 161, 289, and *passim*. Araki shows that this process was initiated by Zhenzong; see *SHY, XJ*, 14.22b, 15.9b, *passim*. Cf. the supplement on examination regulations (*yunlue tiaoshi*) in Ding Du et al., eds., *Libu yunlue*.

issued after the fall of the northern capital of Kaifeng concerned procedural matters. The regulations explained the organization and administration of the examinations and covered the levels and kinds of *jinshi* examinations, the different officials and clerks needed to oversee the examinations and check the papers, the qualifications for examination candidates and for supervisory personnel, and quotas. The regulations also addressed the general structure of each track: the genres to be tested, the number of pieces required for each genre, the order in which the different genres were to be tested, and, in some cases, the sources from which the topics for particular genres were to be drawn.

In 1145, Gaozong issued a milestone decree prescribing the organization of the *jinshi* examinations.[9] The regulations promulgated in 1145 remained in force until the last examination under Song auspices. The structural stability of the Southern Song examinations stands in sharp contrast to the abrupt and frequent changes in examination procedures during the eleventh and early twelfth centuries. The history of the poetry session illustrates the difference between the Northern and Southern Song periods in the rate of policy change. In the century preceding Gaozong's 1145 edict, rules concerning the position of poetry in the examinations changed six times. The ten reforms proposed by Fan Zhongyan stipulated that policy response essays were to be examined in the first session and that the poetic genres were to be tested last. Renzong 仁宗 (r. 1023–63) endorsed these proposals in 1044, but rescinded them the next year. In 1071, Wang Anshi eliminated poetry from the curriculum. In 1086, poetry was again tested in the examinations, only to be removed once the reformers returned to power. In 1127, poetry was reinstated.[10]

---

9. For this and earlier legislation regarding the examinations during the early years of Gaozong's reign, see Chaffee, "Examinations During Dynastic Crisis." The detailed and frequent attention Gaozong and his court devoted to the examinations demonstrates that, by the early twelfth century, the court and officialdom considered the examinations of crucial significance in the relationship between government and the literate elite.

10. Chaffee, *The Thorny Gates*, 71; Lee, *Government Education and Examinations in Sung China*, 152–53; Elman, "An Early Ming Perspective on Song-Jin-Yuan Civil Service Examinations"; idem, *A Cultural History of Civil Examinations*, 730–33.

The regulations issued in 1145 were the result of a compromise between proponents and opponents of the continuation of the use of
poetry in civil service examinations. As explained in the Introduction,
two tracks were established within the *jinshi* examinations, one in poetry
and one in classical scholarship. The prefectural and departmental examinations consisted of three sessions each. Examination candidates
who opted for the poetry track were required to write one regulated
poem and one regulated poetic exposition in the first session, one exposition in the second session, and three policy response essays in the
last session. Those who opted for the "meaning of the Classics" track
wrote three essays on the meaning of their classic of specialization and
one essay each on *The Analects* and *Mencius* for the first part, and then,
like their colleagues in the poetry track, one exposition in the second
session, and, finally, three policy response essays in the final session.[11]
From extant records it is impossible to determine precisely the numbers
of candidates who opted for each track, but court edicts and biographical data on Imperial Library personnel suggests that court policy aimed,
with some success, to balance the number of graduates (and by extension the number of candidates) in each track.[12]

---

11. Chaffee, *The Thorny Gates*, 5; Araki Toshikazu, *Sōdai kakyo seido kenkyū*, 394; *SHY*,
*XJ*, 4.21b–22a, 28b. Zhu Shangshu (*Songdai keju yu wenxue kaolun*, 198, 204) demonstrates
that misgivings about the use of poetry remained. In 1157 students were asked once
more to prepare for examinations in both poetry and essays on the meaning of the
Classics, but this measure was only in effect during the examinations held in 1159.

12. John Chaffee argues that poetry "consistently attracted the most candidates"
(*The Thorny Gates*, 71). Biographical data on Imperial Library personnel suggests that
among *jinshi* degree holders hired into it in the Southern Song period the number of
those who had specialized in poetry roughly equaled the number of those who had majored in one of the Classics. A small number of candidates majored in both streams.
This is based on my own comparison of Wang Yu's ("Nan Song kechang yu Yongjia
xuepai de jueqi," 152) numbers for Classics *jinshi* with the number of poetry *jinshi* mentioned in *juan* 7 and 8 in *Nan Song guan'ge xu lu*. Wang Yu (ibid.) uses the data on the
Classics degree holders to track the popularity of each classic between 1162 (original
mistakenly gives 1062) and 1262. His evidence demonstrates that *The Book of Documents*
remained the most popular classic of specialization, leading *The Spring and Autumn Annals* and the ritual classics. Even though the rankings changed little, his numbers suggest
that there was some fluctuation in the percentage of candidates specializing in particular
classics. *The Spring and Autumn Annals*, for example, appears to have gained in popularity
over the course of the thirteenth century.

The regulations accommodated the demand for the restoration of poetic composition in the examinations, but they also incorporated several of the changes implemented during the decades of reform. Classical scholarship became a well-established option. Wang Anshi's emphasis on government was further reflected in the preservation of another aspect of his examination reforms. In the last stage of the examinations, the palace examination, examinees were required to submit one policy response. Wang had introduced this requirement as a substitute for the test in poetic and exposition writing customary before 1071. The 1145 regulations thus consolidated the impact of Wang's institutional reforms while eliminating the main objections of the conservative opposition.

The Southern Song government's success in ending factional disputes over the examination sessions was not accompanied by any effort to establish an examination curriculum. The official endorsement of Wang Anshi's commentaries and *Explanations of Characters* was revoked. Examination regulations published after the fall of Kaifeng outlined the procedural requirements for each of the genres tested on the examinations but provided little guidance concerning specific texts for the examinations.

The government's retreat from curricular demands was apparent in the publishing world of the twelfth and thirteenth centuries. In the first century of the Northern Song period, editions of the Classics, the histories, and the official rhyme books published by the Directorate of Education dominated the book market.[13] According to Lucille Chia, the Directorate printed all the standard titles used in the examinations in the eleventh century—the Classics, *The Analects* and *Mencius*, the seven dynastic histories, and rhyme books.[14] The Directorate distributed its publications to government schools. Under Wang Anshi, the court used the

---

13. During the Southern Song period, *An Overview of Rhymes by the Ministry of Rites* (*Libu yunlue* 禮部韻略; ca. 1037) became the most frequently used rhyme book among examination candidates. It was printed and distributed by the Directorate of Education and subsequently reprinted by commercial distributors for the use of examination candidates. For examples of Directorate and commercial editions of *Libu yunlue*, see Poon, "Books and Printing in Sung China," 266–67; and Chia, *Printing for Profit*, 111, 130. On its use in the examinations, see Araki Toshikazu, *Sōdai kakyo seido kenkyū*, 98.

14. Chia, *Printing for Profit*, 118.

same tactics to impose the new commentaries and Wang's dictionary as required reading in government schools.[15] After the demise of the reforms, the Directorate increasingly lost ground in the printing world. Chia found that "the records of donations of Directorate editions to schools apply only for the Northern Song."[16] This suggests that the government abandoned the practice of the empire-wide distribution of standard texts in the early Southern Song. The Directorate's role accordingly changed from that of a major supplier of textbooks to that of inspector of more widely available commercially printed textbooks.

Starting in the first decades of the twelfth century, the commercial printing of examination aids burgeoned. Bookstores, private publishers, writers, critics, and editors—for the most part successful or frustrated examination candidates and teachers—issued encyclopedias, style manuals, anthologies of old and contemporary texts, collections of successful examination writings in all genres, annotated editions of the collected works of popular writers, rhyme books, and cheap editions of the Classics, the histories, and some works from the philosophy and literature sections.[17] Those central government institutions like the Directorate of Education and local and regional government institutions involved in printing quickly lost ground in the market for examination materials.

Titles of books that are no longer extant indicate that manuals and anthologies existed for all examination genres in the twelfth and thirteenth centuries. For example, *Guide to Essays on the Meaning of "The Book of Songs"* (*Shiyi zhinan* 詩義指南) was written by Duan Changwu 段昌武, a Southern Song *jinshi* who, it was claimed, ranked first in the meaning of the Classics session in both the local and the departmental examinations. Duan Changwu compiled another examination aid for those specializing in *The Book of Songs*, but is otherwise unknown.[18] For

---

15. Bol, *"This Culture of Ours,"* 232.

16. Chia, *Printing for Profit*, 118–19.

17. Poon, "Books and Printing in Sung China," 100–112. For a comprehensive listing of manuals in various genres, see Zhu Shangshu, *Songdai keju yu wenxue kaolun*, 261–301.

18. This work is mentioned in a seventeenth-century catalogue, Huang Yuji's *The Thousand Acre Studio Catalogue* (*Qianqing tang shumu*), 1.38a; cited in Zhou Yanwen, "Lun lidai shumu zhong de zhiju lei shu," 8. For Duan's other work on *The Book of Songs* and scanty biographical information, see *SKTY*, 15.306; and *SRZJZL*, 1703. For other examples of collections of essays on the meaning of the Classics, see Chen Dang's memorial

those opting for the poetry track, there were anonymous commercial anthologies such as *The Ten Thousand Gem Heights of Poetry* (*Wanbao shi shan* 萬寶詩山)[19] or *Guiding* Fu *with Annotations* (*Zhinan fu jian* 指南 賦箋).[20] Extant editions of writing manuals such as *The Art of the Exposition* (*Lun jue* 論訣) and *Secret Tricks for Responding to Policy Questions* (*Dace mijue* 答策秘訣), and anthologies such as *Standards for the Study of the Exposition* (*Lunxue shengchi* 論學繩尺) and *Standards for the Study of the Policy Response* (*Cexue shengchi* 策學繩尺), provide vivid evidence of how late Southern Song teachers and commercial printers mediated the dissemination of current standards for expositions and policy response essays (see Chapters 3 and 7; Fig. 1).

The Song court changed its strategies for dealing with commercial print shops to accord with the general policy of Great Impartiality. Commercial printers were allowed to market examination manuals as long as they submitted each title for inspection and approval to regional authorities and the Directorate of Education. This policy legalized the participation of commercial printers in the examination field; it also recognized the curricular activities of private teachers specializing in examination preparation. Private teachers supplied the printers with guides and anthologies, as we have seen in the case of the "Yongjia" teachers in Chapter 3. By allowing for the legality of commercially printed examination manuals, the Southern Song court broke with publishing laws issued in the early twelfth century. To judge from Lucille Chia's table of regulations concerning the commercial publication of examination literature, all such injunctions issued between 1101 and 1117 banned the commercial printing of such works.[21] The printing blocks of examination manuals were ordered destroyed, and awards were of-

---

discussed in the section "The Ban on False Learning and the Politics of Anthologizing" below.

19. The Siku editors referred to this work as "the broadest collection of Song dynasty examination poetry." It is mentioned in their review of another collection of examination poetry, *Liu Xuxi/Chenweng's Collection of Poems on the Four Seasons* (*Xuxi sijing shiji* 須溪四景詩集; *SKTY*, 164.3435).

20. This title is mentioned in Chen Zhensun's catalogue; *Zhizhai shulu jieti*, 15.458. Chen noted that this compilation and a companion volume were done by a bookstore and contained poetic expositions written up to the Shaoxi reign (1190–94).

21. Chia, *Printing for Profit*, 121–23.

fered to commoners for turning in printers engaged in the business of examination literature. The official reports and court orders issued between 1145 and 1253, by contrast, banned only commercially printed material unauthorized by local governments and the Directorate of Education. The frequent repetition of these injunctions (nine are listed in Chia's table for a one-hundred-year period) suggests the government's concern over its inability to function as the final authority in the examination manual business.

The government's insistence on official authorization for all publications qualifies the meaning of the policy of Great Impartiality. Great Impartiality expressed the court's decision to keep factional politics out of examination curricula. Not only did it imply a statement of intention on the court's part to allow diversity, but it was also a standard by which the court measured the activities and publications of its subjects. Examination papers and examination manuals could express a variety of opinions as long as the government did not perceive factionalist arguments in them. As the teachings of the Cheng brothers, who had opposed Wang Anshi's reform program, revived in the early decades of the Southern Song, exclusivist claims to orthodoxy began to impact examination writing. This, successive governments perceived as a political threat and a violation of the "broad scholarship" it saw as the standard for examination writing. The next two sections trace the trajectory of the government's response to this challenge.

### *Prohibitive Interventions in the Twelfth Century*

Throughout the twelfth century, the Southern Song government expressed its commitment to Great Impartiality. This policy led to the resurgence of the Learning of the Cheng Brothers (Chengxue 程學), which had been banned in the last decades of the Northern Song. Ironically, the ecumenical spirit fostered by the government's commitment to Great Impartiality also drove its successive campaigns against the adherents of Cheng Yi's teachings. The rhetoric of political and intellectual unity disguised the censorship of pro-war opinion.[22]

---

22. Cf. Hartman, "The Making of a Villain," esp. 89–93. Hartman emphasized the enforcement of political and intellectual uniformity under Gaozong and Qin Gui. Unlike

After the loss of the north to the Jin, the Northern Song reformers of the first decades of the twelfth century were blamed for the debacle. Grand Councilor Cai Jing 蔡京 (1046–1126) and his fellow reformers had made strenuous efforts to silence the voices of their critics;[23] supporters of Cheng Yi fared badly. During Gaozong's reign, those who had suffered were rehabilitated. Memorials called for conciliation and recognition of the strong points of different intellectual orientations such as the teachings of Cheng Yi and Wang Anshi. The court endorsed the call for conciliation.

The restoration of Cheng Yi's Learning did not go unchallenged. As early as 1136, Chen Gongfu 陳公輔 (1076–1141), then remonstrator of the right (*you sijian* 右司諫), launched an attack on those espousing the philosophy of Cheng Yi, accusing them of contempt of the dynastic law of impartiality. Chen submitted a general accusation of those affiliated with the Learning of Cheng Yi and asked for a ban on their theories. He objected to their intellectual genealogy, which claimed that the Cheng brothers had rediscovered the universal moral truths embedded in the Way of the sages after centuries of aberration in literati culture. Chen saw evidence of factionalist tendencies among believers in Cheng Yi's teachings in their exclusivist claims to truth and their unconventional discourse and behavior. There was some basis for the last claim. Contemporaries of Cheng Yi and his disciples noted that they dressed and walked differently from everybody else.[24] The language Cheng Yi developed to convey his philosophical theories sounded jargonistic to

---

Wang Anshi, however, Qin Gui's efforts were directed not so much at the construction of a shared intellectual and political culture, but at the censorship of those critical of his pro-peace stance. Ari Levine's dissertation, "A House in Darkness," provides an in-depth account and analysis of factionalism in the Northern Song period and discusses the activities of and retaliation against Cheng Yi's Luo party (Luo dang 洛黨) in the broader context of factional rhetoric and practice. Whereas histories of factional politics of the Song period tend to focus on the Northern Song period, Shen Songqin's study of the history of factionalism in the Southern Song period, *Nan Song wenren yu dangzheng*, looks at the continuation of factionalism throughout the remainder of Song history. He gives a narrative account of factional struggles, explains the political concepts structuring Southern Song factional politics, and considers the political, intellectual, and literary stakes at play.

23. Chaffee, *The Thorny Gates*, 79.

24. Tillman, *Confucian Discourse and Chu Hsi's Ascendancy*, 21.

contemporaries trained in the classical tradition or in Ancient Prose. Chen drew a parallel between the conformity in speech and behavior characteristic of his Chengist contemporaries and the so-called degeneration of the literati under the absolutist rule of Wang Anshi.[25]

Chen Gongfu's indictment of Cheng Learning was a clear sign that factionalist politics were alive and well despite the court's attempt to put such disputes to rest. Opposition to Cheng Yi's Learning in the 1130s was linked to support for Qin Gui.[26] Qin Gui had been captured by the Jurchen Jin along with 3,000 members of the imperial family and court in 1127. Originally an advocate of resistance, Qin Gui began to espouse a pro-peace stance during his captivity. He escaped from his Jurchen captors and appeared at the Southern Song court in 1130.[27] He advocated peace with the Jurchens at whatever cost. Emperor Gaozong, whose position on the throne depended on the continued captivity of the last Northern Song emperor, Qinzong 欽宗 (r. 1126–27), gradually became convinced that a pro-peace policy would guarantee the consolidation of his rule over the southern territories. He nominated Qin Gui to the post of councilor. Between 1138 (when the peace conditions were announced) and 1155 (the year of Qin Gui's death), peace with the Jurchens was the "court line."[28] Voices transgressing this line were the object of political suppression.

Qin Gui's pro-peace policy generated dissent among officials and the literati. Literati interested in the moral philosophy of Cheng Yi generally adopted a staunch pro-war stand. At court Qin Gui faced opposition from Zhao Ding 趙鼎 (1084–1147), a close adviser to Gaozong since 1127 and a supporter of an aggressive military policy toward the Jin. Blaming Wang Anshi and the reform government for the military debacle, he advocated the rehabilitation of members of the opposition groups, including those adopting the views of Cheng Yi. In the early 1130s, Gaozong continued to favor Zhao Ding. In recognition of Zhao's success in restoring peace to key areas along the Yangzi River,

---

25. Li Xinchuan, *Dao ming lu*, 3.24; *SS*, 28.528.

26. *SB*, 241–47.

27. In contemporary opinion, Qin Gui's alleged escape had been orchestrated by the Jurchen aristocrat and warlord Dalan (Hartman, "The Making of a Villain," 65).

28. Yü Yingshi, *Zhu Xi de lishi shijie*, 1: 373.

he was promoted to the post of second privy councilor (*canzhi zhengshi* 參知政事) in 1134.[29]

In this position, Zhao became the main obstacle to Qin's pro-peace policy efforts. Qin Gui associated the uncooperative attitude of his main rival with the opposition waged by men known as advocates of the Learning of Cheng Yi. The effort to oust the main opponents of Qin's pro-peace policy was therefore presented as an attempt to defeat a conspiracy among the followers of Cheng Learning. A key figure among the advocates of Cheng Learning, Hu Anguo 胡安國 (1074–1138),[30] was a court lecturer during the 1130s. Hu, whose commentary on *The Spring and Autumn Annals* would become the standard interpretation of this classic in Neo-Confucian curricula, wrote a lengthy rebuttal of Chen Gongfu's accusations against Cheng Learning. His defense of the teachings of Cheng Yi was read as a confirmation of the factionalist tendencies among Chengists at court. Subsequent protest by Chen Gongfu and others led to Hu's departure.

In the 1140s and 1150s, Qin Gui launched a series of attacks against "the confined learning" (*zhuanmen zhi xue* 專門之學) of his Chengist critics. He charged Zhao Ding with leading this faction.[31] In the 1140s, Qin Gui grew concerned over what he saw as the use of the civil service examinations to recruit supporters of Cheng Learning, men who were, in his view, *ipso facto* opposed to the peace treaty concluded between Jin and Song in 1142. The conclusion of the peace treaty and the subsequent elimination of Song generals with meritorious service in campaigns against Jin provided a new focus for the opposition. Qin Gui intensified his efforts to silence the opposition and continued to dismiss allies of Zhao Ding and alleged advocates of Cheng Learning from court positions, a process set in motion by Chen Gongfu's allegations and proposed measures.

To prevent the infiltration of new opponents, Qin Gui began to investigate the channels of bureaucratic recruitment. The civil service examinations came under intense scrutiny. In 1144, the court endorsed Palace Censor (*dianzhong shiyushi* 殿中侍御史) Wang Bo's 汪勃 (1088–

---

29. *SB*, 72–82.
30. *SB*, 434–36; van Ess, *Von Ch'eng I zu Chu Hsi*, 201.
31. Schirokauer, "Neo-Confucians Under Attack," 166.

1171) request to exhort examiners to fail all examination papers that "borrowed from confined subversive theories and strayed into aberrant and bizarre spheres."[32] The reference to confined theories and bizarre discourse was a clear indication of the court's intent to keep Cheng Learning out of the examination curriculum and advocates of Cheng Learning out of the bureaucracy.

The anti–"confined learning" tide subsided after Qin Gui's death in 1155. Even though Qin Gui's pro-peace policy remained the dominant policy at court, Qin's measures against Cheng Learning seemed not to have had the effect he desired. Complaints against the use of Chengist discourse in examination writing resurfaced in the 1160s, even as defenders of Cheng Learning passed the examinations in growing numbers. Conrad Schirokauer has shown that in the examinations held during the late 1150s and 1160s the number of successful candidates who would later be persecuted as advocates of False Learning and fit roughly into the earlier category of those supportive of the Learning of Cheng Yi tied the number of graduates who would pose as their persecutors.[33] In memorials from the 1160s and early 1170s, examiners and high officials began to complain about the appearance of Chengist discourse in examination essays. Examiners referred to the distinctive discourse of Cheng Learning in derogatory terms, noting the awkward language, the terminology reminiscent of Chan Buddhism, and the lack of a foundation for the new terminology in the classical tradition. Although no explicit reference was made to the teachings of Cheng Yi, it is likely that these indictments were directed against the infiltration of Chengist discourse into examination essays.[34]

The critics of the 1160s and early 1170s displayed the same concern about the philosophical turn in current examination writing as Chen

---

32. Li Xinchuan, *Dao ming lu*, 4.3b–4a; cited in Yuan Zheng, *Songdai jiaoyu*, 54*n*3.

33. Schirokauer, "Neo-Confucians Under Attack," 167.

34. *SHY, XJ*, 4.34b–35a, 39b, 40b–41a. The memorials date from Shaoxing 31 (1161), Qiandao 5 (1169), and Qiandao 7 (1171). The last memorial was written by Liu Zheng, who does not seem to have had any conflicts with those later branded as Learning of the Way affiliates. Indeed, Liu himself appeared on the list of those advocating False Learning; see Schirokauer, "Neo-Confucians Under Attack," 180; *SB*, 624–28. Cf. Araki Toshikazu, *Sōdai kakyo seido kenkyū*, 395–96. See also Zhu Shangshu, *Songdai keju yu wenxue kaolun*, 242–60.

Gongfu had in his description of Cheng Learning at court. They were concerned with the cultural implications of the rise of philosophical discourse but, unlike the critics under Qin Gui's regime, steered clear of accusations of political factionalism. As advocates of historical scholarship and classical writing, they inferred from the cultural and political history of the Jin 晋 dynasty (265–316) that metaphysical speech led to administrative disaster. One critic, Probationary Imperial Recorder (*shi qijulang* 試起居郎) Liu Zheng, described the rise of philosophical discourse as a problematic trend in the examination field because of its effects on elite culture, but he was careful not to blame the teachings of Cheng Yi for this development.[35] Memories of the reigns of Wang Anshi and Qin Gui taught that direct accusations led to fierce political battles. Liu Zheng instead proposed that examiners and students alike be reminded of the general criteria used in the examinations. He singled out conformity with the prosodic or structural requirements of each genre and the demonstration of broad scholarship (see Chapter 2).

The success of men supportive of Cheng Learning continued in the 1170s and 1180s. In the four examinations held during the 1170s and early 1180s, prospective victims of the campaigns against False Learning did even better, outshining their future opponents sixteen to two.[36] The earliest explicit indictments of proponents of Cheng Yi's teachings for subverting the conventions of examination writing and literati culture date from this period. In 1178, Censor (*shiyushi* 侍御史) Xie Kuoran 謝廓然 (?–1182) demanded that the examiners be admonished not to select scholars on the basis of the teachings of Cheng Yi and Wang Anshi. His complaint was overtly political. He charged the proponents of Cheng Yi's teachings not only with promoting theories with no foundation in the classical tradition, but also with using the examinations to further partisan goals.

---

35. Liu Zheng's intervention at this time is noteworthy in light of his later attempts to mediate factional conflict. Liu Zheng became a councilor in 1186 and remained at the top of the civil administration until 1194. In this capacity he recommended for office Learning of the Way advocates, including Zhu Xi, but also continued to work alongside allies of the former grand councilor Wang Huai discussed below. Zhu Xi criticized him for not making a clean sweep. For Liu's role in government in the 1180s and 1190s, see Yü Yingshi, *Zhu Xi de lishi shijie*, vol. 2, chaps. 10–11.

36. Schirokauer, "Neo-Confucians Under Attack," 167.

The sharper focus of Xie's complaint and the memorials that succeeded it accorded with the growing animosity toward Zhu Xi. In his publications, political acts, and educational activities, starting in the 1160s and lasting through the 1190s, Zhu Xi presented himself as the leading spokesperson of those supportive of Cheng Yi's teachings. Through his efforts, Cheng Learning was transformed into an intellectual movement, which was by the last decades of the twelfth century generally referred to as the Learning of the Way (see the Introduction). Zhu Xi institutionalized the Chengist call for the moral reform of society and created an exclusivist identity for the members of this movement.

Between 1163 and 1177, Zhu Xi completed a set of publications that outlined the genealogy of the Learning of the Way and made the legacy of its teachers available in manuscript or print form. He completed *The Surviving Works of the Cheng Brothers* (*Cheng shi yishu* 程氏遺書) in 1163 and established the status of Zhou Dunyi and Zhang Zai, two other eleventh-century philosophers, as founders of the Learning of the Way by publishing three of their writings in the next two years. In Zhu Xi's view, Zhou Dunyi's *Diagram of the Supreme Ultimate* and *Penetrating the Changes* (*Tongshu* 通書), Zhang Zai's *Western Inscription*, and the collected writings of the Cheng brothers, fully expressed the teachings of the Learning of the Way. Zhu Xi explained the relationship among these thinkers in *Record of the Source of Cheng Learning* (*Yiluo yuanyuan lu* 伊洛淵源錄) in 1173. In this work, Zhu Xi clarified the transmission of the Learning of the Way from the eleventh century through the early twelfth century. *Record of the Source of Cheng Learning* is a collection of biographical records, starting with the founders of Cheng Learning and continuing with the friends and disciples of the Cheng brothers.[37] The genealogy drawn in this work and the publication of the works of the masters in this line of transmission carried great significance.

Zhu Xi's creation of a continuous lineage of masters introduced disciples into a community of transmitters of the Way. They became inductees by attaching themselves to a teacher who could legitimately claim to be a link in the chain that reached back to the masters of the eleventh century. For his part, Zhu Xi declared himself the legitimate

---

37. This work is discussed in Tillman, *Confucian Discourse and Chu Hsi's Ascendancy*, 114–19; Hervouet, *A Sung Bibliography*, 222–23; and Wilson, *Genealogy of the Way*, 160.

present link in the chain not only because of the intellectual genealogy of his teachers,[38] but also because of his efforts to clarify the transmission of the Learning of the Way in voluminous publications. In 1175, Zhu Xi's efforts to reconstitute a community of transmitters of the Learning of the Way culminated in the printing of *A Record for Reflection*.

Zhu Xi took the initiative in the compilation of this anthology, but sought the help of Lü Zuqian, who was familiar with the practice of anthologizing and maintained an interest in the teachings of the Cheng brothers (see Chapter 4).[39] *A Record for Reflection* was intended for beginning students of the Learning of the Way. It introduced them to the teachings of the four eleventh-century masters whose works Zhu Xi had published in the preceding decade. Its organization was based on a set of concepts Zhu Xi perceived as central to the Learning of the Way tradition. The work covered the steps in moral self-cultivation and the relationship of the Learning of the Way to other intellectual traditions and practices. *A Record for Reflection* and its role in the development of Learning of the Way examination materials is discussed in Chapter 6. It is relevant in the present context because the first open accusations of factionalism against Cheng Learning in 1178 and against the Learning of the Way in 1183 followed soon after the printing of this text. The excerpts under each of the headings in *A Record for Reflection* created the impression that the four masters had been working on a common project, a project that was continued in Zhu Xi's efforts to reconstitute the community of Learning of the Way followers.

Zhu Xi alternated work on the lineage and legacy of the eleventh-century masters with work on a new classical canon. In the 1160s and 1170s, he prepared commentaries on the Four Books (*The Analects, Mencius, The Doctrine of the Mean,* and *The Great Learning*). In 1163, he wrote *The Essential Meaning of "The Analects"* (*Lunyu yaoyi* 論語要義), followed

---

38. Zhu Xi's father, Zhu Song 朱松 (1097–1143), was a student of Yang Shi, himself a student of Cheng Yi. He supervised his son's education and, at his death in 1143, entrusted Zhu Xi's education to teachers with a solid grounding in Cheng Learning. Zhu Xi had passed the civil service examinations in 1148, and in the 1150s he attached himself to Li Tong 李侗 (1093–1163), who was, like Zhu Song, a disciple of Yang Shi. See Okada, "Shushi no fu to shi"; Shu Jingnan, *Zhuzi da zhuan*, 1–196; and Tillman, *Confucian Discourse and Chu Hsi's Ascendancy*, 40–41.

39. See Chapter 6, note 67.

by a combined edition of his commentaries on both *Mencius* and *The Analects* in *The Essential Meaning of "The Analects" and "Mencius"* (*Lun Meng jing yi* 論孟精義) in 1172. A printed edition of *The Essential Meaning of "The Analects"* appeared in 1167; four printings of the combined title appeared in three different locations between 1172 and 1180.[40] In 1174 he prepared drafts of his influential works on *The Great Learning* and *The Doctrine of the Mean*: *"The Great Learning" and "The Doctrine of the Mean" in Chapters and Verses* (*Daxue Zhongyong zhangju* 大學中庸章句), and *Questions About "The Great Learning" and "The Doctrine of the Mean"* (*Daxue Zhongyong huo wen* 大學中庸或問). In 1177 he completed drafts of *Collected Commentaries on "The Analects" and "Mencius"* (*Lun Meng ji zhu* 論孟集注) and *Questions About "The Analects" and "Mencius"* (*Lun Meng huo wen* 論孟或問). A commercial printing of the last two titles appeared soon afterward in Jianyang, a printing center famous for its focus on examination preparation publications.[41]

Some of the classical texts in Zhu Xi's combined commentaries had long had a special status in examination curricula. In the twelfth century, *The Analects* and *Mencius* were required reading for candidates in the Classics track. Zhu Xi's grouping of *The Analects* and *Mencius* with *The Doctrine of the Mean* and *The Great Learning*, however, proved controversial. The last two titles were chapters in *The Book of Rites* and did not have an independent status in the classical canon. Zhu Xi's claims that these four works explained the central teachings of the sages of Antiquity fully and more clearly than did the traditional canon of the Five Classics raised concerns among officials and scholars. The attempt to prioritize the Four Books over the Five Classics was read as evidence of the exclusivist and artificial reconstruction of the classical tradition in Cheng Learning.

Zhu Xi's works on the eleventh-century masters and the Four Books in the 1160s and 1170s suggest that by the mid-1170s he had developed a coherent vision of the central beliefs and texts of the Learning of the

---

40. Shu Jingnan, *Zhuzi da zhuan*, 297, 378, 447.

41. Ichiki, *Shu Ki monjin*, 288–89, 401–2. Shu Jingnan (*Zhuzi da zhuan*, 300) dates Zhu Xi's first drafts of *"The Great Learning" and "The Doctrine of the Mean" in Chapters and Verses* to 1172–73. For the commercial editions of *Collected Commentaries on "The Analects" and "Mencius"* and the *Questions*, see Shu, 379.

Way.[42] In his political life, Zhu Xi demonstrated a commitment to these beliefs. In a memorial addressed to Xiaozong in 1163, he exhorted the emperor to improve his personal moral conduct. Even though Zhu Xi supported the eventual recuperation of the northern homeland, he criticized the emperor for his attempts to regain the northern territories in the preceding months and urged the emperor to prioritize moral self-cultivation before waging military campaigns in the north. Throughout the 1160s and 1170s, Zhu Xi declined several appointments. These refusals to serve in the bureaucracy were highly controversial because of their implicit political criticism.[43] Once in office, he proved very tough on matters of bureaucratic morale. In 1182, he indicted the respected scholar and well-connected prefect Tang Zhongyou.[44] Among other things, he charged Tang with the use of government funds for the publication of the works of Xunzi and Yangzi, a charge considered extreme even by those sympathetic to Zhu Xi's cause.[45]

Apart from his busy publication record and controversial political career, Zhu Xi attracted the court's attention for his efforts to institutionalize the teaching of the Learning of the Way in schools and academies. In the 1180s and 1190s, he established or renovated academies in which students discussed and practiced the ethics of the Learning of the Way. The most famous and controversial example was his restoration of the White Deer Grotto Academy (Bailu dong shuyuan 白鹿洞書院) in 1180. Located in Nankang 南康 (Jiangnan East circuit), the White Deer Grotto Academy had achieved empire-wide renown as one of only four academies to have received the support of the early Song court. By 1179, when Zhu Xi became magistrate of Nankang, the acad-

---

42. In his study of Zhu Xi's teaching activity, Ichiki (*Shu Ki monjin*, 211 and *passim*) similarly argues that Zhu Xi's moral philosophy was fully developed by the mid-1170s.

43. Schirokauer, "Chu Hsi's Political Career," *passim*; Yuan Zheng, *Songdai jiaoyu*, 56.

44. Schirokauer, "Neo-Confucians Under Attack," 169; and Haeger, "The Intellectual Context of Neo-Confucian Syncretism," 506, both of which follow the standard account in *SSJSBM*, 80.679. For the conflict between Zhu Xi and Tang Zhongyou, see Tillman, *Confucian Discourse and Chu Hsi's Ascendancy*, 134; Zhu Ruixi, "Songdai Lixuejia Tang Zhongyou," 43–53; and other works mentioned by Tillman, ibid., 280*n2.

45. Zhu Chuanyu, *Songdai xinwen shi*, 217. Chen Liang, still on good terms with Zhu Xi in the early 1180s, disapproved of Zhu Xi's intervention; see Tillman, *Confucian Discourse and Chu Hsi's Ascendancy*, 136, 162.

emy had lost its former glory. Zhu Xi mustered local support to rebuild the academy and designed regulations for it based on the moral philosophy of the Learning of the Way. The renovation was widely publicized. Zhu Xi sought the court's approval for his undertaking and submitted repeated requests to the emperor asking that the court replace the name plaque and the commentaries on the Classics that Taizong had given to the academy but had been lost. Imperial recognition for the academy would have translated into imperial support for Zhu Xi's emerging Learning of the Way movement. In the 1180s and the 1190s, no support was forthcoming, and Zhu Xi received much criticism for his efforts.[46]

In reports about examination writing from the 1180s, accusations against Chengist discourse were discussed in growing detail. In an 1180 memorandum, Palace Library Assistant (*bishulang* 秘書郎) Zhao Yanzhong 趙彥中 (*js.* 1169) expressed great concern over the fact that exclusivist claims for the Chengist interpretation of the Classics were gaining ground:

For the writings for the examinations there are established regulations. Nowadays those who revere the teachings on nature and principle (*xingli zhi shuo*) praise one another for their random words and unfounded theories. It is permissible that scholars who trust in the Way and discipline themselves take the sages and wise men of the Six Classics as their teachers. However, as a result of their setting up a separate "Luo Learning," their indulging in awkwardness and inciting foolishness, the scholarly atmosphere declines day by day and human talent becomes scarcer and scarcer.[47]

科舉之文成式具在. 今乃祖性理之説, 以浮言游詞相高. 士之信道自守, 以六經聖賢爲師, 可矣. 而別爲洛學, 飾怪驚愚, 士風日弊, 人才日偷.

The rhetoric in the memorials of Xie Kuoran and Zhao Yanzhong was reminiscent of that used in the campaign against "confined learning" under Qin Gui in the 1140s. Xie and Zhao wrote that the metaphysical discourse of Cheng Learning ("the teachings on nature and principle") created artificial boundaries within the larger literati com-

---

46. Chaffee, "Chu Hsi and the Revival of the White Deer Grotto Academy," 58–59; idem, "Chu Hsi in Nan-k'ang"; Walton, *Academies and Society*, 25–41.

47. *SSJSBM*, 80.679.

munity. Those participating in Chengist discourse discussed core values (such as "the Way") and core traditions (such as "the Six Classics") shared by all cultured elites, but they glossed their meaning in a discourse that could be accessed only by members of Chengist circles. The examiners portrayed the candidates invoking Chengist explanations as members of a community of discourse, in which "those who revere the teachings on nature and principle praise one another for their random words and unfounded theories." Xie and Zhao noted that this exclusivist discourse was spreading through examination writing and that the effects were becoming more apparent every day.

The renewed apprehension about the "confined" and antagonistic character of Chengist discourse escalated into a straightforward attack against "the Doctrine of the 'True' Way" in 1183 when Censor Chen Jia 陳賈 submitted a memorial to the court in which he questioned the sincerity of the claims of Chengist philosophical discourse. His denunciation of Doctrine of the "True" Way scholars' hypocritical display of sincerity and moral uprightness foreshadowed the campaign against False Learning a decade later.[48] Censor Chen fought the Learning of the Way on its own terms. He undertook an analysis of Chengist discourse and reappropriated terminology that he saw as common to the classical tradition but that had been redefined in Chengist discourse.[49]

He started by defining the meaning of "sincerity" (*cheng* 誠) and "falseness" (*wei* 僞). "Sincerity" was "the matching of exterior and interior"; "falseness," "a mismatch between exterior and interior." Chen's definition of sincerity and the deliberate link to falseness challenged Learning of the Way interpretations of "sincerity." According to Daniel Gardner, Cheng Yi and Zhu Xi understood "sincerity" (he renders *cheng* as "truthfulness") as meaning to be true to one's nature. The individual maintains the link to his endowed nature by keeping his original mind unified and free from distractions.[50] Censor Chen pointed out that sincerity meant something different outside Doctrine of the "True" Way circles. Advocates of that doctrine had laid exclusive claim to passages

---

48. Haeger, "The Intellectual Context of Neo-Confucian Syncretism," 506; Schirokauer, "Neo-Confucians Under Attack," 169–70.

49. *SSJSBM*, 80.869.

50. Gardner, *Learning to Be a Sage*, 89.

from the Classics such as "making the will sincere and rectifying the mind" (*cheng yi zheng xin* 誠意正心). This passage from *The Great Learning* was indeed used by Cheng Yi and Zhu Xi to define the central message of the Learning of the Way. For them, it symbolized the centrality of the cultivation of the mind in the classical tradition. They integrated sincerity into their philosophy of the mind and interpreted it as a modality of self-cultivation. Chen Jia countered that the claim of possessing the correct interpretation of sincerity in Doctrine of the "True" Way discourse did not match the reality of its meaning within the classical tradition at large. Within the larger commentarial tradition, Chen argued, sincerity referred to the agreement between intent and expression and was not linked to the philosophical discussion of self-cultivation.

He continued that the Doctrine of the "True" Way interpretation of sincerity qualified as falseness in the traditional reading of the word. Exterior and interior were at odds in the discussion of sincerity in the Doctrine of the "True" Way. Chen charged that the performance of advocates did not match their claims of sincerity. As censor, Chen bore a responsibility to evaluate the performance of government officials. He concluded that Doctrine of the "True" Way discourse operated as a code that allowed incapable men to obtain power. He recommended that those whose background checks revealed an engagement with this doctrine be removed from office.

Chen Jia's indictment of the Learning of the Way was targeted at Zhu Xi and the growing influence of the movement he inspired. In contrast to the writings of his contemporaries, including those interested in Cheng Learning, references to the Learning of the Way abound in Zhu Xi's work. Chen Jia's critique of the use of the term was thus most likely a response to Zhu Xi's construction of the Learning of the Way as a coherent textual, intellectual, and moral community (see the Introduction). Zhu Xi's political activities contributed to the sense of urgency apparent in Chen Jia's recommendations.

In 1182, Zhu Xi came under close scrutiny after he had indicted Tang Zhongyou. As intendant for granaries, tea, and salt for the Liang Zhe East circuit (*tiju Liang Zhe dong lu changping cha yan gongshi* 提舉兩浙東路常平茶鹽公事), Zhu Xi held supervisory power over Tang Zhongyou, who held the post of prefect of Taizhou, one of the prefectures under Zhu's jurisdiction. Not only was Tang a noted scholar, as a relative of incumbent chancellor Wang Huai 王淮 (1127–89), he had powerful connections at court. Zhu Xi's indictment of Tang elicited immediate sharp reactions at court.[51] Wang Huai grew convinced that Zhu Xi and his allies were staging a political battle and supported Chen Jia's sharp attack on the Doctrine of the "True" Way in 1183.

Following Chen Jia's report, the court increased its efforts to monitor the examinations. The denunciatory tenor of the report had a demonstrable effect on the examinations at both the local and the departmental levels. At the local level, examiners became hesitant to recommend those with Learning of the Way sympathies for the higher-level examinations. The impact was felt even more strongly in the departmental examinations. Following the 1184 departmental examinations, one of the examiners was dismissed for violating the criteria set by the chief examiner. His transgression consisted of support for the Learning of the Way.[52] After the departmental examinations in 1187, the chief examiners—Chen Jia, by then junior grand master of remonstrance (*you jianyi dafu* 右諫議大夫), Hong Mai 洪邁 (1123–1202), and Ge Bi 葛邲—handed in a detailed report about violations of examination standards in the essays they had graded.[53] They pointed out three recurrent problems.

First, candidates had made illicit use of Song official historical sources. Sources such as the ongoing project on the history of the current dynasty (*guoshi* 國史) were not intended for public consumption. Song law forbade the dissemination of archival records as well as their publication. Yet, students referred to these sources, often incorrectly, and even repeated recent stories they had picked up from hearsay.[54]

---

51. Ibid., 6; Shu Jingnan, *Zhuzi da zhuan*, 484–97.

52. Yü Yingshi, *Zhu Xi de lishi shijie*, 2: 107–10.

53. *SHY, XJ*, 5.10a–11a.

54. For prohibitions, see Zhu Chuanyu, *Songdai xinwen shi*, 197–99, 210, 214. Some officials defended references to contemporary history and the use of official sources about contemporary history. See, e.g., the memorial by He Dan 何澹 (*js.* 1166) from 1188 in *SHY, XJ*, 5.11b–12a. Another official asked that examiners clearly relate the circumstances of recent events when making references to them in policy questions so as not to disadvantage students who did not have access to official documents (Zhu Chuanyu, ibid., 200). For an extended discussion of both prohibitions on and access to archival compilations, see De Weerdt, "Byways."

Second, examination papers violated the length requirements mandated in the official regulations.

According to the current regulations, a regulated prose-poem should be no longer than 360 characters; expositions should count 500 characters or less. Nowadays, in an essay on the meaning of the Classics, an exposition, or a policy response essay, some write up to 3,000 words. In the prose parts of the prose-poem they have sentences of up to fifteen or sixteen characters long, coming to a total of 500 to 600 characters per piece.[55]

考之今式, 賦限三百六十字, 論限五百字. 今經義、策論一道, 有至三千言; 賦散句之長者至十五六字, 一篇計五六百言.

Finally, the examiners listed thirteen expressions as illustrations of the intrusion of colloquial terminology derived from "heterodox teachings" into examination writing:

As for what we have called their eccentricities, expressions like "firm vision" (*dingjian*), "power" (*liliang*), "to imagine" (*liaoxiang*), "to divide one's efforts" (*fenliang*), "comes from somebody's collected works" (*laizi ji zhong*), "determined direction" (*dingxiang*), "view" (*yijian*), "appearance" (*xingxian*), "air" (*qixiang*),[56] "system" (*titong*), "closing off the mind" (*guxin*), "every mind has a ruler, still there is a lot of noisy wrangling" (*xinxin you zhu, huihui zhengming*), and "with one trampling one can arrive; while washing the hands it can be accomplished" (*yijiu ke dao, guanshui ke zhi*)—all these are heterodox and vulgar expressions.

所謂怪僻者, 如曰定見, 曰力量, 曰料想, 曰分量, 曰自某中來, 曰定嚮, 曰意見, 曰形見, 曰氣象, 曰體統, 曰錮心, 及心心有主、喿喿爭鳴, 一蹴可到、盥手可致之類, 皆異端鄙俗文辭.

The violations, hinted at in the earlier complaints mentioned above and listed in Chen Jia's report for the first time, are indicative of the informal character of the influence of Learning of the Way discourse on the examinations in the twelfth century. During this period, examination candidates learned about "the teachings of nature and coherence"

---

55. For a slightly different, probably edited, version of this passage, see *SS*, *XJ*, 2.3633; and Nakajima, "*Sōshi*' '*Senkyoshi*' *yakuchū*, 1: 223*n*215. The transmitted *SHY* text has "thirty" (*sanshi* 三十) instead of "three thousand" (*sanqian* 三千).

56. For the significance Zhu Xi attached to comportment in political communication, see Marchal, "Lun Zhu Xi, Lü Zuqian de jian jun sixiang."

in personal exchanges with teachers and from texts recording oral exchanges. Examination papers written by early followers of the Learning of the Way resembled the recorded sayings (*yulu* 語錄), disciples' transcripts of conversations and lectures of prominent teachers. They embodied the discursive break between Learning of the Way interpretations of the classical texts and conventional commentary.[57] Chapter 7 analyzes the transformation of late twelfth-century examination papers and the relevance of recorded conversations in understanding these transformations in more detail. The three violations observed in the 1187 departmental examinations reflect some features of early Learning of the Way examination writing.

The technical terms listed in the 1187 report related to the discussion of mind-cultivation in Learning of the Way discourse. "To divide one's efforts," for example, meant to adopt a gradual approach to mind-cultivation. The use of the technical vocabulary of mind-cultivation is well attested in Zhu Xi's work. Half the terms in the quotation appear in *Master Zhu's Classified Recorded Sayings* (*Zhuzi yulei* 朱子語類), which was based on records of conversations between Zhu Xi and his disciples held for the most part in the 1180s and 1190s, and all but two are attested in his commentaries and other writings.[58] The extension of Learning of the Way philosophical discourse into examination essays in the late twelfth century can be confirmed in expositions from this period. "To divide one's efforts" appears in Zhu Xi's recorded conversations and in an examination essay of one his disciples from around 1193 (Chapter 7).

The cavalier attitude toward the regulations on length and the conventions of prosody described in Chen Jia's report, while not necessar-

---

57. Gardner, "Modes of Thinking and Modes of Discourse in the Sung."

58. According to the *Index of Colloquial Terms in Zhu Xi's Classified Sayings*, *liliang, yijian*, and *qixiang* occur throughout *Zhuzi yulei*; other terms like *fenliang, titong*, and *laizi* appear between seven and four times each (Shiomi, *"Shushi gorui" kōgo goyi sakuin*). *Dingjian, dingxiang, liaoxiang, xingxian*, and *xinxin you zhu* do not appear in *Zhuzi yulei*, but Zhu Xi used them in texts included in his collected works, *Huian ji*. A search of Zhu Xi's work in the electronic version of the Siku quanshu yielded multiple hits for these terms. Some of the terminology listed in the memorial appears with lesser frequency in Zhu Xi's commentaries on the Four Books, *Sishu zhangju*. A search of the online edition at Academia Sinica yielded ten occurrences for *qixiang* and one for *fenliang*.

ily limited to advocates of the Learning of the Way, fit in with the adaptation of philosophical discourse to examination writing. Zhu Xi spoke out against the preoccupation with compositional techniques in examination preparation. He developed a theory of writing as the unmediated result of philosophical understanding and denied it independent status. For Cheng Yi and Zhu Xi, the understanding of philosophical truths enabled the mind to express these truths effortlessly in speech and writing alike. Unmediated speech was a sign of personal moral insight or at least the commitment to attain it; records of the speech of a teacher were aimed at preserving the personal communication of truth without the interference of established literary conventions. The recorded conversations of Cheng Yi and his disciples, three collections of which Zhu Xi edited between 1159 and 1173, provided a model for the transposition of orally and personally communicated truth in written form.[59] Contemporary critics noted that the application of the model of the unrestricted expression of moral truth to examination writing led to violations of conventions of prosody and length. Complaints about the use of esoteric expressions and unrestrained language escalated and during the last decade of the twelfth century led to a demand that books of recorded sayings be destroyed.[60]

Philosophical discourse in Learning of the Way circles also led to the discussion of Song politics. The moral reformation of the bureaucracy was a priority for Zhu Xi. In memorials to the court, he urged the emperor to surround himself with morally superior men, and he backed up this recommendation with criticisms of those in charge. His recorded conversations from the 1180s and 1190s contain negative comments on twelfth-century statesmen such as Qin Gui.[61] Earlier in his career, Zhu Xi encouraged students to evaluate Qin Gui's character and politics in examination questions of his own design (Chapter 7). The stories about Qin Gui may have come to Zhu Xi through hearsay and his

---

59. Even as the recorded conversations broke with extant conventions of written commentary and the literary language more generally, they established new literary conventions. The recording of the conversations was necessarily subject to an editorial process that transformed the original speech.

60. See the memorial by Ye Zhu cited below.

61. *ZZYL*, esp. chap. 131, "The Reigning Dynasty: Individuals from the Establishment of the Southern Song Through the Present Day."

connections at court, but he also relied on draft histories of the reigns of eleventh- and twelfth-century emperors. In conversations with students, Zhu Xi related that he had read the Veritable Records (*shilu* 實錄; digests of court diaries) of the Huizong and Gaozong reigns, and he occasionally referred to them.[62] Zhu Xi's admission was not unusual. Despite the ban on circulation of the archival records of the reigning dynasty, collections of such records appeared on the book market and were updated with regularity.[63] The documented discussion of contemporary politics in Zhu Xi's exchanges with students and its appearance in their examination papers added fodder to the emerging campaign against the Learning of the Way.[64]

The 1187 report confirmed Zhao Yanzhong's feeling that few remained unaffected by Learning of the Way ideology. The examiners stated that they had failed the papers manifesting the symptoms most acutely but had been unable to fail all papers with these deficiencies because the selection quota had to be met. They further admitted that even top-ranking essays were plagued by these problems. Based on their analysis, Chen Jia and the other examiners argued for a radical solution. Earlier memorials had requested that examiners be instructed to fail subversive essays. Chen Jia and the other examiners proposed that examination candidates be directly addressed about the problem of Learning of the Way discourse. The court agreed to forward their report to all prefectural schools through the Directorate of Education and ordered that the list of improprieties be posted at prefectural schools throughout the empire.

Over the next few years, mounting criticism of the Learning of the Way was counterbalanced with significant support from high-ranking court officials. With the dismissal of Grand Councilor Wang Huai in 1188, one of the principal supporters of the attack on the Learning of

---

62. *ZZYL*, 104.2624, 107.2665–66. Ironically, Hong Mai, one of the examiners pointing out the violation of the ban on historiographical records in examination writing, shared a copy of the records of several reign periods in the *History of the Current Dynasty* (*Guoshi* 國史) with Zhu Xi; see Shu Jingnan, *Zhuzi da zhuan*, 665.

63. De Weerdt, "Byways."

64. See the policy questions designed by Zhu Xi and other teachers affiliated with the Learning of the Way discussed in Chapter 7.

the Way left the court. Zhou Bida,[65] grand councilor of the right since 1187 and promoted to grand councilor of the left in 1189, was supportive of the Learning of the Way and recommended Zhu Xi to the court.[66]

Zhu Xi had an audience with Emperor Xiaozong but declined the position he was offered in the Ministry of War.[67] He subsequently submitted a memorial explaining that in order to solve the empire's administrative problems, the emperor needed to concentrate on self-cultivation and surround himself with "a group of morally superior men."[68] Senior officials read his refusal and 1188 memorial as a further attempt to monopolize political power and as confirmation of Chen Jia's ideological analysis of Zhu Xi's program of moral reform. Zhou Bida's efforts to bring reform-minded scholar-officials like Zhu Xi to court escalated the tensions between officials associated with Wang Huai's regime and his politics of the status quo, on one hand, and the alliance of the advocates of reform (which included both Learning of the Way and Yongjia scholars), on the other hand.

The accession of Guangzong in 1190 brought a respite for the Learning of the Way supporters, and they took advantage of the change to push their agenda. They continued to recommend and recruit supporters for top court positions. Their efforts in this regard focused on the Censorate. As the organ that could make and break bureaucratic careers, the Censorate had operated as a nexus of factional struggles throughout

---

65. *SB*, 275–77.

66. Tillman, *Confucian Discourse and Chu Hsi's Ascendancy*, 135–39; Schirokauer, "Neo-Confucians Under Attack," 190; idem, "Chu Hsi's Political Career," 176–79.

67. This appointment has received conflicting interpretations among Zhu Xi's biographers. Shu Jingnan (*Zhuzi da zhuan*, 645) interpreted it as an attempt by the court to sidetrack Zhu Xi. While appearing to give him a court position, this appointment was in this view engineered to sideline Zhu Xi from court politics. As a bureau director in the Ministry of War (*Bingbu langguan* 兵部郎官), Zhu Xi would have been working under Lin Li and in a department dominated by the faction of Councilor Wang Huai. And indeed Lin Li sent in an impeachment letter the day following the appointment. More recently, Yü Yingshi (*Zhu Xi de lishi shijie*, 2: 251), who does not refer to Shu Jingnan's works, read the appointment as evidence of Emperor Xiaozong's support for Zhu Xi and his attempt to bring him and his political allies to power.

68. For a detailed discussion of Zhu's memorial, see Shu Jingnan, *Zhuzi da zhuan*, 706–23.

the twelfth century. Supporters of the Learning of the Way such as Liu Guangzu 劉光祖 (1142–1222) used it to similar effect. In 1190, Liu was promoted to the influential position of palace censor (*dianzhong shiyushi* 殿中侍御史). He submitted letters of impeachment in quick succession, targeting those like Chen Jia who had fired the first salvo in the emerging False Learning campaign as well as others who were undermining the reformers' use of the Censorate in their attempt to overhaul the bureaucracy.[69] Liu Guangzu's work on behalf of the Learning of the Way had a temporary effect. Guangzong endorsed his 1190 memorial requesting that the claims to orthodoxy for the Learning of the Way be granted legitimacy. The Censor asked the emperor to recognize these claims so as to forestall future attacks.[70]

## *The Ban on False Learning and the Politics of Anthologizing*

Guangzong's ratification of Liu Guangzu's defense of the Learning of the Way did not bring about a major shift in its status. The movement continued to expand in the provinces, but the court remained split between those who denied its exclusivist claims and those who tried to defend its program against political caricaturization. Another urgent concern preoccupied court officials in the early 1190s when it became apparent that Emperor Guangzong was proving incapable of performing the imperial functions. When he refused to attend to the funeral of his father, court officials secured his abdication in favor of the crown prince. The principal agents in this intervention were Zhao Ruyu 趙汝愚 (1140–96), administrator of the Bureau of Military Affairs (Shumi yuan 樞密院) and a member of the imperial clan, and Han Tuozhou 韓侂胄 (?–1207).[71] In his capacity of supervisor of the Palace Postern (Gemen 閤門), Han Tuozhou served as a guard at imperial audiences.

Ningzong's succession initially augured well for Learning of the Way advocates. Zhao Ruyu immediately recommended Zhu Xi. Only one

---

69. Yü Yingshi, *Zhu Xi de lishi shijie*, 2: 258–60, 288–93.

70. Li Xinchuan, *Dao ming lu*, 6.8b–12b; cited in Yuan Zheng, *Songdai jiaoyu*, 67.

71. Chaffee, "Chao Ju-yü, Spurious Learning, and Southern Sung Political Culture." On Han Tuozhou, see *SB*, 376–84.

month after Ningzong's enthronement, Zhu was appointed to the position of court lecturer. Zhu's lectures to the new emperor focused on *The Great Learning*, the first and core text in Zhu Xi's new canon of the Four Books. His lectureship was short-lived, however; it lasted for only forty-six days.[72] Zhu Xi's dismissal resulted from the escalating conflict between Zhao Ruyu and Han Tuozhou. The alliance between Zhao and Han broke down soon after Ningzong's ascension to the throne. Han Tuozhou dismissed Zhao Ruyu's appointees and allies. Zhu Xi and Chen Fuliang were among those sent away from court because of their connections to Zhao. In 1195, Han Tuozhou secured the demotion and eventually the exile of Zhao himself. Through these dismissals and the purges that followed, Han maintained control over the court and the bureaucracy during the next decade (1194–1206). For this bureaucratic overhaul, Han relied on the support of officials who continued to espouse the status quo politics of Wang Huai even after the latter's departure from court in 1188. They likewise saw in Han Tuozhou an opportunity to gain access to the emperor and rid themselves of their opponents with imperial support.[73]

The dismissal of Zhao Ruyu and those associated with him aroused a wave of protests from men with Learning of the Way sympathies.[74] Consequently, toward the middle of 1195, court officials defending the political status quo prevailed on Han Tuozhou to direct existent misgivings about the Learning of the Way into a campaign against False Learning (*weixue* 偽學). Arguments over the examinations once more played a prominent role in the justification of the action against the Learning of the Way.

The supervising examiners of the 1196 departmental examinations, Ye Zhu and Liu Dexiu 劉德秀 (?–1208), both of whom had become

---

72. Gardner, *Learning to Be a Sage*, 6. Biographers have traditionally counted either forty or forty-six days depending on how the starting date of his appointment is determined (Yü Yingshi, *Zhu Xi de lishi shijie*, 2: 216–20).

73. For a description of the cooperation between the anti-reformers and Han Tuozhou, as well as the tensions inherent in this relationship between bureaucrats and a court favorite, see Yü Yingshi, *Zhu Xi de lishi shijie*, 2: 374–81.

74. Schirokauer, "Neo-Confucians Under Attack," 177–79; Tillman, *Confucian Discourse and Chu Hsi's Ascendancy*, 140.

staunch opponents of Learning of the Way politics,[75] confirmed the allegations Censor Chen Jia made in 1183. They added to the charges by accusing the leaders of the Learning of the Way of disturbing the social and political order of the empire. In a memorial submitted after the examinations, they charged its leaders with "stealing the power of the sovereign and causing a stir in the empire."[76] The charge of rebellious intent could not be ignored and, if proved, would warrant more thorough measures. The accusers renamed the Learning of the Way "False Learning" and thus branded the movement as inherently subversive. To curb the influence of the movement, the examiners recommended that recorded sayings and similar materials be destroyed.

A few weeks later, Ye Zhu, by then minister of personnel, explained the rationale behind his demand for the ban on recorded conversations:[77]

For the last twenty years scholars have been growing accustomed to False Learning. They have destroyed their innate minds. They find the Six Classics, the histories, and the philosophers insignificant, and official regulations irrelevant. They devote themselves to false teachings in recorded sayings to cover up the baseness of their hollowness and laziness. They mix in Chan words so as to deceive others. With the grading at the Imperial College during the triennial major examinations in mind, adherents of False Learning make special use of awkward expressions and secret codes so that they can be recognized and put in the front ranks. As a consequence, real talents are not selected.

. . . We hope that because of these defects a decree will be issued to the officials to admonish the scholars to take Confucius and Mencius as their teachers and to study the Six Classics, the philosophers, and the histories. Let them no longer transmit recorded sayings, whereby they spread their falseness, misappropriate fame, and deceive the world.

二十年來, 士子狃於偏學, 沮喪良心, 以六經子史爲不足觀, 以刑名度數
爲不足考, 專習語錄詭誕之說, 以蓋其空疏不學之陋, 雜以禪語, 遂可欺

---

75. Ye Zhu was impeached by Liu Guangzu in 1190 (Yü Yingshi, *Zhu Xi de lishi shijie*, 2: 288–89). Liu Dexiu nurtured a deep-seated hatred for the Learning of the Way after several of its supporters had distanced themselves from him at a meeting called by Liu Zheng during the latter's tenure as councilor (1189–94) (ibid., 2: 338–47).

76. *SSJSBM*, 80.873; *XZZTJ*, 726.

77. *SHY, XJ*, 5.17b–18a; cf. *WXTK*, 5.302b. Schirokauer ("Neo-Confucians Under Attack," 180) mistakenly dates this memorial to 1197.

人. 三歲大比, 上庠校定, 爲其徒者專用怪語暗號, 私相識認, 輒寘前列.
遂使眞才實能, 反擯不取.

   ... 欲望因今之弊, 特詔有司, 風論士子, 專以孔孟爲師, 以六經子史爲
習, 毋得複傳語錄, 以滋其盜名欺世之僞.

Examiners had been protesting the infiltration of metaphysical jargon
into examination writing for decades. By advocating a ban on the publication and distribution of recorded conversations, Ye Zhu struck at the
heart of Learning of the Way instruction. Recorded conversations functioned as a major vehicle for the transmission of Learning of the Way
teachings in the twelfth century. They contained the teacher's understanding of the transmitted truth of the Learning of the Way and included the teacher's explanation of passages from the Classics and the
histories, a feature that proved convenient for examination candidates.

   The recorded conversations also embodied the pedagogical ideals of
the Learning of the Way. For disciples interacting with a teacher like
Zhu Xi, recording his conversations testified to the immediacy of their
relationship with the teacher even after they had left. Through their records, these disciples linked themselves as witnesses to a living chain of
transmission. For students who learned about the teacher through the
recorded conversations, the question-and-answer format allowed them
to imagine themselves as students asking questions of the teacher and
listening to his replies.[78] The teacher and his words provided access to
the moral truths of the Learning of the Way more readily than the classical canon.

   The reference to Chan Buddhism in the report echoes the common
twelfth-century perception that the Learning of the Way discourse of
mind-cultivation was based on contemporary Chan Buddhist discussions. The use of philosophical jargon was definitely not limited to the
recorded conversations of Chan or of Learning of the Way origin;
commentaries on the Classics written by Zhu Xi and those thinkers he
included in the Learning of the Way lineage made ample use of it.
However, the prevalence of unrestricted philosophical discussion in the
recorded conversations made them a symbolic target for the campaign.
The recorded conversations were not strictly linked to the established

---

78. Gardner, "Modes of Thinking and Modes of Discourse," 586–88.

sources of tradition such as the Classics and the histories mentioned in the examiners' report. Recorded conversations therefore allowed greater freedom of interpretation than standard commentaries on the Classics. The teacher elaborated on selected phrases or short passages from the Classics at great length and interpreted them in terms of contemporary philosophical discourse.

Ye Zhu recommended the destruction of recorded conversations in order to prevent exclusivist traditions of scholarly opinion from taking over the examination preparation curriculum. Chapter 7 explains how the genre of the recorded conversations functioned as a model for examination writing. In addition to the ban on recorded conversations, Ye Zhu proposed the following countermeasure:

We also ask that all schools, from the Imperial College to the prefectural schools, send in the top three examination writings in the monthly examinations to the Censorate for inspection. In the case of the Imperial College, they should be sent monthly; in the case of [the schools in] all the circuits, these should be sent each term. In the case of the Imperial College, the educational officials can send them directly; in the case of the regions, the superintendents of education should send them. If things remain as before and do not change, the educational officials and the superintendents should be punished.

更乞内自太學, 外自州軍學, 各以月試取到前三名程文, 申禦史台考察. 太學 以月, 諸路以季. 太學則學官徑申, 諸路則提學司類申. 如仍前不改, 則坐學官、提學司之罪.

Ye Zhu's proposed measures—the destruction of the recorded sayings, the collection and inspection of successful examination papers, and the threats against those assigned to implement the measures—illustrate the determination of the persecutors of False Learning. The emperor endorsed the proposals. Half a year later, after renewed allegations that "the power over the examination halls had completely fallen into the hands of the proponents of False Learning,"[79] Han Tuozhou's regime resorted to even more extreme measures. A decree was issued that candidates needed the qualification "is not a man of False Learning" on their family certificates.[80] Students with connections to the

---

79. *SSJSBM*, 80.874.

80. Schirokauer, "Neo-Confucians Under Attack," 180; *XZZTJ*, 727. A family certificate was a form containing biographical information about the candidate; cf. Chaffee,

Learning of the Way were thereby theoretically excluded from taking the examinations.

Throughout the twelfth century, the government continued in its self-proclaimed role of impartial arbiter of examination preparation. It justified its actions against the Learning of the Way as interventions to root out factionalism and preserve impartiality. In the 1190s, the government tried on the additional role of trendsetter in examination preparation. It charged the Directorate with compiling and printing authoritative anthologies, a policy not considered since the doomed years of the late Northern Song. On the occasion of the departmental examination in 1199, the minister of rites, Huang You 黄由 (*js.* 1181, d. 1210), later blacklisted as a proponent of False Learning, called for a more positive involvement of the government.[81] He concluded from the Directorate of Education's surveys of commercial anthologies of model examination writings that current anthologies did not provide adequate models. He proposed that the government solve this problem by stepping into the market of examination materials.

Huang's strategy consisted of two components. First, he suggested that the top twenty papers of the last departmental examination be forwarded to the Directorate of Education. After screening and editing, the Directorate was to print and distribute the essays. This measure built on a departmental communication from 1190 in which the Directorate was charged with the selection and printing of model essays from the past in all genres.[82] Second, the minister recommended that

---

*The Thorny Gates*, 53, 60–61; Araki Toshikazu, *Sōdai kakyo seido kenkyū*, 51, 60, esp. 70; and Hymes, *Statesmen and Gentlemen*, 43–45.

81. *SHY*, *XJ*, 5.21b–22a; note following Chen Dang's memorial from 1199, ibid., 21a–b. Huang You's offense derived not from his advocacy of Learning of the Way teachings but from his recommendation not to set up a blacklist (Deng Guangming and Cheng Yingliu, *Zhongguo lishi da cidian—Song shi*, 423). For Chen's involvement in the persecutions, see *SS*, 474.13773.

82. Peng Guinian, *Zhitang ji*, 1.2. The idea of selecting recent essays as models for examination candidates had been brought up before. In 1171 the Imperial College was ordered to select worthy papers in its examinations so that they could be used as standards throughout the realm (Liu Hsiang-kwang, "Yinshua yu kaoshi," 184; *SHY* [AS], *XJ*, 4.41). This pronouncement differed from the later ones in that it was intended as an admonition for examiners to base their evaluations on traditional standards of accordance with rhyme and broad scholarship rather than as a directive to collect

collections of essays written before the large-scale contamination of examination writing by Learning of the Way discourse be collated and reprinted.[83] This measure adopted a proposal by Chen Dang 陳讜 (*js.* 1163), an active persecutor of Han Tuozhou's targets. Chen Dang's list of legitimate collections included *The Triple Champion's Standard Fu from the Yuanyou Period* (1086–94) (*Sanyuan Yuanyou hengjian fu* 三元元祐衡鑒賦), *The Best Expositions from Around the Shaoxing Period* (1131–62) (*Shaoxing qianhou lun cui* 紹興前後論粹), and *Policy Response Essays Precious as Rhinoceros Horn and Ivory* (*Zhuoxi baxiang ce* 擢犀拔象策). None of these titles are extant. All of them contained essays dating to the period between 1086 and 1162. In his annotated catalog, Chen Zhensun listed *Policy Response Essays Precious as Rhinoceros Horn* and *Policy Response Essays Precious as Ivory* as two separate collections. The first collection contained policy response essays written between 1086 and 1131; the second collection gathered essays from the later years of the Shaoxing period, presumably the 1140s through the 1160s.[84] The resurrection of these older guides in the 1190s was inspired by a desire to return to models predating the two waves of factionalist examination writing most dreaded by the Song court. Essays from the Yuanyou period seemed to avoid the ideology of the reformist regime in the early twelfth century, and those dating before the 1170s had not yet been corrupted by the novel discourse of Learning of the Way.

Private collections of successful and imitation examination essays had been circulating since the Tang dynasty. Some Tang anthologies were still available in Song times, but they were soon superseded by collections of more recent essays.[85] As related in the first section of this

---

and print model essays approved by the Directorate, the court's supervisory organ in education.

83. Chen Dang mentioned these three titles in his discussion of legitimate Song models for the various kinds of examination writings. In addition, he listed *Essays on the Meaning of "The Book of Rites"* (*Liji yi* 禮記義) by Zhou Kui 周葵 (1098–1174) and Chen Songlin 陳宋霖 (*js.* 1135) and *Essays on the Meaning of the "Book of Documents"* (*Shu yi* 書義) by Xu Lü 徐履 (ranked first in the departmental examinations in 1148) (*SHY, XJ*, 5.21a–b).

84. Chen Zhensun, *Zhizhai shulu jieti*, 15.458. Zhu Ruixi, "Song, Yuan de shiwen," 35.

85. In the Tang dynasty, examination papers were freely circulated among friends and appeared in bookstores as well (Twitchett, *Printing and Publishing in Medieval China*, 17; Waley, *Po Chü-i*, 40; Zhou Yanwen, "Lun lidai shumu zhong de zhiju lei shuji," 3).

chapter, the expansion of commercial publishing in the early twelfth century led to a rapid increase in the number of examination anthologies. The Northern Song government responded to its loss of intellectual control by issuing orders on several occasions to have commercial editions of anthologies and other examination guides burnt along with their printing blocks. In 1108, Huainan West circuit Educational Superintendent Su Yue 蘇栻 (*js.* 1100) recommended that the commercial compilations be replaced with government editions of selected examination essays.[86] The court attempted to stop the flow of commercially printed examination manuals by borrowing strategies from the commercial printers. Prior to 1108, the Directorate published standard editions of the Classics, the histories, and officially endorsed commentaries on them. The 1108 proposal to add anthologies of successful essays to its repertoire signaled the government's attempt to take over a crucial sector in the market for examination publications. Anthologies of successful essays in recent examinations were sold and bought as the most reliable indicators of the intellectual and political climate among examiners.

After its retreat to the south, the court redefined its position in the examination market. Commercially published examination materials were allowed as long as they were submitted for approval. The court monitored the market by checking submitted materials and inspecting bookstores, but it abandoned the publication and distribution of Directorate editions as the primary means of controlling examination standards. Huang You's proposal was representative of a wider call for more active government intervention in the examination field among court officials. Starting in the 1190s, repeated proposals were submitted to the court for greater government participation in the market for examination publications. As concern about the position of the Learning of the Way in the examination field climaxed in the 1180s and 1190s, memorialists argued that the government could set standards by issuing anthologies of selected examination papers. Huang You's proposal was not as ambitious as that of Su Yue, his Northern Song predecessor. It

---

86. Poon, "Books and Printing in Sung China," 106–8; Chia, *Printing for Profit*, 121; *SHY* (AS), *XF*, 2.48.

recognized that, if implemented, the government would at best become a strong competitor; it would not obtain a monopoly and substitute for commercial alternatives. His proposal therefore reaffirmed the legitimacy of commercial anthologies and the need for government inspection of them.

The government was not very successful in implementing the proposed measures. Unauthorized commercial compilations continued to be sold in all major commercial centers in the southeast, and no official compilations appear in the records. The private and commercial publication of recent examination policy response essays was a particularly sore thorn in the side of the government. Policy response essays about current affairs became forbidden territory, since they might contain information concerning state security.[87] Citing the sale of essays on border affairs to Jin agents, the court repeatedly reminded local officials of the ban on the publication of all examination materials relating to border affairs. In 1182 and 1190, the court issued prohibitions banning the publication of policy response essays on current affairs in general.[88] The Qingyuan (1195–1200) law code specified banishment as the punishment for printing materials touching on border affairs or printing examination policy response essays on current affairs. The unauthorized publication of other examination papers was punished by the lighter sentence of eighty strokes. The reward for informants was set at thirty strings of cash for tips leading to a conviction on the charge of unauthorized publication and at fifty for those leading to the confiscation of policy response essays.[89] These measures expressed the court's concern over internal as well as external threats. Since the questions on current affairs (*shiwu ce* 時務策) invited students to evaluate past measures and develop proposals, they were a vehicle for the expression of dissent. The policy response collections by Chen Fuliang and Ye Shi discussed in Chapter 3, for example, were banned for their critical attitude toward policies similar to those advocated by the Han Tuozhou regime.

---

87. Poon, "Books and Printing in Sung China," 61.

88. Zhu Chuanyu, *Songdai xinwen shi*, 193–94. For a critical reading of reports on the Song-Jin trade in sensitive material, see De Weerdt, "What Did Su Che See in the North?"

89. Zhu Chuanyu, *Songdai xinwen shi*, 214–15.

Despite the ban, the commercial publication of policy response essays continued unabated. In the 1190 edict, the court admitted that in the commercial publishing center of Jianning (Fujian) the printing of policy response essays persisted despite earlier prohibitions. In 1198, the Directorate reported on continuing violations in the same printing center. One anthology of examination papers contained policy questions and response essays attributed to a certain Guo Mingqing 郭明卿, who allegedly took first place in the 1197 spring term internal examinations held at the Imperial College. Further investigation revealed that the questions listed in the anthology were not those asked during those examinations, nor was Guo Mingqing on record as a student at the college.[90]

The court's attempt to weed out illegitimate printers and to claim a larger share of the examination publication market was unsuccessful for reasons other than the intransigence of commercial printers. Scholars and officials also engaged in the private publication of policy response essays and other unauthorized examination materials. The printing of examination essays continued to serve as a powerful means to protest decisions made by examination officials. In 1214 Zhen Dexiu 眞德秀 (1178–1235), who had held several high court positions earlier in the decade and had become lesser lord of the imperial sacrifices (*taichang shaoqing* 太常少卿) in 1213, financed the commercial publication of the essays of one of his protégés because he ranked lower than expected.[91]

Among scholars and officials there was considerable resistance to the government's plans to become involved in the business of examination anthologies. Peng Guinian 彭龜年 (1142–1206), an ally of Zhao Ruyu and Zhu Xi, listed several objections to the 1190 proposal to have the Directorate select and print model essays in all genres.[92] First, the publication of examination essays was unworthy of the government. He

---

90. *SHY* (AS), *XF*, 2.129. Lucille Chia (*Printing for Profit*, 122) and Zhu Chuanyu (*Songdai xinwen shi*, 166) give 1177 as the year in which this report was sent. Current editions give "Chunxi 4" as the year in which the report was sent. This is a mistake for "Qingyuan 4." The report gives 1197 as the year in which the examination was held. The online edition of the *SHY* lists the report in the correct chronological order without indicating the change made.

91. Poon, "Books and Printing in Sung China," 108.

92. Peng Guinian, *Zhitang ji*, 1.2–4; cited in Poon, "Books and Printing in Sung China," 107.

argued that by encouraging students to imitate contemporary essays, the Directorate, as the central educational institution of the empire, was giving in to the dictates of the examination market and relinquishing its obligation to set high standards. Peng shared the Learning of the Way view that moral conduct and insight into moral principles, not literary prowess or broad scholarship, should be the primary standards for recruiting officials and evaluating essays.

Scholars and officials opposed the government's projected plans not only in principle but also on practical grounds. Peng Guinian pointed out that publishing anthologies of old essays would prove futile under current conditions in the examination market. Students bought anthologies of the most recent essays as a way to keep up with changing standards. He noted that students had become accustomed to the notion that since standards changed over time, arguments made decades earlier could no longer serve as adequate models. Even if the pieces in the proposed government compilations discussed general truths, students would believe that they were written in an old-fashioned style. Another official emphasized the inequities that would result from the implementation of this policy. The proposal would disadvantage students from areas far removed from the capital. Students closer to the location of the departmental examination would continue to have easy access to the results of the most recent examinations and be able to keep abreast of the current tastes of the examination officials, whereas those in the provinces would have recourse only to compilations of old essays.[93]

Huang You responded to these objections by recommending that the court select recent examination essays for distribution. He acknowledged students' need for current models in addition to old ones and demanded that the Directorate select from the essays submitted for the last departmental examinations. The plan to charge the Directorate of Education with the selection, editing, and printing of recent top essays received support in several memorials in the following years. The politics of selection put the government at a definite disadvantage over the commercial publishers. Procedural issues caused immediate difficulties for the plan's implementation. Some proposed to assign the task of se-

---

93. *SHY, XJ*, 6.19b–20b (1214).

lection to educational officials; others proposed to have all the scholars in the Directorate jointly select the pieces, because educational officials would be under too much pressure from the powerful eager to see their essays published.[94]

The measure never got off the ground. No evidence of government anthologies has come to light. All extant anthologies of examination essays as well as those known only from their titles listed in Song catalogs are commercial endeavors. The prohibitions against Learning of the Way publications and the expression of Learning of the Way beliefs in examination papers also proved ineffective. In 1202 the government canceled all discriminatory measures. Over the next few decades, it further recognized the unifying potential of the Learning of the Way in literati culture.

### Changing Standards in the Thirteenth Century

The court's policies toward the Learning of the Way and its impact on examination preparation changed drastically during the first half of the thirteenth century. After the campaign against False Learning in the 1190s, the government made a series of concessions to the demands of Learning of the Way proponents, honoring their founding teachers and recommending their writings to scholars empire-wide. These concessions culminated in the full-scale endorsement of the work of the Northern and Southern Song masters of the Learning of the Way in 1241. The chronology of and the rationale behind the gradual official canonization of the Learning of the Way has been recounted in several recent studies.[95] This section highlights the major turning points and those decisions impacting the position of the Learning of the Way in the examination field.

The first steps in the rehabilitation of the Learning of the Way occurred in the first decade of the thirteenth century. The major targets in Han Tuozhou's campaign, Zhao Ruyu and Zhu Xi, passed away in 1196

---

94. Ibid., 5.31b (1205), 6.10b (1211).

95. James Liu, "How Did a Neo-Confucian School Become the State Orthodoxy?"; Neskar, "The Cult of Worthies," chap. 6; Tillman, *Confucian Discourse and Chu Hsi's Ascendancy*, 231–34; Yuan Zheng, *Songdai jiaoyu*, 70–76.

and 1200, respectively. Han did not want to become mired in an ideological struggle. Following the death of He Dan, the last leading prose-
cutor of the reformers, Han Tuozhou's government announced in 1202
that the "False Learning party" had been dissolved and that "the orientations of the scholar-officials had returned to normalcy."[96] All discriminatory measures were annulled. In a further effort to heal the
wounds of the persecuted, the government reinstated and promoted
some of those who had been blacklisted.[97] The reconciliation effort fit
into Han's larger vision of a reunified empire covering all the Chinese
territories. Han, with his military background, had served on two mis-
sions to the Jin and was intent on the recovery of the north. The end of
the campaign against False Learning in 1202 was a prelude to the beginning of the campaigns against the Jin empire in 1204. The end of the
worst partisan infighting since the defeat of the Song in 1127 was supposed to boost the war against Jin. Conversely, the war was intended to
muster support for Han's regime and erase the factional infighting of its
first decade.

The war effort did not rehabilitate Han's regime, but it did create
further opportunities for the recognition of the Learning of the Way.
The Song armies were unable to score major victories and faced large-
scale desertions.[98] At court, the call for peace negotiations grew stronger. Han Tuozhou was dismissed from office in 1207 and was decapi-
tated immediately. Only when the Song court agreed to send Han
Tuozhou's head north did the Jin court agree to a peace settlement in
1208. Han's demise opened the way for those requesting justice for the
victims of the False Learning campaign. An honorific title was be-
stowed on Zhu Xi in 1209. In the following decades, the court responded piecemeal to the stream of requests for honors for the masters
of the Learning of the Way and for the adoption of Zhu Xi's commentaries on the Classics.

In 1212, Zhu Xi's commentaries on *The Analects* and *Mencius* were
recommended for use in official schools. By this time, some teachers
were already teaching a broader version of the Learning of the Way

---

96. Li Xinchuan, *Dao ming lu*, 7*xia*.89.
97. Schirokauer, "Neo-Confucians Under Attack," 193.
98. H. Franke and Twitchett, *The Cambridge History of China*, 6: 245–50.

canon at local and central official schools. Wu Rousheng 吳柔勝 (1154–1224), director of the Directorate School[99] and later a teacher at the Imperial College, used all Four Books for his lectures and examination questions.[100] Imperial recognition for Zhu Xi's work on all Four Books followed shortly. In 1227 Emperor Lizong 理宗 (r. 1225–64) issued a decree in which he expressed his esteem for Zhu Xi's annotations on *The Analects, Mencius, The Great Learning,* and *The Doctrine of the Mean* and called his commentary a useful contribution to "ordering the world."[101] The occasion for this decree was the enfeoffment of Zhu Xi as Lord of Xin in the first month of 1227; the emperor's praise for Zhu Xi's program of learning at this time did not have direct implications for the curricula at official schools.[102]

Indirectly, the official recognition of Zhu Xi and his commentaries provided impetus for a shift in curricula and evaluation criteria. The imperial concessions granted legitimacy to arguments and evaluations invoking Learning of the Way ideology. Examiners' reports no longer complained about the intrusion of the distinctive discourse of Learning of the Way in examination papers. On the contrary, in 1219 and 1220 the court ratified requests by the director of studies at the Directorate of Education, Wang Fei 王棐 (*js.* 1199), and by Hu Wei 胡衛 (fl. 1220), a palace censor, demanding that the criteria for the examinations be based on moral reasoning rather than literary skill.[103]

---

99. The Directorate School was part of the Directorate of Education. Whereas students at the Imperial College came from various backgrounds, the Directorate School was intended for sons of higher officials. It is not clear whether, at this point, the school functioned as a separate school or was joined to the Imperial College (Lee, *Government Education and Examinations in Song China,* 48; Chaffee, *The Thorny Gates,* 64; Zhu Ruixi, "Guozi sheng," in *Zhongguo lishi da cidian—Song shi,* 281).

100. *SS,* 400.12148; cited in Yuan Zheng, *Songdai jiaoyu,* 71; and Chen Wenyi, *You guanxue dao shuyuan,* 187. Wu Rousheng had been prosecuted during the Campaign against False Learning; cf. *SB,* 1213–14.

101. *Song shi quanwen,* 31.32b.

102. According to Yuan Zheng (*Songdai jiaoyu,* 72), this decree signifies the formal establishment of Zhu Xi's Learning of Coherence as state ideology. I argue below that the major breakthrough in the establishment of the Learning of the Way as official ideology did not happen until 1241.

103. *SHY, XJ,* 6.32b–33a (1219), 40a–41a (1220).

The sources from which such criteria derived were officially canonized in the last four decades of the Song reign. After the failed attempt to recapture the northern capitals of Kaifeng and Luoyang from the crumbling Jin empire in 1234–35, and faced with the new political and ideological threat of the emerging Mongol empire, the Song government grew more introspective. Partisan politics continued to plague the central government, but the emperor and many of his advisers now trusted that the moral and political philosophy of Zhu Xi's Learning of the Way would work as a catalyst in healing and reinvigorating the Song state.

In the first month of 1241, Lizong decreed that the central figures of Zhu Xi's Learning of the Way—Zhou Dunyi, Zhang Zai, Cheng Hao, and Cheng Yi—as well as Zhu Xi himself be enshrined in the Confucian temples attached to the Directorate of Education and local schools and academies. This implied that the masters would be worshipped in the ceremonies attended by the headmasters and students in schools all over the empire. Key figures in Zhu Xi's lineage of the Way had been upheld as models for the literati in local shrines since the twelfth century. Lizong now accepted their claims to orthodoxy. The decree called not only for the induction of the masters of the Learning of the Way but also for the removal of Wang Anshi from the temples.

The government had resisted the removal of Wang Anshi from the temple[104] as long as it remained committed to Gaozong's policy of impartiality to all Confucian intellectual formations. With his removal, the government relinquished its self-assumed role of impartial arbiter and became a trendsetter in literati culture. Apart from the canonization of Learning of the Way masters in Confucian shrines, the Southern Song government bolstered the propagation of the Learning of the Way through its active support for Learning of the Way academy education. Days after the 1241 decree, Emperor Lizong sent a copy of Zhu Xi's "Instructions for the White Deer Grotto Academy" ("Bailu dong xue-

---

104. *SS*, 42.821. For the history of Wang Anshi in the Confucian temple, see Neskar, "The Cult of Worthies," 275–301. She attributes Wang's continued presence in the temple to Qin Gui and "the continued appeal of institutional reform until 1241" (ibid., 296–98). I do not find this explanation very persuasive. I have seen no evidence that Wang's institutional reforms retained much appeal in the first decades of the thirteenth century; it is also not proven that there was less interest in them after 1241.

gui" 白鹿洞學規; originally, "Bailudong shuyuan jieshi" 白鹿洞書院
揭示) in his own calligraphy to the Imperial College.[105] He thereby expressed his support for Zhu Xi's pedagogical ideals. Emperors Lizong
and Duzong 度宗 (r. 1265–74) also recognized the authority of the
Learning of the Way texts, which had become established as the core
curriculum at Learning of the Way academies (Chapter 6). In 1241
Lizong recommended Zhu Xi's commentaries on all Four Books and
the writings of the five masters in general. In 1270 Duzong declared,
"All scholars in the empire should study [Zhou Dunyi's] *Explanations of
'The Diagram of the Supreme Ultimate*,' [Zhang Zai's] *The Western Inscription*,
[Cheng Yi's] 'Preface to *Commentary on the Changes*,' and 'Preface to
*Commentary on the Spring and Autumn Annals*.'"[106]

By then the work of Zhou Dunyi, Zhang Zai, the Cheng brothers,
and Zhu Xi had become standard in academies and government
schools, as well as in the book market. The government's adoption of
Zhu Xi's Learning of the Way was shaped by a century of Learning of
the Way activity in education (Chapter 6). The full-scale endorsement
of the work of the Northern and Southern Song masters of the Learning of the Way in 1241 consolidated the position of authority it had
gradually obtained in the examination field (Chapter 7). The impact of
the 1241 decision is evident in the shift in examination standards discussed in the following section.

## CHANGING STANDARDS IN EXAMINERS'
## COMMENTS ON EXPOSITIONS

Examination graders and supervisors often bemoaned the huge piles
of papers they had to grade within a span of only a few weeks. The
workload was enormous.[107] Ten examination officials for the local

---

105. Liu Yue 劉燁 (1144–1216) had asked that the "regulations" be posted in the Imperial College in the 1210s. Cf. *SS*, 401.12171; and Yuan Zheng, *Songdai jiaoyu*, 74. For the
role of the White Deer Grotto regulations in the history of academies, see Chaffee,
"Chu Hsi and the Revival of the White Deer Grotto Academy"; Chen Wenyi, *You
guanxue dao shuyuan*, chap. 2.

106. *SS*, 46.905; cited in Yuan Zheng, *Songdai jiaoyu*, 74.

107. Araki Toshikazu, *Sōdai kakyo seido kenkyū*, 182–219. The procedure for local examinations and palace examinations was similar. Supervisors of the local examinations

examination in Fuzhou in 1186 were responsible for the papers of 14,000 to 15,000 candidates.[108] Fuzhou was an extreme case. As the most successful prefecture in the examinations, it was also one of the most competitive, with large numbers of participants contesting the small number of slots every three years. The pressure to grade large numbers of papers with a small staff within a limited time span was felt by examiners at all levels across the empire and intensified as the number of candidates grew over the course of the twelfth and thirteenth centuries. Examiners reacted to the growing pressure by economizing on the time spent on each paper. Scholars and officials protested that papers did not receive due attention. In 1213 one official complained that examiners just skimmed through papers to select "awaiting entrance" candidates for the Imperial College examinations. The court responded by requiring graders to write more detailed comments on all papers, indicating the positive and negative aspects of each piece.[109]

The court's order for more detailed comments was more than an attempt to satisfy the demands of disgruntled candidates. Underlying the order was a tension between the examiners and the court concerning examination standards. Examiners had a great deal of autonomy in setting evaluation criteria. The court monitored the examiners by developing standardized ranking systems and requiring more detailed comments. This section first describes the ambivalent relationship between the court and the examiners. It then argues that the court's decision to endorse the Learning of the Way in the thirteenth century enhanced the authority of the court and weakened the position of the examiners in

---

in the Southern Song were assigned several graders based on the number of candidates (ibid., 18–42, 321–32). Examination papers first went through the hands of examination hall clerks and lesser examination graders. The examination hall personnel included guards, proctors, clerks to seal original examination papers, copyists, and clerks to check the copies handed in against the original examination papers. After having been sealed and copied, papers went to the first grader (*dianjianguan* 點檢官 or *chukaoguan* 初考官), who counted the typos, marked papers down for sloppiness, and assigned them a grade. From there they went to a second grader (*canxiangguan* 參詳官 or *fukaoguan* 覆考官), who checked and ranked them independently. The final decision was left to the supervisors, sometimes after another consultation with the graders in case of major differences in their evaluations.

108. Araki Toshikazu, *Sōdai kakyo seido kenkyū*, 40–41.

109. *SHY, XJ*, 6.17a–b. On "awaiting entrance," see Chaffee, *The Thorny Gates*, 104.

the examination field. This decision initiated a process of standardization in examiner comments.

In the twelfth century, examiners enjoyed a large measure of autonomy in grading papers. The court's adoption of the policy of Great Impartiality abrogated the curricular standards set by Wang Anshi's reform policies. The authority of examiners in the examination field was also a result of the particularities of the organization of the civil service examinations. Examiners were in charge of all aspects of the examinations they were assigned to supervise. The supervising examiners (*zhiju* 知舉) designed the questions in consultation with the vice supervisors (*tong zhiju* 同知舉) and other examination officials (*kaoguan* 考官). They interviewed the mutual responsibility teams of candidates sent from the prefectures, determined the list of successful candidates, and investigated irregularities in the examination halls.[110] At the local level, the supervising examiner enjoyed similar authority in designing the questions and selecting the graduates. Throughout the Song period, and during the remainder of imperial history, the authority of the examiners was premised on the absence of standardized testing. The court did not develop standardized questions; examiners designed questions for the examinations to which they were assigned.

The Song court instituted regulations to circumscribe the power of the examiners. The title of examiner was a temporary one and did not match a regular position in the bureaucracy. Examiners were appointed ad hoc. Supervisors and vice supervisors for the departmental examinations were selected from among the highest court officials: ministers (*shangshu* 尚書) of the six ministries, Hanlin academicians (*Hanlin xueshi* 翰林學士), vice directors (*shilang* 侍郎), and supervisory secretaries (*jishizhong* 給事中) in the Chancellery. Departmental examiners were, with only one exception,[111] never appointed to supervise consecutive examinations. In the case of the local examinations, which were until 1171 regularly supervised by the prefect, the constant rotation of officials ensured that the examination questions in a given prefecture were never designed by the same person twice in a row. After 1171, examiners

---

110. Araki Toshikazu, *Sōdai kakyo seido kenkyū*, 182–219. The procedure for local examinations and palace examinations was similar (ibid., 18–42, 321–32).

111. Cheng Bi 程珌 (1164–1242) supervised the examinations of 1223 and 1226.

were chosen by the Regional Fiscal Commission (Zhuanyun si 轉運司)
from various categories of local officials other than the prefect and the
vice-prefect.[112] To prevent disclosure of the names of the examiners,
appointments were announced secretly about one month before the examinations. Examiners were then immediately escorted to the examination compound and spent the remainder of the time before the examination in isolation.[113] Such regulations limited the long-term impact of
individual examiners on examination standards. At the same time, they
introduced a measure of unpredictability. Students frequently complained about the unpredictability of examiners' standards. Given the
factional politics of the Song court and bureaucracy and the court's
unwillingness to endorse a standard curriculum, the rotation in examiner positions created uncertainty about examination standards.

In the examination regulations issued throughout the Song period,
the court included provisions to guide the examiners in their evaluations. Examiners were required to abide by a standardized ranking
scheme. During the twelfth and thirteenth centuries, the definition of
the criteria for the rankings changed in accordance with the court's
shifting role in the examination field. As the court moved from impartial arbiter to a position of endorser of orthodoxy, examiners' criteria
began to reflect close adherence to the ideological standards endorsed
by the court.

According to the examination regulations issued in 1011, papers were
to be ranked in one of five classes: (1) excellent scholarship, flawless
composition (*cili* 詞理), (2) richness of thought, smooth composition
(*wenli* 文理), (3) clear composition, (4) average composition, or (5) crude
composition.[114] In this ranking scheme, compositional criteria were central. The scheme was based on the assumption that there were well-
established compositional requirements pertaining to each genre tested
in the examinations. "Wording and pattern" (*cili*) or "phrasing and pattern" (*wenli*), both translated as "composition" above, were binomials
referring to the paper's reproduction of the compositional requirements
of the genre.

---

112. Araki Toshikazu, *Sōdai kakyo seido kenkyū*, 18–36.

113. Ibid., 200–201.

114. *SS, XJ*, 1.3610; cf. Nakajima, *"Sōshi" 'Senkyoshi' yakuchū*, 1: 57.

"Pattern" (*li* 理), a term designating the natural pattern ingrained in things, referred to the natural coherence of the text. The analogy with cosmic and natural processes was assumed, but it applied only to writing in this usage. Each text had a body with mutually dependent components. Although all genres shared this feature of preordained organization, the natural coherence of a text, like the pattern inherent in each thing, was genre-specific. Expositions, discussed in Chapter 2, were constituted of sections comparable to the "head, neck, heart, belly, waist, and tail" of a body, each with appropriate qualities, dimensions, and functions. The compositional requirements for expositions were different from those for the regulated poem. As a general criterion, "phrasing and pattern" was used to refer to the structure and interconnectedness of the text.

This ranking scheme reappeared in proclamations issued during the Southern Song period. In 1135 an instruction to examiners for the departmental examinations stipulated: "Excelling in composition (*wenli*) will be considered according with the standards [i.e., a passing grade]."[115] The revival of criteria in use before Wang Anshi's examination reforms was motivated by the court's intent to steer away from partisan curricular debates. The emphasis on compositional requirements replaced the priority assigned to familiarity with the new commentaries and agreement with the reforms. Compositional requirements were considered less controversial. As in the 1011 regulations, the court promoted broad scholarship as the distinguishing feature of work meriting the highest ranking. Like compositional requirements, the court perceived broad scholarship as a nonpartisan criterion. It measured a candidate's versatility in the classical tradition broadly conceived, a tradition that the court considered opposed to the more narrowly defined curricula advocated by proponents of Cheng Learning or Wang Anshi's reforms.

The court monitored examiners' grading criteria. Departmental examiners were required to report questions and results to the Ministry of Rites. Memorials submitted between the 1170s and 1190s frequently mentioned violations of the rule that compositional requirements and broad scholarship were the criteria for grading. The examiners and

---

115. *SS, XJ*, 2.3628; Nakajima, *"Sōshi" 'Senkyoshi' yakuchū*, 1: 173.

censors quoted in the previous sections reported that papers espousing the Learning of the Way violated compositional requirements and abandoned broad scholarship in favor of theories not grounded in classical learning. Such papers passed nevertheless. Some examiners placed them in the top ranks; others mentioned that they could not avoid passing papers influenced by the Learning of the Way. In the second half of the twelfth century, examiners passing these papers disregarded criteria promoted by the court and responded to the growing appeal of the Learning of the Way in literati culture. In the last decades of Song rule, in contrast, examiners came under increasing pressure from the court to use familiarity with the Learning of the Way canon as a criterion in paper grading.

The impact of the court's about-face in 1241 is evident in the comments of examiners preserved in *Standards for the Study of the Exposition*. In this anthology of examination expositions (discussed in Chapters 2 and 3), all expositions were prefaced by summary notes of appreciation. In about a fifth of the essays, thirty in total, the comments were explicitly attributed to the presiding examination officials. The comments of teachers and literary critics such as Feng Yi 馮椅 and Ouyang Shoudao 歐陽守道 (1209–after 1267; *js.* 1241) were cited for other essays. Anonymous comments may have been written by the compilers, critics, teachers, or examination officials.[116]

In comments on essays from the twelfth century, literary style and versatility in an undefined literary tradition were the main concerns of the critics. The appreciative comment on the exposition by Huang Huai read: "Totally a natural creation, without artificial elaboration. Indestructible writing for all ages."[117] An essay by Wei Zhen 危稹 (*js.* 1187), whose writing was admired by eminent writers of the time like Yang Wanli and Hong Mai,[118] "smelt meaning and cast phrases, [so that the

---

116. I counted thirty essays with examiners' comments. In a few cases, the summary statements were not attributed to the examiner, even though the examiner was cited in interlinear comments. Commercial printing houses were known to publish examination essay anthologies with fake official comments. I have found no indications that this is the case with *Standards for the Study of the Exposition*.

117. *LXSC*, 2.89a. I discuss Huang's essay in Chapter 2.

118. *SRZJZL*, 643.

text appeared] perfectly rounded [linked] and direct."[119] In both cases the commentator found that the greatness of these pieces lay in their phrasing and composition.

In the thirteenth century, examiners and critics increasingly focused on the truth of the arguments proposed in the essays. Truth was measured by the candidate's ability to explicate the operation of moral coherence (*li* 理) in events. One examiner commented on an exposition written for a local avoidance examination:

[This candidate has gained] clear insight into the Learning of Coherence (*lixue*). His perspective is grand. We can truly say that his argument that those who know from having seen it, know this Way, and those who know by having heard about it, also know this Way, matches the mind of the sages across the span of hundreds of years.[120]

理學玲瓏; 地位開闊. 說見而知者, 此道; 聞而知者, 亦此道; 真足以契數
聖人之心於千百載之上.

The examiner commended the author's insight into the Learning of Coherence. (The Learning of Coherence, frequently translated "the Learning of Principle," became increasingly common as a replacement for "Learning of the Way" in the thirteenth century, probably because of the tainted history of the latter term in the twelfth century.) He continued that the author's mental insight matched that of the sages of Antiquity. According to the masters of the Learning of the Way, truth was communicated through the mind; writing was the articulation of the personal understanding of moral truth. The examiner's comment on the personal understanding of coherence reflected his endorsement of twelfth-century Learning of the Way discourse; it also implied disregard for the traditional criteria of broad scholarship and composition.

After the official canonization of the masters of the Learning of the Way in 1241, examiners increasingly measured candidates' understanding of the Learning of Coherence by their ability to ground it in the textual legacy of its leading thinkers. This legacy was by and large the result

---

119. *LXSC*, 4.95a. For a systematic introduction to critical terminology and a ranking of the qualities of good prose writing, see Chen Yizeng, *Wenzhang ouye* (1332).

120. *LXSC*, 2.64b.

of Zhu Xi's editorial and commentarial work.[121] Zhu Xi was recognized as the central figure in the transmission of the Learning of the Way. Between the 1240s and the 1270s, his commentaries and teachings became the primary standard for examination writing. The following comments were attached to two expositions for the departmental and Imperial College examinations held in 1268:

He bases his argument on Master Zhu's explanations on *The Doctrine of the Mean* and brings in his own judgment. His arguments are well founded, and his writing brings clarification. Among examination essays, this piece ranks as one of the very best.[122]

本朱子中庸之説. 而參以己意. 議論有根據; 文理有發明. 此時文中之冠冕者.

He advocates Huian's [Zhu Xi's] teaching of sequence and gradual self-cultivation. He sees through the defects of Xiangshan's [Lu Jiuyuan's] direct shortcut. He has a very good understanding of the basic intent of the passage from *Mencius*. This will be of great benefit to future students. His writing style is mature, the structure is integrated. This is an outstanding exposition.

主晦庵循序漸進之説. 破象山直詣徑造之病. 深得孟子此章本旨. 有功於後學多矣. 筆力老蒼, 文脈貫通. 論中之巨擘也.[123]

These two comments attest to the transformation of Learning of the Way discourse in thirteenth-century examination writing. Learning of the Way discourse became centered around the written work of Zhu Xi. As an official ideology, the Learning of the Way commanded adherence to the letter of Zhu Xi's commentaries as well as to the spirit of his larger heritage. In the first comment, the candidate was lauded for his apt use of Zhu Xi's commentaries on one of the Four Books. He was also complemented for voicing his "own judgment." The voicing of personal judgment was an application of one of the examination reforms Zhu Xi had suggested in 1195. In "Private Opinion on Schools and

---

121. Kojima Tsuyoshi ("Shushigaku no hatten to insatsu bunka"; "Shisō dentatsu baitai to shite no shomotsu") attributes the ascendancy of Zhu Xi's Learning of the Way in literati culture to Zhu Xi's textual productivity and to his disciples' use of printing technology to continue its textual transmission.

122. *LXSC*, 9.1b.

123. Ibid., 10.64b.

Selection" ("Xuexiao gongju siyi" 學校貢舉私議), Zhu Xi had set out a new format for examination papers on the Classics. He proposed that students be required to round off their summary of the commentaries and the explanations of various scholars with a personal evaluation of their pros and cons.[124] He designed this format in reaction to "those who randomly bring in their own opinion" and "the trend for everybody to set up their own interpretive school."[125] Personal judgment according to Zhu Xi's reform proposal required of every candidate the personal confirmation of the truth of the Learning of the Way.

This expression of personal commitment to the Learning of the Way is also a core evaluation criterion in the second comment. The paper's refutation of Lu Jiuyuan's interpretation of a passage on self-cultivation in *Mencius* underscored the examinee's commitment to Zhu Xi's interpretation. Zhu Xi had articulated his differences from Lu Jiuyuan, and in conversations with disciples, he admonished students to avoid Lu's mistakes. In those conversations, as well as in later examination papers, the articulation of right and wrong, agreement and disagreement, reinforced the exclusive truth of Zhu Xi's Learning of the Way.

In thirteenth-century evaluations, the compositional features of examination writing were of secondary importance. Many examiners and critics adopted Zhu Xi's perspective on writing. Writing was valuable to the extent that it was based on the expression of a cultivated mind, a mind that clarified moral coherence in writing. Because composition could, according to Zhu Xi, not be evaluated apart from the author's understanding of moral coherence, it was evaluated by the same standards as content. Clarity was the standard for the examinee's understanding of the Learning of the Way as well as the language in which it was expressed (see the first comment from 1268).

At the same time, Learning of the Way values were expressed in ways better suited to the conventions of examination writing as they began to dominate the examination field in the last decades of Song rule. Examiners and critics noted that the selective use of conventional rhetorical devices conveyed the message of the Learning of the Way more persuasively.

------

124. Zhu Xi, *Zhu Xi ji*, 69.3638–40.
125. Ibid., 69.3639.

His interpretation derives from Nanxuan [Zhang Shi]. The wording he gets from Donglai [Lü Zuqian]. His interpretation brings things to light and explains them thoroughly. His writing uses the methods of "turning things over and over" (*fanfu*) and "opposing and elevating" (*yiyang*). When you read it, it makes you feel as if you are purified through personally receiving "the teaching that resembles the timely rain."[126]

意脈本之南軒. 字面得之東萊. 發越透徹, 反覆抑揚. 讀之使人洒然如親承時雨之敎.

In this evaluation, the examinee's versatility in the Learning of the Way canon ranked first. Zhu Xi recognized Zhang Shi as a legitimate transmitter of the Learning of the Way and edited his work in 1184. The critic highlighted the examinee's effective use of rhetorical strategies. Late Southern Song examiners were responsive to the preoccupation with composition in examination preparation. They had an eye for the skillful use of writing techniques, provided such skills were subordinate to the clarification of the Learning of Coherence. The techniques of alternating perspectives to judge a person or an event (*fanfu, yiyang*) were characteristic of Ancient Writing. Lü Zuqian, whose work had influenced the examinee, was well known for his courses on composition (Chapter 4). In Zhu Xi's estimation, Lü's writing manuals were detrimental to moral self-cultivation. However, as subsequent chapters demonstrate, Learning of the Way advocates in the thirteenth century saw training in composition as a requisite, if subordinate, part of their curricula.

The court's endorsement in 1219 and 1220 of requests to substitute moral reasoning for literary skill as the main criterion in examination grading and the canonization of the masters and the writings of the Learning of the Way in 1241 had an effect on examiner evaluations. With the imposition of an official canon, examiners lost some of the latitude inherent in the earlier, purposely vague, criteria of broad scholarship and adherence to compositional requirements. Whatever their authority, however, examiners have never been credited with bringing about major curricular changes in the twelfth and thirteenth centuries. By contrast, the next two chapters clarify the crucial role of teachers in the rise of the Learning of the Way in the examination field.

---

126. *LXSC*, 10.44b. "The teaching that resembles the timely rain" refers to the topic. The source for the quote is *Mengzi*, VIIA:40; see Legge, *The Chinese Classics*, vol. 2, *The Book of Mencius*, 473.

# Part IV

—

*The Learning of
the Way Movement in the
Examination Field*

# 6

———

## Preparing for the Examinations (ca. 1150–1274): Developing the Learning of the Way Curriculum

The Learning of the Way's tortuous path from persecution to official endorsement traced in Chapter 5 was paved by its curricular efforts. Teachers in the Learning of the Way tradition developed curricula and textbooks that both contested and adapted existing conventions in the examination field. As the Learning of the Way teachers gradually displaced other teachers in the examination field in the thirteenth century, strategies of contestation gave way to those of adaptation.

Throughout the second half of the twelfth century, the Learning of the Way adopted the position of an opposition group in politics as well as in the examination field. In curricular matters, teachers of the Learning of the Way launched a comprehensive critique of current examination standards. The first section in this chapter demonstrates that this appraisal, as articulated by the leader of the Learning of the Way, Zhu Xi, moved beyond earlier criticisms of the civil service examinations in the Neo-Confucian tradition. Zhu Xi's aim was to expose the curricular weaknesses of the masters of "Yongjia," the most influential authorities in the twelfth-century examination field. At the same time, he advocated changes in examination procedures and curricula that would have aligned examination standards more closely with Learning of the Way goals.

In the early stages of their involvement in the examination field, Learning of the Way teachers offered a comprehensive critique of prevailing examination standards and preparation practices as well as a radical alternative. Zhu Xi insinuated several times that examination essays could be written in accordance with the principles of the Learning of the Way. Yet, in his private teaching practice, he kept the explanation and practice of moral principles separate from examination preparation, which he did not teach. Some of Zhu Xi's disciples adopted a more aggressive stance in the examination field. For Chen Chun 陳淳 (1159–1223), one of Zhu Xi's most prominent students, examination writing was a weapon to be used in the Learning of the Way's campaign for the reformation of literati culture. Examination preparation was no longer an extracurricular activity for students of the Learning of the Way. Chen Chun encouraged his students and peers to learn how to write examination essays in the language of the Learning of the Way. The second section of this chapter therefore examines the early production of Learning of the Way textbooks. Chen Chun's well-known introduction to Learning of the Way concepts embodied the radical stance of early Learning of the Way advocates and exemplified the effort to formulate a Learning of the Way alternative in examination preparation. His work and the models for it also support my larger contention that the Learning of the Way program for intellectual and cultural reformation was refracted through the conventions of the examination field.

The curricular offensive launched by Zhu Xi and his followers effected a major change in examination preparation in the thirteenth century. As the Learning of the Way canon designed by Zhu Xi and his immediate disciples gained official support, Learning of the Way ideology moved from the periphery to the center of the Southern Song examination field. This shift in authority transformed Learning of the Way ideology. The third and fourth sections discuss the final stages in the adaptation of the Learning of the Way to the conventions of the twelfth- and thirteenth-century examination field. Whereas Learning of the Way teachers adopted the layout of examination textbooks early on, it was not until the 1230s that examination teachers and commercial printers reconciled Learning of the Way moral philosophy with the more traditional examination fields of history, government, and composition. By the mid-thirteenth century, policy response manuals and Ancient Prose anthologies were disseminating Learning of the Way values. The ideology foregrounded in these manuals was, however, no

longer marked by the critical and antagonistic stance of Zhu Xi and
Chen Chun. In official Learning of the Way ideology as represented in
mid-thirteenth-century curricula, Zhu Xi's legacy served as the focal
point around which diverse currents in literati culture were unified.

## Zhu Xi's Critique of Examination Preparation and the "Yongjia" Curriculum

Between 1150 and 1200, Zhu Xi incessantly discussed the problems of
examination preparation and examination writing and established a
reputation as the most vociferous critic of contemporary examination
preparation. In conversations and letters, he censured students and
peers for focusing on composition and applying utilitarian methods of
analysis to historical and administrative problems. He singled out the
teaching and the manuals of the masters of "Yongjia" for particular
criticism. In official and pseudo-official memorials, he proposed solutions to the problems he perceived in examination preparation and advocated the use of selection criteria in agreement with the reform
agenda of the Learning of the Way.

Despite his fierce attacks on examination preparation, Zhu Xi supported the use of examinations for the selection of officials as well as in
schools. In his support for, or at the very least his acceptance of the
unavoidability of, the civil service examinations, Zhu Xi differed from
Cheng Hao, one of his principal sources of inspiration. Cheng Hao,
along with other eleventh-century reformers, believed that the examinations should be replaced with a system of recommendation through
schools; in contrast, Zhu Xi and most leading Southern Song intellectuals acknowledged that schools would never replace the examinations.[1]
They criticized declining standards in education and the atmosphere of
careerism among contemporary students but attributed these problems
to sociopsychological rather than institutional reasons: superficial
scholarship was promoted by the prevailing scholarly ethos, but the examinations themselves were a neutral means of selection.

---

1. For a discussion of the differences in emphasis of Cheng's and Zhu's proposals,
see Chen Wenyi, *You guanxue dao shuyuan*, 85–88, 218.

Zhu Xi's critique of the examinations was based on a conviction that the pedagogical problems associated with the civil service examinations could not be avoided by alternative pedagogies outside examination preparation and had to be addressed by incorporating alternative pedagogies into the examination regimen. This section argues that Zhu Xi's intent was to challenge the authority of the "Yongjia" masters in the examination field and to formulate an alternative to what contemporaries perceived as "Yongjia" scholarship's main strengths. In letters and conversations related to examination preparation, he targeted "Yongjia" scholarship's reading and writing methods and its emphasis on administrative reasoning. In a pseudo-memorial on examination reform, "Private Opinion on Schools and Selection" ("Xuexiao gongju siyi" 學校貢舉私議; 1195), he outlined the Learning of the Way alternative to "Yongjia" reading and writing methods and presented the evaluation of moral insight as an alternative to an assessment of a capacity for administrative reasoning.[2] Below, I discuss each of these three dimensions of Zhu Xi's criticism of "Yongjia" scholarship (administrative reasoning, reading, and writing), together with the Learning of the Way alternative proposed in Zhu Xi's "Private Opinion."

Zhu Xi observed that "utilitarian discourse" (*shigong* 事功) began to dominate literati culture in the 1160s.[3] In his reading of successful policy response essays dating from the 1160s to the 1190s, a concern with results had supplanted the explanation of moral norms:

Because there are only so many current affairs in the world, all of them have equally been discussed before; people can't find anything [new] to say. . . . They can't find anything [new] to say, but they are also concerned only with novelty. The most harmful thing is that these essays make light of moral behavior, destroy integrity, uphold cleverness, and honor deceit. Reading them is painful and infuriating. I don't know what kind of generational transformation

---

2. For a detailed discussion of the proposal and the history of its reception, see De Weerdt, "Changing Minds Through Examinations"; and Zhu Xi, *Zhu Xi ji*, 69.3632–43. For the dating of this piece, see Shu Jingnan, *Zhuzi da zhuan*, 947–48; and *ZZYL*, 109.2698–99. Ning Huiru ("Zhu Xi lun keju," 136) gives 1187 as the date of writing but provides no evidence for this earlier date.

3. Hoyt Tillman discusses the use of this term and a major representative of the utilitarian way of thinking in the twelfth century in *Utilitarian Confucianism* and *Ch'en Liang on Public Interest and the Law*.

has brought this on. It is frightening! These are all inauspicious signs. Before the Longxing period [1163–64], it was not like this. Since the Longxing period, there has been a lot of debate on the recovery [of the north]. It is all about achievements and fame (*gongming*); nothing has come out of it. Looking back, this has, on the contrary, consumed a lot of energy. Those of us in the Confucian school do not [usually] discuss Guanzi, but even so he said [something of value]: "Propriety, righteousness, integrity, and a sense of shame, these are the four cardinal virtues."[4] Nowadays, propriety, righteousness, integrity, and a sense of shame have all been swept away to make room for results (*shigong*)![5]

緣世上只有許多時事, 已前一齊話了, 自無可得說. 既無可得話, 又只管要新. 最切害處, 是輕德行, 毀名節, 崇智術, 尚變詐, 讀之使人痛心疾首. 不知是甚世變到這裏. 可畏! 可畏! 這都是不祥之兆. 隆興以來不愊地. 自隆興以後有恢復之說, 都要來說功名, 初不曾濟得些事. 今看來, 反把許多元氣都耗卻. 管子、孔門所不道, 而(此)[其]言猶曰『禮義廉恥, 是謂四維』. 如今將禮義廉恥一切埽除了, 卻來說事功!

Zhu Xi justified his reaction against the prominent status of utilitarian thinking in literati culture with a comparison to Confucius' reaction to a similar intellectual and political trend in the seventh through the fifth centuries BCE. *The Book of Master Guan* (*Guanzi* 管子) was attributed to Guan Zhong 管仲 (d. ca. 645 BCE), minister of the powerful state of Qi 齊 in the seventh century BCE.[6] Guan Zhong was also the chief architect of the Duke of Qi's hegemonic power over several other states occupying the Chinese territories. Confucius occasionally expressed misgivings about Guan Zhong's legendary efforts to expand state power and his alleged willingness to bypass moral principles in order to obtain results beneficial to the maintenance or expansion of state power. Zhu Xi emphasized that the cultural change wrought by Guan Zhong's political philosophy, generally considered to have steered the states on a collision course during the following period of the Warring States, was minor relative to the changes occurring in the second half of

---

4. *Guanzi*, 1.2; Rickett, *Guanzi*, 53.

5. *ZZYL*, 109.2701.

6. For a discussion of the textual history of *Guanzi*, see W. Allyn Rickett's article in Loewe, *Early Chinese Texts*, 244–51. For a discussion of Confucius' attitude toward Guan Zhong as seen in *The Analects*, see Schwartz, *The World of Thought in Ancient China*, 109–10, 162, 386.

the twelfth century—at least Guan Zhong recognized the importance of morality.

The blame for the shift to utilitarian thinking Zhu Xi assigned, tacitly in the quote above and overtly in passages cited below, to the most prominent advocates of administrative reasoning in the twelfth century, the "Yongjia" teachers. The "learning of results" (*shigong zhi xue* 事功之學) and "the learning of results and benefit" (*gongli zhi xue* 功利之學) were commonly used appellations for the learning of intellectuals associated with Yongjia, either those teaching in Yongjia, such as Chen Fuliang and Ye Shi, or more broadly those in the East Zhe area with similar intellectual leanings, such as Chen Liang. The rubric epitomized the value attached to the evaluation of the likely outcomes of competing measures in administrative decision making. The label "utilitarian" is appropriate insofar as it reflects the "Yongjia" belief that the best solution for administrative questions is the one that achieves maximum benefit for society and the state. The basis for determining maximum benefit was primarily the calculation of the positive and negative consequences of the measures considered and not their strict accordance with a set of moral principles (Chapter 4).

Zhu Xi attributed the sudden rise in the popularity of this mode of thinking to the perilous geopolitical situation of the Southern Song dynasty. The military threat from its northern neighbors and the desire to recover the northern homeland were major issues facing twelfth- and thirteenth-century intellectuals. The Jin court's military initiatives and the Song army's subsequent victory in the Battle of Caishi 采石 in 1161 raised the prospect of a new chapter in Song-Jin relations. Emperor Xiaozong toughened the Song stance on Jin demands and made no secret of his interest in recapturing the northern territories. In their research and teaching, the "Yongjia" scholars proposed a vision and program of reform. Their reform proposals were predicated on the methodological premise that historical analysis and administrative reasoning were indispensable means to arrive at solutions for current political problems (Chapters 3 and 4). The "Yongjia" manuals and treatises arguing for administrative reform and illustrating the methodology of historical analysis and administrative reasoning received imperial attention and scored high sales among examination candidates throughout the second half of the twelfth century. Chen Fuliang, for example, presented his *Discourse on "The Rites* of Zhou" (*Zhouli shuo* 周禮説) to Xiaozong in 1192—*The Rites of Zhou* had long been a source of inspiration for many interested in institutions and governmental reform.[7] This publication was, like Chen's *Awaiting Reception*, another collection of his political writings (see Chapter 3), a best-seller among examination candidates.[8]

In Zhu Xi's view, the negative effects of "Yongjia" research and teaching on history and government were twofold. First, he found fault with the "Yongjia" approach to the study of institutional history. Zhu Xi agreed that history, in various forms including institutional history, was a worthwhile subject. He recommended the reading of history texts to his students, and his graduated reading program included history, albeit in the later stages. In "Private Opinion," he proposed a short list of historical texts for use in the examinations, as they would be held under his reform plan. Zhu Xi's appreciation of historical study was based on a logic that ran counter to the "Yongjia" approach to history. Reading history was useful insofar as it allowed students to detect the foundational laws of morality. History did not provide lessons of its own; rather, it supplied illustrations of the analyses of human nature and human relationships contained in the classical texts of the Learning of the Way canon.

For this reason, students were to approach history only after being introduced to philosophical texts that explained the laws of morality directly. Students writing on historical subjects were to demonstrate moral judgment by applying the unchanging laws of human nature and demonstrating insight into the patterns that structured the cosmos as a moral order. Zhu Xi contended that "Yongjia" historical studies lacked

---

7. *SB*, 106. The work is no longer extant. According to Sun Yirang 孫詒讓 (1848–1908), Chen Fuliang may have written two works with the same title: one in three volumes presented to the throne, and one in twelve volumes, which was combined with another study on Zhou government by Xu Yuande 徐元德 (1139–1201), also associated with the Yongjia school, into *The Essence of the Institutional System in "Zhou Officials"* ("*Zhou guan" zhidu jinghua* 周官制度精華—*Zhou guan* is a common variant of *Zhou li*); see Zhou Mengjiang, *Ye Shi yu Yongjia xuepai*, 88, 90, 290–91. Cf. Xu Gui and Zhou Mengjiang, "Chen Fuliang de zhuzuo ji qi shigong sixiang shulue," 10. Zhu Xi (*ZZYL*, 123.296) complained that Chen Fuliang was obsessed with *The Rites of Zhou*. For a discussion of Chen Fuliang's interpretation of *The Rites of Zhou* based on extant fragments, see Lo Wing Kwai, "Chen Fuliang yanjiu," chap. 7 and appendix 2.

8. *WXTK*, 181.1558.2.

direction because of their disregard for the moral laws that guided the course of history and the conduct of government.

Zhu Xi perceived the lack of direction as a trait of "Yongjia" scholarship more generally. "Yongjia" treatises and essays principally concerned the analysis of the positive and negative effects of measures and institutions and adopted a situational, historicist approach to events. Zhu Xi's focus on questions of approach, rather than on the larger reform package of which the methodology was part, led him to conclude that there was no "Yongjia" system of thought—just a lot of talk, "without head or tail."[9]

Second, Zhu Xi argued that "Yongjia" administrative reasoning subverted the purpose of the examinations—as he defined it. For Zhu Xi the legitimacy of the examination system depended on its capacity to cultivate and select "morally superior men."

Someone remarked, "Human talent in the reigning dynasty surpasses that of the Han and Tang periods, yet our administrative achievements fall short. This is because the Han and Tang governments did not actively attack the morally inferior (*xiaoren*), whereas the current administration has made it its special mission to drive away the morally inferior." The Master retorted, "When you talk like this, do you think that men in the past were talking nonsense when they referred to 'appoint the morally superior, oust the morally inferior'?[10] I don't know what kind of argument this [statement of yours] is! I am afraid that [your argument] derives from the focus in Yongjia scholarship on calculating benefits and harms."[11]

或言:「本朝人才過於漢唐, 而治效不及者, 緣漢唐不去攻小人, 本朝專要去小人, 所以如此.」曰:「如此說, 所謂『內君子, 外小人』, 古人且胡亂恁地說, 不知何等議論! 永嘉學問專去利害上計較, 恐出此.」

"Yongjia" administrative reasoning expressly denied the value of the distinction between morally superior and morally inferior statesmen. The administrative principles discussed in *To the Point in All Cases* and applied in twelfth-century expositions included the principle that employing men with blemished records might prove most beneficial to the sovereign (see Chapter 3). The reasoning behind this view was that a

---

9. *ZZYL*, 45.1149, 55.1311, 86.2207, 114.2758, 122.2951, 123.2961.

10. *Yijing*, hexagram *Tai*; Wilhelm, *The I Ching*, 441–42.

11. *ZZYL*, 37.987–88.

man with a blemished record has an incentive to prove himself; recruitment in such cases was based on psychological considerations and calculations of benefit rather than moral worth. Furthermore, through their teaching the "Yongjia" masters instilled the belief that anyone who could write persuasively about government was worthy of an examination degree.

Zhu Xi's criticism of "Yongjia" historical studies and administrative reasoning was thus motivated by his advocacy of the use of moral criteria in determining success in the civil service examinations and government service. In his proposal on examination reform, Zhu Xi outlined several measures intended to establish moral character as a selection criterion. The most conspicuous was his proposal for the creation of a morality track and the conversion of half the quota of local examination graduates in the regular *jinshi* track into moral conduct track graduates. County magistrates would be required to investigate and send a fixed number of candidates for the moral conduct track to the prefecture every year examinations were held. After further checks at the prefectural level, the prefect was supposed to send the successful candidates to the Ministry of Rites—the trajectory followed by the moral conduct candidates would thus be similar to that of the regular *jinshi* examination candidates.

Once at the capital, moral conduct candidates would receive special treatment. They would automatically be enrolled in the Imperial College and would be exempted from the monthly and other examinations held at the college. College administrators would evaluate these candidates based on personal interactions. In their second year at the college, they would be given internships in government offices. Those who performed well would be awarded government positions during their third year. Those who did not qualify for immediate appointment would be given the opportunity to sit for the next departmental examinations.

The reduction in the number of regular *jinshi* graduates and the creation of the moral conduct track suggest that Zhu Xi believed that moral conduct could most effectively be promoted by making it a determining factor in the examinations. In "Private Opinion," he further expressed the hope that the introduction and promotion of the moral conduct track would affect all engaged in the process of examination preparation and transform education generally. The appointment of moral conduct examination graduates to teaching positions in government schools, another item in the list of proposals in "Private Opinion," would have contributed toward this goal.

Apart from utilitarian modes of thinking applied to government and history, Zhu Xi's critique also addressed the role of classical scholarship in twelfth-century examination preparation. Instruction in the Classics was a particularly important matter to Zhu Xi. His program of learning was predicated on the idea that a careful reading of the Classics was fundamental to the cultivation of scholar-officials. In the area of training in classical scholarship, as in other areas of examination preparation, the "Yongjia" masters presented a major challenge to Zhu Xi's pedagogical goals. When Ye Weidao 葉味道 (*js.* 1220), a disciple from Wenzhou prefecture who studied with Zhu Xi between 1191 and 1200,[12] told Zhu that he was planning to specialize in *The Spring and Autumn Annals* in the upcoming examinations, Zhu Xi exclaimed, "Mister Chen [Fuliang] and Mister Cai [Youxue][13] from your hometown have scrutinized *The Annals* exhaustively. Current examination essays on the Classics are getting craftier and more farfetched; it is worst in the case of *The Annals*. We may say that the whole empire has been changed by your hometown."[14] Both Chen Fuliang and Cai Youxue hailed from Ruian county in Wenzhou and had gained a reputation at the Imperial College for their mastery of *The Annals*. After obtaining *jinshi* degrees in 1172, they continued to teach *The Annals* while awaiting civil service appointments. Students of Chen Fuliang continued the tradition of "Yongjia" scholarship on *The Annals* after his death.[15]

Zhu Xi's concern was not so much the "Yongjia" scholars' interpretations of specific passages in *The Annals* or any of the other Classics.

---

12. For biographical information on Ye, see Wing-tsit Chan, *Zhuzi menren*, 279–80; and Morohashi and Yasuoka, *Shushigaku taikei*, 6: 503.

13. *SB*, 1035–37; Zhou Mengjiang, *Ye Shi yu Yongjia xuepai*, 297.

14. *ZZYL*, 114.2761.

15. Chen's *Chunqiu houzhuan* is extant. For an overview of Chen Fuliang's work on *The Annals* and that of other Song scholars, see Song Dingzong, *Chunqiu Song xue fawei*. Chen's students Cai Youxue and Zhou Mian 周勉 continued his research on *The Annals* (Zhou Mengjiang, *Ye Shi yu Yongjia xuepai*, 104). For Chen's examination writing on *The Annals*, see Wang Yu, "Nan Song kechang yu Yongjia xuepai de jueqi."

He pointed out their misreadings in conversations with his students but usually attributed these mistakes to more general methodological problems with "Yongjia" scholarship. The classical commentaries of the "Yongjia" scholars appealed to twelfth-century readers because they generated novel readings, but, in Zhu Xi's estimation, the novelty of "Yongjia" interpretations owed much to historical imagination. "Yongjia" scholars generated new meanings by constructing scenes and narratives around passages and individual characters:

[Zhu Xi] further asked, "How did he [Chen Fuliang] discuss *The Annals?*" Sheng responded, "Junju [Chen Fuliang] said, 'People of our generation suspect that Zuo Qiuming's differentiating between good and evil is different from that of the Sage. They say that the affairs he recorded are often at odds with the classic. There are grounds for this. For instance, in the case of the flight of Xian Mao [first half 7th c. BCE] from Jin, people say only that Xian Mao fled to Qin. Now in this case, Xian Mao had not yet settled the succession; therefore, it says "fled" to show disapproval.'"[16] The Master [Zhu Xi] said, "What kind of talk is this! Xian Mao really fled to Qin; how could the text not say 'fled'? Now, if writing 'fled to Qin' is explained as 'showing disapproval,' I don't see why not writing 'fled' would carry a positive connotation. Yesterday I was discussing this with our friends; now this is precisely what we mean by getting stuck in the phase of exploration (*bo*) and not returning to the stage of synthesis (*yue*). This is adding minute investigations to minute investigations; I am afraid this will lead students of the coming generations astray."[17]

又問:「春秋如何説?」滕云:「君舉云:「世人疑左丘明好惡不與聖人同, 謂其所載事多與經異, 此則有説. 且如晉先蔑奔, 人但謂先蔑奔秦耳. 此 乃先蔑立嗣不定, 故書「奔」以示貶.』」曰:「是何言語! 先蔑實是奔秦, 如何不書『奔』? 且書『奔秦』, 謂之『示貶』; 不書奔, 則此事自不見, 何以爲襃? 昨説與吾友, 所謂專於博上求之, 不反於約, 乃謂此耳. 是乃於 穿鑿上益加穿鑿, 疑誤後學.」

---

16. *Chunqiu*, Wengong, 7th year (619 BCE); for a translation of Zuo's commentary, see Legge, *The Chinese Classics*, vol. 5, *The Ch'un Ts'ew with the Tso Chuen*, 248–49. According to Zuo's commentary, Xian Mao went on a mission to Qin to welcome Yong as the rightful successor of Duke Xiang of Jin after the latter's death. In the meantime, a child was set up as the successor of Duke Xiang, and Jin invaded Qin. Thus, Xian Mao saw no other alternative but to seek refuge in Qin.

17. *ZZYL*, 123.2959.

In Yongjia [learning], when they read texts, they don't read those passages that are plain; they go to the small characters in the commentaries and look for details and regard that as sophistication (*bo*).[18]

永嘉看文字, 文字平白處都不看, 偏要去注疏小字中, 尋節目以爲博.

"Yongjia" teachers encouraged wide-ranging explorations of the possible meanings of classical texts. The kind of creativity demonstrated in their close analyses of short passages of texts appealed to examination candidates preparing for essays on the Classics. Questions on the meaning of the Classics (*jingyi* 經義) or expositions on classical texts (*lun* 論) took the shape of short excerpts from the Classics and were typically only five to ten characters long. Zhu Xi was critical of the interpretive range of the "Yongjia" masters and resented the impact of their work on the teaching of the Classics. As demonstrated in the two quotations translated above, the main issue for Zhu Xi was his suspicion that the imaginative exploration of detail (*bo* 博) prevented an integrated understanding of the original text (*yue* 約). In his view, "Yongjia" scholarship approached the Classics as a disconnected set of anecdotes. His reading method was designed to teach the student how to approach classical texts as cohesive statements and, ultimately, to understand the moral philosophy that underlay them. Paraphrasing Confucius, Zhu Xi called for a return to the fundamental message of the Confucian tradition (*yue*). Confucius' favorite disciple, Yan Yuan, credited Confucius with "broadening (*bo*) me with the patterns of the past, and bringing me back to the fundamentals (*yue*) with ritual."[19]

Reading was to Zhu Xi more than an act aimed at understanding the text and the philosophy that sustained it; reading, especially the reading of classical texts, had to result in a transformation, or better a reformation, of the reader. In this regard as well, Zhu Xi contrasted the efficacy of his reading method with that of the "Yongjia" teachers:

As for *The Songs*, if you explain [this book] as you just did, it is clear in and of itself and easy to understand. You just have to immerse yourself deeply and recite it, ponder the moral principles, savor its flavor, and you will reap the bene-

---

18. *ZZYL*, 2964.
19. *Lunyu*, IX:10.2.

fits. You can skim through the whole book in just two or three days, but you won't get the real taste, you won't be able to remember it; this is all useless. The ancients said that *The Songs* can arouse.[20] You have to have been aroused when reading it—that counts as reading *The Songs*. If you can't be aroused, this does not count as reading *The Songs*. *Therefore I say that Yongjia learning boils down to trying to develop clever arguments. It is picking up things here and there in a short time, putting odd things together, and then constructing an argument. Even if the argument is valid, it is simply to no avail. Not to mention if it is also wrong.*[21]

詩, 如今怎地注解了, 自是分曉, 易理會. 但須是沉潛諷誦, 玩味義理, 咀嚼滋味, 方有所益. 若是草草看過一部詩, 只兩三日可了. 但不得滋味, 也記不得, 全不濟事. 古人説「詩可以興」, 須是讀了有興起處, 方是讀詩. 若不能興起, 便不是讀詩. 因説, 永嘉之學, 只是要立新巧之説, 少間指摘東西, 湊零碎, 便立説去. 縱説得是, 也只無益, 莫道又未是.

In opposition to current conventions in the examination field, which, Zhu Xi alleged, programmed students to skim the text and use it for making arguments, he called for a slow and engaged reading of the whole classic. Pondering the moral principles in *The Songs*, or any other text, "arouses" the reader; in Zhu Xi's reading, a correct understanding of the moral principles involved motivates one to perform one's proper moral duties. In order to reach this level of understanding, Zhu Xi recommended that students recite each poem from *The Songs* about one hundred times. Not until they fully understood the poem should they move on to the next one.[22]

Such a reading method could in Zhu Xi's view be promoted through the examinations. In his early teaching career, he designed examination questions that tested students' ability to reflect on the experience of reading all of a classic. In his policy questions, he asked students at the district school of Tongan 同安 (Quanzhou 泉州, Fujian) to reflect on their experience in reading *The Analects*. He scrupulously avoided asking questions about details (Chapter 7).

In "Private Opinion," Zhu Xi proposed curricular changes that would have resulted in the institutionalization of Learning of the Way

---

20. Ibid., XVII:10; Legge, *The Chinese Classics*, vol. 1, *Confucian Analects*, 323.
21. *ZZYL*, 80.2086.
22. Ibid., 2087.

pedagogical ideals in examination preparation. The general curriculum proposed in "Private Opinion" corresponded with two of the main principles of the reading methods Zhu Xi recommended to his students.[23] The first stipulated that reading should involve a limited set of texts—an idea that ran counter to the habits of Song elites. In order to allow students more time to immerse themselves in a text and reflect on its relationship to the process of moral self-cultivation, Zhu Xi proposed to redistribute the burden of the textual tradition. Only *The Great Learning, The Analects, The Doctrine of the Mean*, and *Mencius* were to be covered in every triennial examination. All other recommended materials were to be covered in a cycle covering four examination periods. The Classics were divided into three groups: *The Changes, The Documents*, and *The Songs* would be examined every first and seventh year; *The Rites of Zhou, The Book of Rites*, and *The Book of Etiquette* every fourth year; *The Spring and Autumn Annals* and its commentaries every tenth year. The philosophical, historical, and administrative texts would similarly be distributed over the course of four examinations. Sources and topics for the next cycle of examinations would be announced right after the end of the palace examination so that students could focus for about three years on careful reading of a limited set of texts.[24] This approach was preferable to the established practice of leaving the curriculum wide open, Zhu Xi argued, because even those who failed would have learned how to read carefully and responsibly.[25]

The curricular proposals in "Private Opinion" envisioned not only the institutionalization of the Learning of the Way reading method but also the diffusion of the doctrinal beliefs of the Learning of the Way as synthesized by Zhu Xi. The selection of a core curriculum and the definition of sets of recommended commentaries for each of the Classics contributed toward this end. As mentioned above, only *The Great Learning, Analects, The Doctrine of the Mean*, and *Mencius* were to be covered in every triennial examination. These Four Books were first pub-

---

23. For a discussion of Zhu Xi's reading methods, see Gardner, *Learning to Be a Sage*, esp. 35–56.

24. Zhu Xi, *Zhu Xi ji*, 69.3632–43; ZZYL, 109.2699.

25. ZZYL, 109.2699.

lished as a set by Zhu Xi in 1182 and formed the core of the developing Learning of the Way canon (Chapter 1). *The Great Learning*, the first text in the set, had been established as the catechism of the Learning of the Way by the mid-twelfth century. Zhu Xi maintained that *The Great Learning* contained an outline of the Learning of the Way and introduced students to the steps of self-cultivation that integrated the individual, his family, his community, and eventually the empire in the moral order of the cosmos. Once students had understood the central message of *The Great Learning*, they could apply it to the other texts on the reading list. Doctrinal belief fashioned Zhu Xi's reading method.

The Four Books was intertwined with a second principle of Zhu Xi's reading method: students should read the limited set of texts in the curriculum in sequence. Students started with the Four Books. Within the Four Books, they began with easier texts (*The Great Learning*) and then moved on to the more difficult texts. Reading in sequence allowed students to see different texts as part of a coherent whole and as all centered on moral self-cultivation. Only after students had read through the Four Books should they move on to the Five Classics and then the histories and finally to other texts. The Four Books provided the criteria for judging all other literature.

Besides selecting classical texts and foregrounding these texts in reading, Zhu Xi prepared commentaries that posed an even more pronounced challenge to the authority of the "Yongjia" masters in the examination field. In 1195, the year in which "Private Opinion" was written, Zhu Xi's analyses of the Four Books would have been the only set of commentaries available on the core texts of his proposed curriculum.

The partisan agenda of the reforms in "Private Opinion" is evident in the prioritization of commentaries affiliated with the Learning of the Way. Commentaries of those affiliated with the Learning of the Way figured prominently on the list of works deemed most suitable in guiding students' reading of the Classics. Cheng Yi, Zhu Xi's chosen predecessor in the Learning of the Way, was the only commentator whose works were listed under all Classics. The works of others included in Zhu Xi's lineage of the Learning of the Way movement such as Zhang Zai and Yang Shi 楊時 also figured prominently in the selected list of commentaries. Other historians have stressed the inclusiveness of Zhu

Xi's list of recommended commentaries.[26] Among the scholars whose commentaries were to be included in the curriculum, Zhu Xi listed Su Shi and Wang Anshi, whose works Zhu Xi criticized throughout his life.[27]

The list was, however, marked by exclusions that would have been obvious to twelfth- and thirteenth-century examination candidates, notably the commentaries of the masters of "Yongjia." The teachers of "Yongjia" had gained a reputation in the instruction of two classics in particular, *The Spring and Autumn Annals* and *The Rites of Zhou*. Zhu Xi's students were familiar with the "Yongjia" interpretations and frequently asked him questions about them. As in the example discussed above, Zhu Xi typically dismissed the "Yongjia" commentaries, because he considered their reading of the Classics antithetical to the reading methods he hoped to foster among students.[28] Judging from the list of recommended commentaries, Zhu Xi intended to limit students' exposure to the masters of "Yongjia" to only one title, Xue Jixuan's commentary on *The Book of Documents*.

The status of composition as a core field of instruction was the third major focus in Zhu Xi's critique of twelfth-century examination preparation. The problems in historical studies and classical scholarship were in his view exacerbated by teachers' and students' preoccupation with the rhetorical analysis of successful examination papers.

Somebody inquired, "Nowadays, in schools, they deal with nothing except volumes of examination papers (*shiwen*) printed in Masha." The Master [Zhu Xi] replied, "Small wonder! Those in charge have been teaching nothing else. Has it never struck them! As far as examination writing (*shiwen*) is concerned, students, of themselves, pay careful attention to it. What further need is there for you to teach that! Rather, when you establish a school, it should be to teach them to understand their proper duties."[29]

---

26. Nivison, "Protest Against Conventions and Conventions of Protest," 239, 243, 247; de Bary, *The Liberal Tradition in China*, 40–42; Elman, *A Cultural History of Civil Examinations*, 27–29.

27. For a discussion of Zhu Xi's ambivalent attitude toward both men, see Bol, "Chu Hsi's Redefinition of Literati Learning."

28. For a discussion of the commentaries of Yongjia scholars, see Zhou Mengjiang, *Ye Shi yu Yongjia xuepai*, passim.

29. ZZYL, 109.2700.

問:「今之學校, 自麻沙時文冊子之外, 其他未嘗過而問焉.」曰:「怪它不得, 上之所以教者不過如此. 然上之人曾不思量, 時文一件, 學子自是著急, 何用更要你教! 你設學校, 却好教他理會本分事業.」

Literally, *shiwen* means "contemporary writing," "current writing." Even though the term does not refer to a particular genre or a particular style, since the Northern Song period, the term has been associated with the civil service examinations. It was used to refer to examination papers, particularly passing papers from recent examination sessions. Such papers were valued not only for their interpretations of historical, policy, and exegetical questions but also as indicators of the writing styles examiners favored and as examples of the effective application of rhetorical techniques taught in examination preparation courses and manuals. "Contemporary writing" frequently carried a pejorative connotation because of the transient and limited value attached to most examination papers. Most examinations papers were relevant only to candidates preparing for the examinations for a limited time after the session during which they had been submitted.

Zhu Xi's criticism of the preoccupation with "contemporary writing" went beyond the often-cited comments from contemporaries regarding the low quality of examination papers. The comment translated above illustrates his opposition to the establishment of composition as a separate field of instruction. In this exchange, he suggested that students acquire the technical aspects of writing by themselves. On a more philosophical level, he explained in other conversations that a mind that understood the truths elucidated in the Learning of the Way would have the ability to express such truths clearly in writing. Writing was to follow the dictates of the mind trained in self-cultivation; courses in composition by contrast compelled the mind to follow the dictates of rhetorical schemata.

Zhu Xi's criticism addressed two factors that contributed significantly to the priority assigned to compositional skills in twelfth-century examination preparation: the standardization of the layout of examination essays and the authoritative position of Ancient Prose in literati culture. After reviewing both factors, I will turn to Zhu Xi's attempt to formulate a Learning of the Way alternative that was adjusted to examination requirements but cured the preoccupation with rhetorical skill.

By the mid-twelfth century, most candidates followed the same standards in organizing examination expositions. As discussed in Chapter 2, the exposition was compared to a body composed of distinct parts, each of which had a distinct place, size, and function. Parallelism had become the dominant rhetorical strategy in the development of an argument. The same criteria were applied to the essay on the meaning of the Classics tested in the first session of the examinations. Examiners defended the use of such formal standards because the standardization of the format facilitated grading. Teachers specializing in examination preparation, including the "Yongjia" teachers, promoted standardization because it provided a very reliable indicator of the criteria used by examiners and a template from which to teach composition skills.

Zhu Xi observed that the standardization of the formal layout of examination genres had pedagogical consequences and showed how indoctrination in a particular essay format steered students' mental processing of the texts on the curriculum. Current conventions resulted in peculiar ways of reading:

In general, no matter what the scope or the length of the question, the text is always divided into two parts, and people write two contrasting sentences to broach the topic (*poti*). They are furthermore required to use other words to match the characters in the topic. One just has to excel in artistry. As for the remaining 2,000 or 3,000 characters, at most they do not add any meaning but only elaborate back and forth on the argument made in the two sentences in the *poti*. Thus, not only does this not constitute classical scholarship, it also does not constitute proper writing.[30]

大抵不問題之大小長短, 而必欲分為兩段, 仍作兩句對偶破題. 又須借用他語以暗貼題中之字. 必極於工巧而後已. 其後多者三二千言別無他意, 不過止是反復敷衍破題兩句之説而已. 如此不惟不成經學亦復不成文字.

The dualistic pattern conditioned students to divide passages from the Classics into two parts regardless of their internal structure. From Zhu Xi's perspective, this impeded a careful analysis of the text and a holistic understanding of its segments.[31] Students limited themselves to an

---

30. Zhu Xi, *Zhu Xi ji*, 69.3640.
31. *ZZYL*, 21.494.

extended paraphrase of the opening argument (*poti*), by adding disparate anecdotes to substantiate the argument,[32] but neglected to probe and explain the deeper meaning of and the connections between the words of the classical text.

The emphases in reading and writing instruction on the structure of texts and eloquence of diction were in Zhu Xi's estimation aggravated by the authoritative position of Ancient Prose in the instruction of composition. The historical and political essays of Su Xun, Su Shi, and Su Che, including several examination essays, were popular among twelfth-century literati. The Ancient Prose of the Su family circulated separately and figured prominently in anthologies (Chapter 4).[33] A student of Zhu Xi confessed that his study time was divided between a close reading of *The Analects* to probe the meaning of moral principles and the study of the prose of the Su family to learn how to write examination essays.[34] Zhu Xi was especially dismayed by the central position of Su Shi's writing in contemporary elite culture and censured the interest in Su's work.[35]

On several occasions, Zhu Xi proclaimed that he saw the preoccupation with composition and the impact of the learning of the Sus as a worse problem for the Learning of the Way than the damage caused by the Learning of Wang Anshi.[36] He argued that, in contrast to the general awareness of the harms caused by Wang's reformist legacy, Buddhism, and Daoism, literati remained unaware of the spiritual pollution caused by imitating the writing of authors like Su Shi. He opposed the celebration of Su's work, arguing that it did not uncover the true meaning of events. Rather, in his essays Su Shi reconstructed historical scenes and appeared to derive situational truths from the reconstructed narrative.

---

32. Ibid., 83.2157.

33. Bol, "Reading Su Shi in Southern Song Wuzhou."

34. *ZZYL*, 118.2859.

35. Bol, "Chu Hsi's Redefinition of Literati Learning," esp. 171–83; Gardner, *Learning to Be a Sage*, 57–81; He Jipeng, "Zhuzi de wenlun," 1229–32; He (ibid., 1232*m*19) cites evidence that several of Zhu Xi's contemporaries thought his appreciation of Su Shi too critical. On Zhu's censure of Lü Zuqian's interest in Su Shi's writing, see Tillman, *Confucian Discourse and Chu Hsi's Ascendancy*, 129–30; and Lin Sufen, "Lü Zuqian de cizhang zhi xue," 156–57.

36. Gardner, *Learning to Be a Sage*, 70–71.

Su Shi's legacy was most apparent in the work and the success of the masters of "Yongjia." In Zhu Xi's eyes, Chen Fuliang was in the same category as Su Shi: "As soon as he [Chen Fuliang] has the slightest understanding, he insists on bringing it to the fore, even though he can't explain things clearly. It is like Dongpo [Su Shi] and Ziyou [Su Che]: when they saw a truth (*daoli* 道理), even if it did not constitute a truth, they wanted to brag about it right away, allegedly to let others understand."[37] For Zhu Xi, instruction in Ancient Prose was not simply a technical matter; it presented a major intellectual challenge to the Learning of the Way.

On this issue, Zhu Xi disagreed with Lü Zuqian. Lü's teaching contributed significantly to the high profile of Ancient Prose in examination preparation and, as discussed in Chapter 4, to the formation of the Ancient Prose canon in the late twelfth and thirteenth centuries. Zhu Xi criticized Lü's Ancient Prose anthologies in several letters and warned his disciples about their impact on literati culture. His criticism targeted the technical, rhetorical approach in Lü's analyses of Ancient Prose:

In a discussion about Bogong's [Lü Zuqian's] criticism of texts, [Master Zhu Xi] said, "Writing flows on and on in endless transformations. How can he limit it like this?"

Someone added, "Professor Lu[38] has said that Bogong works with a textual frame. As soon as he starts composing, he inserts his compositions in the frame. The text's flow of energy cannot run its course." The Master said, "He has insight, he has seen through it."[39]

因説伯恭所批文, 曰:「文章流轉變化無窮, 豈可限以如此?」

　某因説:「陸教授謂伯恭有箇文字腔子, 才作文字時, 便將來入箇腔子做, 文字氣脈不長.」先生曰:「他便是眼高, 見得破.」

Lü's "frame" for the composition of different genres of prose texts derived from his reading of Ancient Prose, the work of Han Yu, Liu Zongyuan, Ouyang Xiu, and the Su family in particular. In *The Key to Ancient Prose*, discussed in Chapter 4, his annotations and instructions

---

37. *ZZYL*, 118.2960.

38. This may refer to Lu Jiuyuan; the passage does not contain any further indications as to the identity of this person.

39. *ZZYL*, 139.3321.

taught the reader to acquire the patterns of organization and the figures of speech deployed in the work of the masters of Ancient Prose. Lü Zuqian defended his Ancient Prose composition courses and manuals against Zhu Xi's criticisms by claiming that content and presentation were separable.[40] Lü believed that Ancient Prose was useful to examination candidates because it taught them rhetorical strategies for formulating persuasive answers to examination questions; the content of the answers need not be based on the arguments made in the selected texts. Zhu Xi countered that the separation between composition and content was a fallacy.[41]

The standardization of the layout of examination essays and the authoritative position of Ancient Prose in late twelfth-century literati culture were the principal reasons, in Zhu Xi's view, for the literati preoccupation with rhetorical skill. Zhu Xi did not see the preoccupation with rhetorical skill as an inevitable byproduct of the use of written examinations. In "Private Opinion," he argued that a change in format could change examination essays from meaningless exercises into a reliable measure of the student's depth of understanding of the source material. The change Zhu Xi proposed was intended to alter evaluation criteria. The current format was, in Zhu's view, based on purely rhetorical considerations. The new format reflected Zhu Xi's ambition to design examinations that incorporated Learning of the Way pedagogy.

Zhu Xi's proposed format for essays on the meaning of the Classics consisted of three parts: citation of the context of the original quotation, discussion of the two or more commentarial traditions of choice, and, lastly, the student's personal understanding of the passage. This format replicated the three stages Zhu Xi recommended for reading classical texts. Zhu Xi considered commentaries a major source of preconceived notions. Therefore, he recommended that students first confront the original text, without the help of commentaries and other study aids (stage one).

---

40. Bol, "Reading Su Shi in Wuzhou"; Lin Sufen, "Lü Zuqian de cizhang zhi xue," 149; De Weerdt, "Content and Composition."

41. Zhang Shi made the same point; see Chu Ping-tzu, "Tradition Building and Cultural Competition in Southern Song China," chap. 3; Tillman, *Confucian Discourse and Chu Hsi's Ascendancy*, 129–30.

Commentaries played a crucial role in Zhu Xi's model of classical education, albeit a secondary and ancillary role. The original text came first; commentaries were to be consulted only after a prolonged personal confrontation with the integral original text. The subsidiary role of commentaries was reflected in the way readers were to approach them. Commentaries were to be read critically. Zhu Xi held that all commentaries were flawed; the critical evaluation of glosses and interpretations was intended to restore a direct relationship between the reader and the original text (stage two).

The personal understanding of the text, the concluding part of the essay, was the principal goal of Zhu Xi's reading method (stage three). In "Private Opinion," Zhu Xi defined the personal understanding of the text as a discussion of "the original intention of the sages and moral worthies and the actual applicability thereof."[42] Personal understanding of the text was not intended as a call for a critical evaluation of the original text; nor was it intended as an invitation to express innovative readings. The goal in reading the Classics was to experience them personally and identify with them. Students should read and write about the Classics as if the ideas embedded in the Classics seemed to come from their own minds.[43]

Zhu Xi's critique of examination writing was based on an alert appreciation of it. He read and wrote examination essays. He dismissed most of them, but he also mentioned a few examples of "good" examination questions and essays, those that were "not harmful of the Way."[44] He perceived the rise of utilitarian modes of thinking, the decline in classical scholarship, and the preoccupation with rhetorical skill as recent developments related to the authoritative position of the teachers of "Yongjia" in the examination field. His critique suggested that those developments could be countered, and he insinuated that Learning of the Way examination writing was not a contradiction:

---

42. Zhu Xi, *Zhu Xi ji*, 69.3640.

43. Gardner, *Learning to Be a Sage*, 42–43.

44. For discussions of specific texts, see *ZZYL*, 43.1105, 79.2062, 122.2953–54, 133.3200, 135.3226.

Tan Xiong asked about composing examination essays (*shiwen*). The Master replied, "*Generally follow the standard format, but rectify it based on ultimate principle (zhili).*"[45]

譚兄問作時文. 曰:「略用體式, 而櫽括以至理.」

Even though Zhu Xi resisted publication of "Private Opinion" during his lifetime,[46] some of his most prominent disciples continued his initiatives in the transformation of examination preparation and examination writing.

### Moral Philosophy: Chen Chun's Examination Guide for Learning of the Way Believers

Zhu Xi's critique of twelfth-century examination pedagogy left an ambivalent legacy. His criticisms of current examination standards and preparation practices attracted those for whom examination success seemed out of reach. Several of his students gave up on preparing for the examinations after committing themselves to the transmission of the Learning of the Way. On the other hand, many others, including some of the disciples whom Zhu Xi held in the highest regard, continued to participate in the examinations, with and without success.[47]

---

45. *ZZYL*, 13.247. For other examples, see Chen Wenyi, *You guanxue dao shuyuan*, 216.

46. I discuss possible reasons for Zhu Xi's reluctance to publish "Private Opinion" in "Changing Minds Through Examinations."

47. Tanaka Kenji ("Shu Ki to kakyo") outlines Zhu Xi's ambivalent reactions to examination preparation and participation in letters sent to students and acquaintances. He concludes from this mixed record that for Zhu Xi "learning for one's self" (self-cultivation) and participation in the examinations were compatible and that opting out was not in and of itself a recommended path of action. For a brief overview of the different career patterns of Zhu Xi's students, see the table in Ichiki, *Shu Ki monjin*, 389. In the discussion that precedes this table, Ichiki concludes that these students shared Zhu Xi's critical view of the examinations. Zhu Xi's later disciples were in Ichiki's view less likely to spread Learning of the Way ideology through the examinations and more likely to be active in building community at the local level (ibid., 369–93, 423–24). Chen Chun's example discussed here, as well as those of twelfth- and early thirteenth-century examination candidates discussed in Chapter 7, demonstrates that the Learning of the Way attracted students not only as an alternative to examination learning but also as an alternative way of preparing for the examinations.

Whatever their decision about the examinations, Zhu Xi's students were equally frustrated over the conflicting demands of the pursuit of moral self-cultivation, on one hand, and examination preparation, on the other hand. When students raised such problems in conversations or letters, Zhu Xi typically shifted to a conciliatory tone. He admitted that the practice of morality and examination preparation could be pursued simultaneously if students did not get caught up in the latter. He advised students to spend most of their time on the Learning of the Way and prepare for the examinations on the side.[48] In his private teaching practice, however, Zhu Xi did not mix the Learning of the Way and examination preparation. The explanation and practice of moral principles were separate from examination preparation.[49]

Some of Zhu Xi's disciples integrated the Learning of the Way and examination preparation in their teaching practice. For Chen Chun,[50] whose introductory text to the Learning of the Way is the focus of this section, examination preparation was no longer an extracurricular activity for students of the Learning of the Way; examination writing was a weapon that could be used in its campaign for the reformation of literati culture.

In writing, our master [Zhu Xi] has established a style. The right speed of the inexhaustible flow of his words and the right sharpness of their incisiveness are more than sufficient to face the enemy. *The only purpose of examination writing is to oppose the enemy.* Its strengths and weaknesses depend upon this. If we strive for more artistry on top of this and make calculations for gain, we will be misled. Principle and rightness reside in our bodies and minds, and we cannot be without them for even a single day. If we abandon them for a single day, we will lead a meaningless life, we will become like lost people, like ordinary men and vulgar fellows. We will bring about our own decline outside the walls of the house of Confucius and Mencius.[51]

---

48. Gardner, *Learning to Be a Sage*, 19–20.

49. Judging from his examination questions from the 1150s, it appears that Zhu Xi tried to transform examination learning in accordance with Learning of the Way principles. This project was, however, short-lived. See Chapter 7.

50. I am adopting the dates given in Satō, "Shin Jon no gakumon to shisō—Shu Ki jūgaku izen." For a discussion of the controversies surrounding Chen Chun's dates, see ibid., 49*n*1.

51. Chen Chun, *Beixi daquan ji*, 34.7a.

吾子於文已成一機軸. 詞源之正馭, 詞鋒之正銳, 其於對敵有餘也. 科舉
之文足以對敵則已. 其得失有命焉. 若於其上求之益工爲必得之計, 則惑
矣. 理義在吾身心, 不可一日關者. 一日而舍去, 則醉生夢死, 爲迷途中人,
爲庸夫俗子, 爲自暴自棄於孔孟門墙之外.

Chen Chun wrote these words in a letter to a friend facing the moral dilemma that examination preparation posed for followers of the Learning of the Way. Chen Chun shared Zhu Xi's critical appraisal of contemporary examination preparation and examination writing and reiterated Zhu's observations on the reading methods fostered in examination preparation courses and manuals. In his view, the focus on composition encouraged a fragmented reading of the literary tradition, including the Classics, the histories, and the philosophers. Students were taught to memorize short passages selected for them in courses, anthologies, and encyclopedias and taught to reconfigure such textual evidence to create novel arguments; to learn this skill, they relied mostly on the model essays of Ancient Prose authors or examination teachers.

Chen emphasized that even though this kind of learning was incompatible with the Learning of the Way, the Learning of the Way was fully compatible with examination preparation. Examination preparation could in his view be transformed into a site to promote the study methods of the Learning of the Way. He upheld Zhu Xi's reading method, with its emphases on an integral reading of the text and the personal application of universal moral standards to the text, as the only viable alternative to current examination preparation conventions. The preceding quotation illustrates his argument that Learning of the Way followers could legitimately participate in the examinations only if they applied the same principles in examination preparation as in their daily practice of moral self-cultivation. Reading was an essential part of the practice of self-cultivation in Zhu Xi's view. The proposals in "Private Opinion," if implemented, would have effected a transformation of the examinations in line with his pedagogy. Chen Chun continued Zhu Xi's steps in this direction and made Zhu's reading method the basis of examination preparation for followers of the Learning of the Way.

The disjunction between the Learning of the Way reading method and the composition skills tested in the examinations was a major source of frustration among examination candidates following the Learning of the Way. Examination teachers successfully promoted Ancient Prose, and twelfth-century masters of Ancient Prose such as Chen Fuliang achieved stellar success in the examinations. Throughout his career, Zhu Xi remained critical of the contributions of Su Shi, the most celebrated model of Ancient Prose, and censured student interest in his work. In the course of the thirteenth century, Zhu Xi's disciples implemented strategies that extended Learning of the Way influence over all fields of examination preparation including composition. Chen Chun's efforts marked a first step in this process.

Chen Chun advised examination candidates that composition was at once a byproduct and a measure of the consistent practice of moral self-cultivation. In the letter cited above, he wrote that if integrating the Learning of the Way into examination writing proved to be a problem, the student's insufficient insight, rather than the standards of examination officials, was to be blamed. A true understanding of the Learning of the Way could not be disengaged from writing:

As determined by the existence of Heaven and Earth, there is this coherence. As determined by their existence, human beings have the five constants in their heart-and-minds, attached to their bodies they have the five positions,[52] and in their daily activities there are the ten thousand affairs. The Way operates within all of this; it cannot be separated from these things. Explaining this Way is learning. Practicing this Way is moral conduct. Truly getting this Way is virtue. Upholding and instituting this Way all over the empire is accomplishment. *Developing it in words is composition.* Thus, the Way and writing are not two different things.[53]

自有天地, 則有此理. 有生人, 則在心所具有五常, 在身所接有五品, 在日用動靜有萬事. 而道行乎其間; 不能與之相離. 講明是道, 則爲學. 實踐是道, 則爲行. 實得是道, 則爲德舉. 而措之天下則爲事業. 而發達於言詞則爲文章. 故道與文非二物也.

---

52. The "five constants" refers here to the five virtues, which are, in embryonic form, inherent in the human mind, namely, humanity, rightness, propriety, wisdom, and trustworthiness. The statement about the five positions means that human beings cannot stand outside the five basic human relationships; they occupy one or more positions in the following pairs: father/son, husband/wife, elder brother / younger brother, friend/friend, ruler/minister.

53. Chen Chun, *Beixi daquan ji*, 34.4b.

Zhu Xi had posited a correspondence between understanding the Way and writing. He, too, explained to students that true language reflected the speaker's understanding of moral coherence in a transparent fashion.[54] The acceptance of this argument, first among Learning of the Way teachers and during the first half of the thirteenth century among examination teachers, effected the subordination of the instruction of composition to the teaching of the Learning of the Way. Among examiners it led to the substitution of standards measuring the candidate's understanding of Learning of the Way beliefs for criteria based on composition (Chapter 5).

Chen Chun went further than Zhu Xi. In his teaching and published work, he offered students a model for the expression of Learning of the Way truths in writing. As the first quotation translated above suggests, Chen Chun recommended the written legacy of Zhu Xi as both the embodiment of truth in writing and a weapon to oppose the enemy in the examination field. For him, the language of the eleventh- and twelfth-century canonical sources of the Learning of the Way was the point of departure for replacing existing examination conventions with those of the Learning of the Way. His introduction to the Learning of the Way foregrounded the written legacy of Zhu Xi and offered students analytical tools for use in examination writing.

## Guides to Learning of the Way Concepts and Texts

Chen Chun became one of Zhu Xi's most trusted students in the last decade of Zhu Xi's life. He studied twice with Zhu Xi, first in 1190 and then again in 1199, for a total of about seven months.[55] In the interval between their first and last meetings, Zhu Xi and Chen Chun frequently corresponded on questions regarding the Learning of the Way.

Chen had been exposed to the Learning of the Way long before he first met Zhu Xi in person. By his own admission, he was converted at the age of twenty-two after reading a copy of *A Record for Reflection*. In his official biography in *The History of the Song Dynasty* (*Song shi* 宋史),

---

54. He Jipeng, "Zhuzi de wen lun"; Ichiki, "Shu Ki no Rikuchō hyō."
55. Wing-tsit Chan, *Zhuzi menren*, 221.

this admission was represented as a simultaneous renunciation of examination success. Chen Chun wrote that he had become satiated with the "language of examination preparation" (*juye yuyan* 舉業語言) between the ages of fifteen and twenty.[56] The "language of examination preparation" refers to the commentaries on classical texts and the composition manuals used in examination preparation. The renunciation of examination success became a trope in biographies of men associated with the Learning of the Way, but what Chen was renouncing was the conventional sources of examination learning, not examination success per se. Despite his early dissatisfaction with examination preparation and his interest in *A Record for Reflection*, Chen Chun continued to participate in the examinations. He passed the prefectural examinations in Zhangzhou 漳州 (southern Fujian) at the age of thirty-two but failed in the following departmental examinations. In 1216, toward the end of his life, he again traveled to Lin'an to participate in a special examination, again without success.

Chen Chun's early conversion to the Learning of the Way evolved into a lifelong commitment. Like most of Zhu Xi's later disciples, Chen Chun became familiar with the Learning of the Way through Zhu Xi's publications. After reading *A Record for Reflection*, he studied Zhu Xi's 1172 and 1177 commentaries on *The Analects* and *Mencius*, his 1163 edition of the recorded conversations of the Cheng brothers (*The Surviving Writings of the Cheng Brothers*), his editions of the seminal works of Zhou Dunyi and Zhang Zai, and his interpretations of *The Great Learning* and *The Doctrine of the Mean*.[57] By the time he accomplished his lifelong dream of meeting with Zhu Xi, Chen Chun had been studying Zhu Xi's work for twenty years. His reading list during this time period suggests that his attention was drawn exclusively to those sources Zhu Xi had marked as central to the transmission of the Learning of the Way.

By reading the emerging canon of the Learning of the Way, Chen Chun nourished the belief that he was called on to become involved in the transmission of the Learning of the Way. His contributions to the spread of the Learning of the Way were recognized during and after his

---

56. Satō, "Shin Jon no gakumon to shisō—Shu Ki jūgaku izen," 48.
57. Ibid., 56.

lifetime.[58] He impressed Zhu Xi, whom he regarded as "the leader of the alliance" of the Learning of the Way. During Zhu Xi's tenure in Chen's hometown in Zhangzhou prefecture in 1190 and 1191, Chen engaged the master in frequent conversations. After his premature departure and return home to Jianyang in northern Fujian, Zhu Xi remarked that he was delighted to have found Chen Chun as a collaborator in southern Fujian.

Chen Chun's most influential contributions to the dissemination of the Learning of the Way dated from the period after his first encounter with Zhu Xi. He compiled an extensive collection of notes recording his interactions with Zhu Xi. His records of conversations with the master (*yulu* 語錄) constitute a substantial portion of the current edition of *Master Zhu's Classified Recorded Sayings* (1270); he ranks third in the list of contributors to that collection. Chen Chun was also the author of several primers on the Learning of the Way, each directed at different audiences. He wrote instructions for children and women and introductory texts for adult male students. In the last group are three collections of lectures: *Oral Explanations of the Four Books* (*Yu Meng Daxue Zhongyong kouyi* 語孟大學中庸口義), *The Lectures at Yanling Prefecture* (*Yanling jiangyi* 嚴陵講義), and *The Correct Meaning of Terms by Chen Chun* (*Beixi ziyi* 北溪字義).[59]

The prominence of the lecture format in Chen Chun's legacy was no coincidence. The lectures bore witness to his lifelong career as a teacher. He supported his family through professional teaching in Zhangzhou, and, in response to a petition, the court awarded him the position of magistrate in a nearby county in recompense for his years of teaching service (he passed away before he could take office). Chen's preference for the lecture format further reflected the centrality of oral discourse in the transmission of the Learning of the Way. The place of oral discourse was evident in the importance attached to direct transmission

---

58. Ichiki ("Shin Jun ron josetsu") focuses on the later use of Chen Chun's lexicon among students preparing for the examinations. The time frame for its expanding use is unclear.

59. Wing-tsit Chan, trans. and ed., *Neo-Confucian Terms Explained*, 5, 12–27; idem, *Zhuzi menren*, 220–21; Honma, "Yomigaeru Shu Ki," 1–17. For an overview of the most important contributors to *Classified Sayings of Master Zhu*, see Ichiki, *Shu Ki monjin*, 365–66.

from teacher to disciple as well as in the recording of direct interactions between master and disciples.

Records of conversation became emblematic of the impact of the Learning of the Way on examination writing, as attested in the requests for their prohibition in the reports of examiners from the 1190s (Chapter 5). The recorded conversations explained Learning of the Way concepts in more detail than commentaries and did so in a language closer to the vernacular. The interacting voices of master and disciples inspired readers to imitate the model and speak in a personal voice when responding to questions. The language of the recorded conversations thus presented an alternative to students searching for ways to translate their commitment to the Learning of the Way in written answers to examination questions. This rhetorical mode was prevalent in examination writing during the first phase of expansion of Learning of the Way discourse in examination writing in the late twelfth century. During this phase, examination candidates articulated their understanding of Learning of the Way teachings in an unconventional language of philosophical debate and adopted an antagonistic stance toward competing traditions of learning (Chapter 7). This stance characterized Zhu Xi's intellectual and teaching career and shaped the thinking of his first-generation disciples.

Chen Chun's recorded lectures embodied the features of this first phase in the ascendancy of the Learning of the Way in the examination field. His collections of lectures bore the stamp of his involvement in their publication. Unlike the "Yongjia" manuals discussed in Chapter 4, which probably resulted from the collaboration between anonymous note takers and commercial printers, Chen Chun's lectures were recorded by identifiable disciples. In at least two cases, he revised their notes before publication.[60]

*The Correct Meaning of Terms*, printed between 1219 and 1223 shortly before his death, best exemplifies his effort "to oppose the enemy" in the examination field. This lexicon provides relatively short definitions of key terms in Learning of the Way discourse. Each entry resulted from lecture notes on the topic and consists of a discussion of the term

---

60. Wing-tsit Chan, trans. and ed., *Neo-Confucian Terms Explained*, 175*n*, 239.

in the colloquial style of the recorded conversations. In the rest of this section, I first show that in its organization and mode of explanation, *The Correct Meaning of Terms* went back to Learning of the Way textbook models. I then argue that it did so deliberately, in reaction to other efforts aimed at repackaging the Learning of the Way for the use of examination candidates.

*The Correct Meaning of Terms* adopted a textbook format that must have been familiar to twelfth-century students. The presentation of the material in twenty-five or twenty-six categories[61] closely resembled the use of topical categories in "category books" (*leishu* 類書), which combined features of both encyclopedias and anthologies. The compilation of category books was a popular occupation among teachers and underemployed scholars and a lucrative enterprise for booksellers. Yue Ke 岳珂 (1183–1240), a contemporary observer, wrote that bookstores in Jianyang were constantly issuing new such titles and revised editions of old titles.[62] Category books for use in the examinations served as digests of classical, historical, and literary sources and provided background information about the variety of subjects covered in examination questions and supplied quotations candidates could use in their essays.

In contrast to standard examination category books, *The Correct Meaning of Terms* was limited to a systematic overview of Learning of the Way moral philosophy. The topics were selected from the core classical texts of the Learning of the Way tradition, especially *The Doctrine of the Mean*. The quotations discussed under each of the entries were drawn from the works of the eleventh- and twelfth-century transmitters of the Learning of the Way, with Zhu Xi's work receiving particular attention. As the title suggests, *The Correct Meaning of Terms* was intended as an index of terms derived from the Classics but defined exclusively by the

---

61. There are twenty-five categories in the oldest surviving edition (Yuan) and twenty-six in the more widely available edition (Qing). Honma ("Yomigaeru Shu Ki," 14) tabulates the differences between the two editions and argues that both may date to Song editions. Wing-tsit Chan discusses editions and translates prefaces from the Song through the Qing periods in *Neo-Confucian Terms Explained*, appendices B–N. See also Inoue, "*Hokki jigi* hanpon kō"; and Zhang Jiacai, "*Beixi ziyi* banben yuanliu lice."

62. For a short discussion of the characteristics of *leishu* and developments in the publication of such works in the Southern Song, see De Weerdt, "Aspects of Song Intellectual Life," 3–5; and idem, "The Encyclopedia as Textbook."

interpretations of the masters of the Learning of the Way, who claimed to have rediscovered their true meaning.

The arrangement of the topics carried significance. It embodied the message that the topics were to be read not as independent entries but as connected elements in a coherent moral philosophy and as steps in a program of learning that joined understanding and moral action. The sequence of the twenty-five topics in *The Correct Meaning of Terms* was based on the opening sentences of *The Doctrine of the Mean*:[63]

> What Heaven endows is called the nature.
> Following the nature is called the Way. Cultivating the Way is called teaching.
> 天命之謂性.
> 率性之謂道, 修道之謂教.

These sentences defined key concepts in Learning of the Way metaphysics (Heaven's endowment, the nature, and the Way) and traced systematic relationships between them. The last sentence links the philosophy of the Way to its practice in teaching and learning.

*The Correct Meaning of Terms* took readers along the same trajectory. The first chapter opened with a discussion of Heaven's endowment and the nature and analyzed other subjects relating to the endowed characteristics of human nature. The second chapter elaborated on the next two sentences. It first defined the Way and discussed concepts explaining both its genesis and its operation such as the "Supreme Ultimate" and "coherence." The last five entries covered subjects that were either encouraged or discouraged in the practice of the Learning of the Way: ritual, music, ancestor worship, and propriety were encouraged, whereas profit and the practice of uncanonical sacrifices, Buddhism, and Daoism were censured.

Chen Chun thus based the selection of Learning of the Way terms and their arrangement in *The Correct Meaning of Terms* on the last of Zhu Xi's Four Books. The superimposition of a canonical source on the philosophy of the masters of the Learning of the Way highlighted the continuity these philosophers claimed with the Way of the Sages represented in the Classics. The strategy of encapsulating Learning of the

---

63. My interpretation is based on Honma, "Yomigaeru Shu Ki," 14–15.

Way moral philosophy in particular passages of the Four Books had been tried before. Chen Chun modeled *The Correct Meaning of Terms* on his favorite Learning of the Way text, *A Record for Reflection*, the text that had introduced him to the Learning of the Way.[64] He accorded great significance to it in his teaching. A comment attributed to Zhu Xi, that whereas the Four Books were the ladder to the Six Classics, *A Record for Reflection* was the ladder to the Four Books, was recorded by Chen Chun.[65] One of Chen's students, Ye Cai 葉采 (fl. 1248), published the first full-scale commentary on *A Record for Reflection*.[66]

Zhu Xi intended *A Record for Reflection* as a concise and step-by-step introduction to the path of moral self-cultivation discussed in great detail in the works of the eleventh-century Masters of the Learning of the Way. In order to "let students know where to start," he selected passages, with Lü Zuqian's help,[67] from the work of Zhou Dunyi, the

---

64. Wing-tsit Chan, trans. and ed., *Neo-Confucian Terms Explained*, 2.

65. *ZZYL*, 105.2629.

66. Wing-tsit Chan, trans., *Reflections on Things at Hand*, 338–39, and *passim*.

67. Du Haijun ("Lü Zuqian yu *Jinsi lu* de bianzuan") has argued that Zhu Xi's contribution to the volume was insignificant compared to Lü Zuqian's. Du's research demonstrates overlap between Lü's ideas and those expressed in the anthology, as well as Zhu Xi's attention to Lü's input. Extant records do not directly discuss who was mainly responsible for the anthology's contents and layout. However, the fact that Zhu Xi claimed it, and that it was part of a series of works in which he discussed the legacy of the Four Masters, lend credence to the older conclusion (voiced by Wing-tsit Chan and others) that the concept behind this work mainly embodied Zhu Xi's interpretation of the tradition of Cheng Learning. Zhu Xi recommended reading *A Record for Reflection* to his students. Wing-tsit Chan translated about ten conversations between Zhu Xi and students in *Reflections*, XL–XLI. Du Haijun relies predominantly on conversations with students in which Zhu Xi comments critically on items Lü Zuqian had either selected or rejected. Such critical comments on intellectuals associated with Cheng Learning were typical of the technique of metacriticism Zhu Xi employed in conversations with students (Chapter 7). I also read them as an indication that Zhu Xi may have come to regret the implications of his collaboration with Lü on this project in particular instances, and not as an admission that Lü was the leading force behind the project.

The postscripts to the anthology provide further evidence of the centrality of Zhu Xi in the project. In his postscript to the compilation, Zhu Xi explained the rationale of the compilation and its general layout. He saw it as an introduction to the work of the Four Masters and the philosophy of moral practice he had discovered in their work. Lü Zuqian's postscript, written one year after Zhu Xi's, on the other hand, only comments on Lü's recommendation to place the chapter on the metaphysical foundations of self-

Cheng brothers, and Zhang Zai and organized them in a sequence inspired by the process of moral growth outlined in *The Great Learning*: from the essentials of learning and self-cultivation to the handling of sociopolitical affairs. Chapters 3 through 8 trace the extension of moral knowledge from "examining things and investigating coherence" to "preserving the mind," "correcting mistakes and doing good, and disciplining the self and returning to ritual," "regulating the family," "accepting and declining office," through "ordering the state and bringing peace to the world." The philosophy of the eleventh-century masters was thus represented through one of the most celebrated sentences in the first of the Four Books:

It is only when things are investigated that knowledge is extended; when knowledge is extended that thoughts become sincere; when thoughts become sincere that the mind is rectified; when the mind is rectified that the person is cultivated; when the person is cultivated that order is brought to the family; when order is brought to the family that the state is well governed; when the state is well governed that peace is brought to the world.[68]

物格而後知至, 知至而後意誠, 意誠而後心正, 心正而後身脩, 身脩而後家齊, 家齊而後國治, 國治而後天下平.

By adopting this sentence from *The Great Learning* as an organizational template, Zhu Xi presented the diverse legacy of the masters of the Learning of the Way as one coherent philosophy reviving the central message of the Classics.[69] As the categories listed above suggest, *A Record for Reflection* was intended as a guide to moral practice. According to Zhu Xi, the work differentiated steps in the extension of moral self-cultivation for pedagogical reasons. In actuality, the steps in the process were organically linked, but the truth of their coincidence could be real-

---

cultivation in the front. In contrast to Zhu Xi, Lü Zuqian does not appear to have recommended the book to his students. For translations of both postscripts, see Wing-tsit Chan, trans., *Reflections on Things at Hand*, 1–4. Cf. Xu Ruzong, *Wuxue zhi zong*, 186–88.

68. The translation is taken from de Bary, ed., *Sources of Chinese Tradition*, 1: 331; cf. Gardner, *Chu Hsi and the Ta-Hsueh*, 93–94.

69. For a discussion of the connection between *The Great Learning* and *A Record for Reflection*, see Honma, "Yomigaeru Shu Ki," 24–27; and Tucker, "An Onto-hermeneutic and Historico-hermeneutic Analysis of Chu Hsi's Political Philosophy," 2–3.

ized only in moral practice. *A Record for Reflection* was intended to motivate students to discover the effects of moral self-cultivation.

The question of the relationship between Learning of the Way moral philosophy and moral practice lay at the core of *A Record for Reflection* and came up in editorial discussions. In the case of *A Record for Reflection*, as in compilations modeled on it like *The Correct Meaning of Terms*, the debate about the priority of theory and practice was resolved in favor of the former. *A Record for Reflection* starts with a section on "the constitution of the Way" (*daoti* 道體) in which Zhu Xi and Lü Zuqian debated the pedagogical merits of placing the most theoretical chapter at the very beginning of the compilation. Zhu Xi ultimately agreed with Lü's argument that some exposure to the cosmology of the Way would motivate beginners to engage themselves in the concrete steps of daily practice and would allow them to gain a fuller understanding of the metaphysical foundations for moral practice.

The discussion of the origins of the Way in the first section of *A Record for Reflection* was based primarily on Zhou Dunyi's *Diagram of the Supreme Ultimate*. Zhu Xi's canon of the Learning of the Way included Zhou's work because it offered the most systematic account of the origins of the cosmos as a moral order. Even though beginners were unlikely to understand the cosmology of the Way on first reading this chapter, Lü and Zhu reckoned that placing it before the other chapters would ensure that readers read the chapters on moral practice with the opening chapter in mind. It would thus ensure that readers took seriously the claim of the Learning of the Way masters that their program of learning was grounded in the order of the cosmos.[70]

Through the selection and organization of concepts in *The Correct Meaning of Terms*, Chen Chun presented his work as a direct continuation of the work of earlier transmitters in the genealogy of the Learning of the Way. The layout of his work was based on two prominent texts in the canon of the Learning of the Way promoted by his teacher Zhu Xi. *A Record for Reflection* exemplified the use of an axiom from one of the Four Books for capturing the unity and continuity of the Learning of the Way. *The Doctrine of the Mean* provided the conceptual framework

---

70. See the prefaces by Zhu Xi (1175) and Lü Zuqian (1176) in Wing-tsit Chan, trans., *Reflections on Things at Hand*, 1–3.

for Chen's work. Apart from selection and organization, Chen Chun connected *The Correct Meaning of Terms* with the Learning of the Way canon through a shared mode of explanation.

Twelfth-century Learning of the Way proselytizing texts (including commentaries, recorded conversations, and anthologies) taught readers not only a vocabulary of moral philosophy but also modes of explaining transmitted truths. Zhu Xi's commentaries on *The Great Learning* and *The Doctrine of the Mean* provided two models for the explanation of transmitted texts. In his edition of and commentary on *The Great Learning*, Zhu Xi reconstructed the received text. He interpreted the eight steps in the extension of moral self-cultivation translated above as the central message of *The Great Learning* and divided the rest of the text up into sections that corresponded to and elaborated on each of the eight steps.[71] The eight steps became for Zhu Xi not only a list of topics but also a way of reading texts on moral self-cultivation. The superimposition of the eight steps on the written legacy of the eleventh-century Masters of the Learning of the Way in *A Record for Reflection* was therefore the result of a way of reading and explaining that Zhu Xi had already applied to a classical text.

In his edition of and commentary on *The Doctrine of the Mean*, Zhu Xi used a different mode of explanation, which was also extended to the interpretation of the words of contemporary transmitters of the Learning of the Way. Zhu Xi uncovered three different layers in the classical text. He argued that in the first part Zi Si, grandson of Confucius, noted down the core text transmitted to him. In subsequent sections, Zi Si explained the core text by quoting Confucius' words. In these sections Confucius spoke, even though his words had been gathered from different contexts, to elucidate the meaning of the core text. In the last sections, Zi Si explained in his own words what he had learned from Confucius, in order to clarify the meaning of the passages he had quoted in the previous sections.[72]

For Zhu Xi and his disciples, Zi Si symbolized the virtues of a transmitter of the Learning of the Way. He noted down and transmit-

---

71. For discussions of Zhu Xi's reorganization of the transmitted text, see Gardner, *Chu Hsi and the Ta-Hsueh*, 27–45, and *passim*; and Honma, "Yomigaeru Shu Ki," 31–32.

72. Honma, "Yomigaeru Shu Ki," 32–36.

ted his master's words and explained their meaning in his own words. His explanations were furthermore based on what he had learned through personal communication with his master. Chen Chun's explanations in *The Correct Meaning of Terms* followed this model.

Chen Chun clarified the meaning of the terms in *The Correct Meaning of Terms* by quoting and explaining statements from the masters of the Learning of the Way. He quoted verbatim the written work of Zhu Xi's eleventh-century predecessors and inserted passages selected from Zhu Xi's vast written legacy as well as from personal communications with him. Such passages conveyed authority of both the masters and the transmitter. Their invocation confirmed the authority of the masters; the validity of their sayings was never questioned. At the same time, the quotation of the masters' writings and sayings conveyed authority to the transmitter. The transmitter clarified Learning of the Way philosophy by recalling appropriate passages from the tradition and explaining their meaning, either through additional quotations or through personal explications. Through his ability to connect the transmitted wisdom of the masters with his personal understanding of Learning of the Way moral philosophy, the transmitter demonstrated authority over the Learning of the Way tradition:

I personally heard Duke Wen [Zhu Xi] when he said, "When the thearch is greatly infuriated,"[73] it is simply that according to the principle of coherence, it should be so. Nothing in the world is honored more highly than coherence, and therefore we call it the thearch."[74] From this we can understand that Heaven is coherence. Thus the blue sky is the body of Heaven. The body of Heaven is spoken of in terms of material force (*qi*), and the operation of Heaven is spoken of in terms of coherence (*li*).[75]

又嘗親炙文公說: 上帝震怒也, 只是其理如此. 天下莫尊於理, 故以帝名之. 觀此亦可見矣. 故上而蒼蒼者, 天之體也. 上天之體以氣言; 上天之載以理言.

---

73. *The Book of Documents*, "Hong Fan," 3. Cf. Wing-tsit Chan, trans. and ed., *Neo-Confucian Terms Explained*, 45*n*27.

74. ZZYL, 4.63.

75. I am largely adopting Wing-tsit Chan's translation in *Neo-Confucian Terms Explained*, 45. Chen Chun, *Beixi ziyi* (SKQS), shang, 7b.

In this passage, Chen Chun was responding to a question regarding Heaven's endowment, the first topic in *The Correct Meaning of Terms*. The question probed the attributes of Heaven as a source providing direction to all things: "When Heaven endows, is there really something above that arranges and orders that?" Chen Chun responded that Heaven stands for coherence, the pattern that is unique in each and every thing; yet, by virtue of its prescribing every thing's proper development and roles, it ties them all together in one cosmic order. He justified his answer to this question by invoking the words of Cheng Yi, citing passages from Zhu Xi's commentaries, and, as demonstrated in the passage translated above, bearing personal testimony to Zhu Xi's words. The passages from the canonical authors of the Learning of the Way carry normative meaning and thus provide explanatory power, but the transmission of the Learning of the Way also requires the personal articulation of the moral philosophy that underlies the variety of canonical passages. Chen Chun deduced the truth of the equation of Heaven and coherence from the masters' legacy and clarifies that this equation holds only as far as the operation, and not the material substance, of Heaven is concerned.

The parallel between the role Zhu Xi assigned to Zi Si and the role Chen Chun assumed in lectures and lecture notes illustrates the continuity in modes of explanation between Zhu Xi's twelfth-century commentary on *The Doctrine of the Mean* and Chen Chun's thirteenth-century effort to produce a comprehensive introductory text of Learning of the Way moral philosophy. Chen Chun's lectures and textbook familiarized students with a mode of explanation that allowed them to become participants in the transmission of the Learning of the Way. As he mentioned in the letter to the desperate examination candidate excerpted above, Chen Chun recommended that examination students learn from Zhu Xi's style. He advocated that they speak in the same authoritative voice; he did not advocate that they restrict their responses to citation of Learning of the Way commentaries corresponding to the passages referred to in the examination question. Chen Chun's strategy was based on Learning of the Way models. Like Zhu Xi's commentaries, the recorded conversations, and *A Record for Reflection*, *The Correct Meaning of Terms* represented the strident voices of advocates of the Learning of the Way in the twelfth and early thirteenth centuries. Unlike those early models, it also offered a deliberate alternative to other strategies aimed at promoting the use of Learning of the Way ideology in examination preparation.

As the Learning of the Way gained more clout among local elites and at court in the thirteenth century, practices current in the examination field were adopted for its promotion. Zhu Xi's redefinition of the genealogy of the Way, in particular his alignment of the eleventh-century transmitters of the Way, resulted in the unofficial canonization of the work of Zhou Dunyi, the Cheng brothers, and Zhang Zai and of *A Record for Reflection*, the source that had brought these thinkers' legacies together. Chen Chun and other Learning of the Way affiliates gave priority to these sources and the commentaries and recorded conversations of Zhu Xi in their curricula. In the course of the thirteenth century, new editions and compilations of Learning of the Way sources occupied a growing share of examination preparation manuals in commercial printing centers, an indication of the Learning of the Way's shift from the periphery to the center of the examination field. The publication of another category of work signaled how important knowledge of the Learning of the Way had become for examination candidates. By the 1220s digests of the works of its leading thinkers, in the format of the examination "category books," appeared on the market.

Chen Chun wrote a scathing review of a compendium in which the editor had gathered quotations on concepts identical to those covered in Chen's *The Correct Meaning of Terms*. He acknowledged that the words of the masters had been turned into tokens that could be cashed in for examination success. For Chen concepts such as "the Way and virtue," "humanity and compassion," and "nature and emotions" fit into a larger philosophy of moral self-cultivation and should be explained only with reference to this broader framework.[76] He objected to the reification of the terms for these concepts into categories, whereby each term served as a semantic container for its divergent occurrences in the works of authoritative thinkers. In his view, category books indexing moral concepts undermined the power of the masters' sayings; such books turned them into words disassociated from speech acts. Students

---

76. Chen Chun, "On Reading '*Scrutinizing What Is Right from the Gao Studio*' (*Gaozhai Shenshi ji*)," in *Beixi daquan ji*, 14.9a–b.

using such tools replicated the words of the masters but without the masters' understanding and intent.

By contrast, Chen Chun advised examination candidates to turn their use of the words of past masters into performative statements. By personally selecting appropriate sayings and articulating a personal understanding of the Learning of the Way, a student demonstrated his commitment to the transmission of the Learning of the Way. This commitment, which was at once a commitment to the moral transformation of society and polity, was in Chen Chun's as in Zhu Xi's eyes the standard for the selection of officials.

Even as Chen Chun advocated examination writing that remained true to the modes of transmission of the Learning of the Way and was uncontaminated by the conventions of the examination field, the process of the adaptation of Learning of the Way ideology to such conventions continued unabated. Commercial printers targeting examination candidates marketed Learning of the Way manuals that cast the textual tradition of the masters in formats traditionally associated with examination preparation. The Zeng family of commercial publishers from Jian'an printed a supplement to *A Record for Reflection* and a special edition of the original, *A Record for Reflection, Classified for Use in the Examinations* (*Wenchang ziyong fenmen "Jinsi lu"* 文場資用分門近思錄).[77] The materials Zhu Xi had organized into fourteen categories were reclassified under 121 subject headings. The inflation of the number of categories was congruent with the growing influence of the Learning of the Way in examination preparation.

The detailed categories in this compilation were not an extension of classification schemes indigenous to the Learning of the Way. The classification schemes used in Learning of the Way texts such as *A Record for Reflection, The Correct Meaning of Terms*, or classified editions of Zhu Xi's recorded sayings were based on a relatively small number of cate-

---

77. I saw the copy at the National Library in Taibei. *Guoli zhongyang tushuguan Songben tulu* (167–68) gives the late Southern Song period as the approximate date of appearance. For the Zeng family business, see Poon, "Books and Printing in Sung China," 164, 314. Poon (ibid., 145) argues that even though some printers used "family school" as a designation for their business, they were, in fact, commercial printers. Cf. Chia, *Printing for Profit*, 129.

gories, ranging between fourteen and twenty-six.[78] A large number of detailed categories was characteristic of examination category books. Compilers and printers producing such manuals advertised the advantage of direct access to quotes directly relevant to the user's topic of interest.[79] A user interested in Zhu Xi's ideas on reading and writing, for

---

78. The earliest known classified compilation of recorded conversations, from 1219, consisted of twenty-six categories. The current edition, compiled by Li Jingde in 1270, consists of an equally small number of major categories. For a comparison of both editions, see Honma, "Yomigaeru Shu Ki," 21–22; cf. Hervouet, *A Sung Bibliography*, 225. A condensation of the growing corpus of recorded conversations, *Huian xiansheng yulu da gangling* (with supplements on some of the major debates in which Zhu Xi was involved), published in the mid-thirteenth century, also used twenty-six categories. I saw a Song edition of this work at Beijing Library (see also Honma, "Yomigaeru Shu Ki," 23).

79. In his typology of "Confucian anthologies," Thomas Wilson (*Genealogy of the Way*, chap. 4, esp., 151–67) differentiates between the "Reflections Type" and the "Nature and Principle Type" in the class of categorized anthologies. The latter he associates with the examination system. He sees the great compilation projects under Ming emperor Yongle, especially *The Great Collection on Nature and Principle* (*Xingli da quan*), as the beginning of categorized Learning of the Way compendia for the examinations. He further argues that the prototype of this subgenre is Li Jingde's *Zhuzi yulei*.

Categorized Learning of the Way handbooks were circulating in the late Southern Song and the Yuan periods. Besides *Classified Reflections for the Examinations* and the work censured by Chen Chun, there is also *A Classified Compilation of Master Zhu on Government, Providing a Standard for Writing* (*Zhuzi jingji wenheng leibian* 朱子經濟文衡類編). This work, in three series consisting of sixty-four, seventy-five, and fifty-two subjects, respectively, offers an index to Zhu Xi's collected works and the recorded conversations. The compilation is attributed to a disciple of Zhu Xi, Teng Gong 藤珙 (*js.* 1187). The earliest edition to which I have seen a reference is a Yuan edition from 1324 (Liu Lin and Shen Zhihong, *Xiancun Song ren zhushu zonglu*, 98–99; cf. Hervouet, *A Sung Bibliography*, 224).

The inspiration for such a classified edition of someone's writings comes from editorial methods current in the Song examination field. The work of Yang Wanli, whose writing, including his examination essay writing, gained wide acclaim in the last decades of the twelfth century and after, for example, was cut up in the same way in *Excerpts from Yang Wanli, Annotated and Classified* (*Pidian fenlei Chengzhai xiansheng wenkuai* 批點分類誠齊先生文膾). In a preface dated 1259, Fang Fengchen, who was also involved in the compilation of examination essays by Chen Fuliang (Chapter 3), wrote that Li Chengfu 李誠父 from Jian'an had selected those excerpts from Yang's work that would be most useful to examination candidates and organized them in a detailed set of subjects. Li's compilation appeared in two installments. This preface appears in the Yuan edition preserved at the Palace Museum Library in Taibei.

example, could refer to the categories "On Learning" and "On Writing" in *Master Zhu's Classified Recorded Sayings*; in *A Record for Reflection, Classified for the Examinations*, he would find entries on "Contemplating Books," "Reading History," "Doubts About the Classics," "Composition," and "Examination Preparation." In the competitive environment of the Song civil service examinations, printed category books were routinely smuggled into the examination halls, and *A Record for Reflection, Classified for the Examinations* would have enjoyed great appeal.

Chen Chun bemoaned the compilation of examination handbooks that merely categorized the words of wise men, even if the categories were related to the Learning of the Way. His lectures and, after his death, his published talks outlined an alternative approach to examination writing. A personal understanding of Learning of the Way truths would, in his view, lead to the personal explanation of these truths in ideologically pure examination writing. Although Learning of the Way affiliates continued to publish alternatives to current instructional materials in the late Southern Song period (discussed below), practices common in the Southern Song examination field, such as the compilation of category books, were increasingly adopted to sell Learning of the Way ideas in a format more familiar to thirteenth-century examination candidates. The next section shows that the adaptation of the Learning of the Way to the conventions of the examination field went beyond the mere reorganization of its core texts.

### *History and Government: Ideological Reconciliation in Commercial Encyclopedias*

Before the 1230s, Learning of the Way instructional publications were concerned principally with moral philosophy. Institutional history and government, key areas of interest among the "Yongjia" teachers and essential fields of study in preparing for the policy essay section of the examinations, received less attention among Learning of the Way teachers, who subordinated the study of history and government to an engagement with moral philosophy. Government, like composition, was thought of as an extension of the process of self-cultivation; separate instruction in administrative or compositional techniques was therefore considered incompatible with the goal of moral education.

History illustrated the axiom that government was an extension of moral self-cultivation. Zhu Xi's two historical projects, *Outline of the Comprehensive Mirror* (*Tongjian gangmu* 通鑑綱目) and *Records of the Words and Deeds of Famous Ministers* (*Mingchen yanxing lu* 名臣言行錄), were aimed at aiding the student in analyzing the impact of historical actors' practice or neglect of moral cultivation on the course of history. Learning of the Way teachers thus rejected the historical studies and administrative encyclopedias of their contemporaries and created alternatives that underscored the dominance of moral philosophy over other aspects of the curriculum.

In the first decades of the thirteenth century, teachers specializing in examination preparation forged a different relationship between Learning of the Way moral philosophy, on one hand, and history and government, on the other. Starting in the 1210s, as Zhu Xi and the other masters of the Learning of the Way received imperial recognition, their work became recommended reading at the Imperial College and gained positive attention in the examinations. Examination teachers and printers responded to the changing appreciation of Learning of the Way ideas and texts by incorporating them into courses and manuals. This section investigates the process of conciliation between the Learning of the Way curriculum and standards of examination teaching associated with the "Yongjia" curriculum. It first examines how thirteenth-century examination teachers and commercial printers represented the Learning of the Way and "Yongjia" examination teaching in their courses and manuals. The combination of these two modes of learning in thirteenth-century examination manuals and the changes in their representation were indicative of the transformation of Learning of the Way ideology. This transformation, the formation of an ideology of conciliation, is also discussed in the next section and in Chapter 7. This section traces the transformation of the Learning of the Way to examination teachers upholding the Learning of the Way as a symbol for unity in Song intellectual and political culture.

## THE ARCHEOLOGY OF AN ENCYCLOPEDIA

On a first reading, Lin Jiong's 林駉 (fl. 1210s) *Ultimate Essays on Origins and Developments (from the Past to the Present)—(Newly Annotated for Examination Success)* ([*Xinjian jueke*] [*Gujin*] *yuanliu zhilun* 新箋決科古今源流

至論), published in three installments and completed by the late 1220s or early 1230s, appears to be an expanded edition of the Learning of the Way manuals discussed in the previous section. The topics covered and the argumentation in many of the essays under each topic draw heavily on the Learning of the Way tradition as synthesized by Zhu Xi.

Lin Jiong completed the first three installments at different times between the 1210s and 1230s. The commercial printer of the text produced each installment in ten chapters, with each chapter encompassing from three to ten entries in the form of essays. In most current editions, the first four chapters in the first installment are concerned with learning. The series opens with entries on two central sources of the Learning of the Way, *The Diagram of the Supreme Ultimate* and *The Western Inscription*. Zhu Xi selected these sources for inclusion in the Learning of the Way canon in the 1160s and 1170s and especially valued *The Diagram*'s explanation of the metaphysical foundations of the Learning of the Way and *The Western Inscription*'s poetic description of the cosmic dimensions of moral self-cultivation. Neither had ever been included as individual entries in prior encyclopedias. The first installment also includes essays on the significance of Learning of the Way interpretations of *The Doctrine of the Mean*, *The Great Learning*, *The Analects*, and *Mencius*. These four titles were gradually attaining the status of classics, a transformation in the classical canon due entirely to the ascendancy of the Learning of the Way in literati culture. Besides the core texts of the Learning of the Way tradition, the first installment also includes essays on key Learning of the Way concepts such as "The Learning of Human Nature," "The Learning of the Mind," "Centrality," "Humanity," and "Finding Joy in the Way." These concepts were also covered in Chen Chun's *The Correct Meaning of Terms*.

The second installment begins with essays on "The Learning of the Way" and "The Learning of Investigating Things," a concept used to explain the core canonical content of the Learning of the Way. Learning of the Way ideology also dominates a fourth installment compiled by Huang Lüweng 黃履翁 (*js.* 1232) around 1233. The first essays in Huang's supplement praise the history compiled under Zhu Xi's direction (*Outline of the Comprehensive Mirror*), his commentaries on the Four Books, and Zhou Dunyi's commentary on *The Changes* (*Penetrating the Changes*). Other entries such as "integrity" (*cheng* 誠) and "rightness and profit" (*yili* 義利) overlap with those included as entries in the core Learning of the Way encyclopedias *A Record for Reflection* and *The Correct Meaning of Terms*.

Learning of the Way thinkers, sources, and concepts were also introduced in two other thirteenth-century examination encyclopedias. *The Epitome of Eminent Men Responding at the Imperial College* (*Bishui qunying daiwen huiyuan* 璧水群英待問會元; ca. 1245) featured the "Learning of the Way" and "Nature and Coherence" among its sixteen main headings. Its compiler, Liu Dake 劉達可, cited the works of the masters of the Learning of the Way at length under these categories. *A Net to Unite and Order the Massive Amounts of Information in All Books* (*Qunshu huiyuan jie jiang wang* 群書會元截江網) included entries on the tradition of Yiluo 伊洛 (the learning centered on the intellectual legacy of the brothers Cheng) and Zhu Xi. The textual legacy of the masters received brief reviews in this encyclopedia, an abridged and updated version of *The Epitome* published around 1250.[80]

The organization of these encyclopedias was based on a logic different from that of the Learning of the Way manuals discussed in the previous section. The structure of *A Record for Reflection* and *The Correct Meaning of Terms* was based on *The Great Learning* and *The Doctrine of the Mean*, both chapters in *The Book of Rites*. Both manuals focused on the main concepts in Learning of the Way moral philosophy, but their editors legitimated the selection of terms and the coherence of the conceptual framework on the grounds that they agreed with the classical canon. In thirteenth-century examination encyclopedias, by contrast, the Learning of the Way was represented by a new set of canonical texts authored by eleventh- and twelfth-century Learning of the Way masters and cast in the format of the examination encyclopedia. The thirteenth-century representation of the Learning of the Way was a sign of the broader acceptance of Zhu Xi's construction of the genealogy of the Learning of the Way.

The prominent place accorded the Learning of the Way in thirteenth-century examination literature was the result of the unofficial and official canonization of the work of the masters of the Learning of the Way in the late twelfth and early thirteenth centuries. The canon

---

80. *Bishui*, chaps. 39, 44–45; *Jie jiang wang*, chaps. 31, 33.

took shape unofficially in private and local schools and was disseminated through private and local government printers. Zhu Xi's disciples and other teachers advocating the Learning of the Way taught Zhu Xi's commentaries and his editions of the collected writings of his chosen predecessors. The official recognition of the authoritative figures of the Learning of the Way and their written legacy encouraged the widespread use of Learning of the Way sources in examination questions and answers (Chapter 7). In edicts endorsing the Learning of the Way in 1212 and 1227, Emperors Ningzong and Lizong highlighted the value of Zhu Xi's commentarial legacy. In 1241, Lizong commended the work of the Five Masters of the Learning of the Way, Zhou Dunyi, Zhang Zai, the Cheng brothers, and Zhu Xi. Thirteenth-century encyclopedias reflect the high profile of the Learning of the Way in elite education and the gradual official endorsement of its leading thinkers and texts.

The canonical status of the sayings and writings of the Learning of the Way masters found expression in the interpretive essays included in the encyclopedias. The authors of the essays on learning in *Ultimate Essays* regularly quoted the central figures of the Learning of the Way. Such quotations indicated that the author's argument was anchored in Learning of the Way canonical texts. In an essay on Zhou Dunyi's *Penetrating the Changes*, for example, Huang Lüweng established his main argument by quoting Zhang Shi and Zhu Xi:

In the past, Zhang Nanxuan [Zhang Shi] described Lianxi's [Zhou Dunyi's] learning as follows, "His work originates in the concept of the supreme ultimate in *The Changes* and in the concept of integrity in *The Doctrine of the Mean*."[81] Thereby we know that the supreme ultimate and integrity are the profound source of self-realization. Zhu Huiweng [Zhu Xi] explained Lianxi's book in the following words, "The integrity of the sage is equivalent to what is referred to as the supreme ultimate."[82] Thereby we know that integrity and the supreme ultimate stand for the highest refinement.[83]

---

81. Zhang Shi, *Zhang Nanxuan ji*, 4.17b.

82. Zhou Dunyi, *Zhou Lianxi xiansheng quanji*, 5, *Tongshu*, 1.2b. For a discussion of Zhu Xi's commentary on *Penetrating the Changes*, see Qian Mu, *Zhuzi xin xuean*, 3: 53, 65–68.

83. *Yuanliu zhilun*, bieji, 1.6a–b.

昔張南軒記濂溪之學曰: 本乎易之太極, 中庸之誠. 是知太極與誠乃自得
之蘊也. 朱晦翁釋濂溪之書曰: 聖人之誠, 即所謂太極. 是知誠與太極乃
無間之妙也.

The voices of Zhang Shi, whose collected work Zhu Xi had edited in 1184, and Zhu Xi authorized the author's argument that integrity was the central concept in Zhou Dunyi's *Penetrating the Changes*. This argument could not be deduced directly from Zhou's original text. The author derived it from Zhang Shi's and Zhu Xi's interpretation of Zhou's work. Zhu Xi justified his inclusion of Zhou's work in the Learning of the Way canon by identifying its core message with that of *The Doctrine of the Mean*. Zhu Xi saw integrity, which he defined as being true to one's endowed nature, as a modality of moral self-cultivation and relied on *The Doctrine of the Mean* as his main classical source for this interpretation. He equated the supreme ultimate, a concept Zhou used in his cosmogony, with the highest standard of integrity. This interpretation underscored Zhu Xi's belief that the process of self-cultivation restored a human being's original endowment, an endowment that ultimately derived from the supreme ultimate. Zhu Xi's interpretation of Zhou's work connected Learning of the Way moral philosophy with an elaborate cosmogonic theory and simultaneously rendered this theory coherent with one of the core classical texts of the Learning of the Way.

The interpretive imperative of coherence shaped Huang's explanation of Zhu Xi's interpretation of Zhou's work. Huang's references to the work of both Zhu Xi and Zhang Shi highlighted the consensus on this interpretation. In essays relating to the Learning of the Way, the authors typically offered multiple quotations, often from different masters; the effect of this explanatory technique was to highlight the agreement among the masters' divergent works. This technique conformed to Chen Chun's explanations of Learning of the Way concepts in *The Correct Meaning of Terms*.

The thirteenth-century encyclopedias were more than indices of the core concepts and texts of the Learning of the Way tradition. They also reflected the impact of Learning of the Way ideology on examination preparation courses and its adaptation to the conventions of the thirteenth-century examination field. Learning of the Way topics were inserted into organizational schemes that preserved the model of conventional policy essay manuals and were presented to readers through pedagogical formats rejected by the central figures of the Learning of the Way because of their association with examination preparation.

The difference in layout between Chen's manual and the thirteenth-century encyclopedias reflects the transformation of Learning of the Way ideology between the 1220s and 1250s. Whereas *The Correct Meaning of Terms* was the work of a disciple of Zhu Xi explaining the meaning of the Learning of the Way in his own words and with the help of his master's personally conveyed understanding as well as his written work, the thirteenth-century encyclopedias were the work of teachers, un-affiliated with the Learning of the Way, providing examples of policy essays on current topics and demonstrating argumentative techniques that had become standardized in examination writing.

The process of adaptation is recorded in the publication history of Lin Jiong's *Ultimate Essays*. The archeology of the text uncovers different layers in its genesis. The encyclopedia appeared in different installments; three were attributed to Lin Jiong and one to Huang Lüweng, who also wrote a preface to a combined edition of Lin Jiong's three installments. In most current editions (including the eighteenth-century Siku edition) the first installment begins with a discussion of Learning of the Way canonical texts and concepts. This arrangement of the installments dates to 1367; in all extant earlier editions the first and third installments appeared in reverse order—what is now the third installment was the first, and what is now the first was the third. The third installment in current editions represents the oldest stratum of *Ultimate Essays*. This conclusion is further supported by internal evidence. The latest time references in the essays appeared in what became the first installment. The reversal of the first and third installments created the impression that *Ultimate Essays* was an encyclopedia of the Learning of the Way and obliterated the traces of the process by which Learning of the Way categories had been superimposed on other layers of literati culture. The legacy of the encyclopedias focusing on institutional history produced by the "Yongjia teachers" receded in the background.

## The "Yongjia" Legacy

Teachers and students between the 1220s and 1250s continued to devote much effort to institutional history and government. The work of the "Yongjia" teachers retained its appeal for those preparing for the policy essay session. The legacy of the "Yongjia" masters shaped the organization, the presentation, and the source materials of thirteenth-century policy essay manuals.

Lin Jiong, who was teaching in a location distant from Lü Zuqian's hometown, copied passages from Lü's *Detailed Explanations of Institutions Throughout the Ages* in the sections on institutional history and integrated selections from "Donglai's [Lü Zuqian's] work" in the essays on the regulations for the production and sale of alcohol and salt and indicated their source in footnotes following each citation.[84] This suggests that teachers used *Detailed Explanations* as a source text. Thirteenth-century manuals also incorporated the political thought of other "Yongjia" scholars. The editors of *The Epitome* and *A Net* quoted the writings of Chen Fuliang, Ye Shi, and Chen Liang regularly in the sections on administrative subjects.[85]

The thematic organization of *Ultimate Essays* was similar to that of *Detailed Explanations*. The essays in the original first installment dated from the 1210s and covered four major subject areas: military organization, financial and fiscal affairs, the civil service, and educational institutions. The topics covered in Lin Jiong's first collection reflected the material covered in his examination preparation classes of the 1210s. Lin Jiong made a career out of private teaching in Longxi 龍溪 (Zhangzhou, Fujian circuit). Just before the publication of the first installment of *Ultimate Essays*, he wrote another manual for students preparing for the policy essay session of the examinations. In a preface to the collection, he explained that it was the result of material he had collected and notes he had taken in preparation for classes on Song institutions and government.[86] He finished the project in 1216; an admirer suggested the title *The Imperial Mirror* (*Huangjian* 皇鑑).

---

84. Ibid., xuji, 4.1a–5b, citing *Xiangshuo*, 6.2a–3a; 4.16a–b, citing *Xiangshuo* 5.5a–6b.

85. In *Jie jiang wang* (5.18b–19b, 13.19a–b, 14.16a–b), for example, Chen Liang was quoted in the entries on grain storage, the government forces, and the local militia. Ye Shi was cited in the sections on accounting, currency, paper money, and the local militia (ibid., 9.12b–13b, 11.14a–15a, 12.9a–b, 14.15b–16a).

86. Lin Jiong explained that he designed lectures on Song history in response to student demand. For Lü Zhong 呂中 (*js.* 1247), another Song teacher publishing lectures on Song history, see Hartman, "The Making of a Villain," 81.

Lin Jiong's *Imperial Mirror* consisted of thirteen main categories: (1) the ruler's virtue, (2) the ruler's government, (3) the organization of officialdom, (4) the selection of officials, (5) examination degrees, (6) personnel administration, (7) the way of the minister, (8) literati learning, (9) military organization, (10) taxation and corvée, (11) government spending, (12) disaster relief, and (13) the vices of our times. Like the "Yongjia" manuals, it provided an index to the areas of administration that were of major concern to the Song government. The compilers of *The Epitome* and *A Net* likewise aimed to train students in "the discussion of techniques of government." The areas of administration, excepting those categories concerning the Learning of the Way, roughly coincided with those covered in *To the Point in All Cases* and *Detailed Explanations* (see Table 5 in Appendix B).

Lin's interest in institutional history and government continued throughout the remainder of his teaching career. Categories on bureaucratic organization and principles of administration were featured in subsequent installments, albeit in a different configuration. In the second and third installments, published in the 1220s and 1230s, respectively, entries on the appropriate sources for literati learning, principally the Learning of the Way tradition as indicated above, took precedence.

The editors and printers of thirteenth-century encyclopedias also adopted the methods of instruction used in "Yongjia" examination preparation in the layout of individual entries and stuck to formats used in twelfth-century policy essay manuals. The combination of primary source texts and explanatory essays in *Detailed Explanations* became a basic organizational principle in thirteenth-century manuals. It suggests that encyclopedias were intended to train students in both the reading of primary source material and the composition of interpretive essays based on these materials.

The essays that constituted the entries in *Ultimate Essays* were footnoted with quotations from and explanations of past and present primary source materials and thus reproduced the division between primary sources and interpretive texts, albeit in a slightly different format. The subordination of primary texts in small-font notes expresses the ancillary function teachers assigned to literal quotations from primary sources.

Lin's expositions on institutional history were based on historical and documentary works such as Li Tao's 李濤 (1115–84) *Extensive Collection for the Continuation of "The Comprehensive Mirror for Aid in Government"* (*Xu zizhi tongjian changbian* 續資治通鑑長編; 1183), Sima Guang's *The Comprehensive Mirror for Aid in Government*, Du You's *The Comprehensive Statutes*, various monographs on institutions, and Song archival sources such as *Sagely Imperial Government* (*Shengzheng* 聖政), *Collected Essentials*, *Precious Instructions* (*Baoxun* 寶訓) and memorials. Bookstores published collections of excerpts of official documents for examination candidates for use in the policy essay session.[87] Lectures and essays by teachers like Lü Zuqian and Lin Jiong provided classified digests of these sources and applied the information in sample arguments.

Lin Jiong's sample expositions generally adopted the same structure as Lü Zuqian's in *Detailed Explanations*. The essays moved from a short general introduction to a chronological overview of the history of the subject (with special attention to the Han and Tang dynasties) to a discussion of policies in the Northern and Southern Song. Lin's essays probed the reasons behind the success or failure of particular measures and the relative advantages and disadvantages of different regulations; they typically ended with brief policy proposals.

*The Epitome* and *A Net* demonstrate how the layout of individual entries was further modified in an effort to facilitate access to the large body of primary materials included (see Table 5B in Appendix B). In the commercialized examination field of the twelfth and thirteenth centuries, printers attempted to gain a competitive advantage by modifying the format of examination manuals. Like *Detailed Explanations*, the entries in these two encyclopedias were divided into two main sections, primary sources and elucidating essays. The primary sources excerpted in the documentary sections of *The Epitome* were subdivided by type of source. In this part of the encyclopedia, students found quotations from the Classics, the histories, Song compilations of official

---

87. Chen Zhensun (*Zhizhai shulu jieti*, 167–69) lists collections with excerpts from the *Shengzheng* for the reigns of Gaozong and Xiaozong, and another compilation of official documents for use in the policy essay session, *The Essentials of the Affairs of Our Dynasty Classified* (*Huangchao shilei shuyao* 皇朝事類樞要). Cf. Hartman, "The Making of a Villain," 83. I discuss these and other examples in De Weerdt, "Byways."

documents, memorials, and the argumentative prose (*yilun*) of many famous Song intellectuals and some pre-Song authors. In a preface dated 1245, Chen Zihe 陳子和 (*js.* 1244), a *jinshi* graduate from Jian'an who had obtained his degree one year earlier, assured the buyer that *The Epitome* was the most complete of all the examination encyclopedias flooding the book market.

The essays featured in *The Epitome* and *A Net* differed from those included in Lü Zuqian's and Lin Jiong's examination encyclopedias. They were not model essays crafted by a teacher, but, in all likelihood, essays written by students sitting for examinations at the Imperial College. Both encyclopedias originally carried a reference to the Imperial College in their title. According to the Siku editors, the full title of *A Net* read *A Net to Unite and Order the Massive Amounts of Information in All Books—An Enlarged Edition from the Imperial College* (*Taixue zengxiu qunshu huiyuan jiejiang wang* 太學增修群書截江網).[88] "Jade Pond" (*Bishui* 璧水) in the title of *The Epitome* was a common metonymic appellation for the Imperial College.[89] This suggests that the selections were made from essays written for the periodic examinations at the Imperial College.[90]

As the highest institution of learning, the Imperial College was supposed to set the standards for the rest of the empire. The currency of Imperial College essays in the market of examination preparation guides indicates that this institution was perceived as a conveyor of standards and as a barometer for current policy debates and fashionable writing styles. Students in the provinces preparing for the examinations tried to keep abreast of changing tastes in the capital by perusing collections of Imperial College essays.

Examination manuals like these were not published by the Directorate of Education, which oversaw the Imperial College and had its own printing office. Commercial printers procured essays from students at the Imperial College for reproduction. Collections of Imperial College essays sold so well that some printers manufactured fake essays and attributed them to students scoring high on recent Imperial College

---

88. *SKTY*, 26.2802.

89. Wu Zimu, *Meng liang lu*, 15.132.

90. Different types of examinations at the Imperial College are listed in Table 2 in Appendix B.

examinations.[91] In addition, the compilers selected essays from successful candidates in other examinations. Papers from local level and departmental examinations were included in *A Net* (see the Appendix).

The organizational improvements in *The Epitome* and *A Net* illustrated the growing impact of marketing considerations in thirteenth-century commercial printing.[92] In addition, the printers tagged texts in the model essay sections for ease of reference. All essays were coded according to the standard divisions of thirteenth-century policy essay writing: the introduction, which stated the main argument; the argumentation supported by evidence from previous dynasties and from earlier in the Song, the discussion of contemporary policy, and the conclusion (Table 5B).[93] Additional examples of examination essays on the subject were listed by the gist of the argument under two rubrics: "the head of the policy essay" and "paragraphs from the policy essay." A third section, "factual materials," cited primary sources relevant to the argument. This layout allowed readers to skim through essays with ease and to navigate the large collection efficiently.

Commercial printers used special editorial features to enhance the appeal of their encyclopedias. Each subdivision in the archival and explanatory sections was generally preceded by a short heading indicating the theme of the essay passage or the primary source material. In a Song edition of *The Epitome*, the theme title was set off from the rest of the text by a large circle.[94] This allowed students to skim through the collection in search of suitable arguments. The parallel couplets and triplets (*ouju* 偶句) interspersed throughout the archival and

---

91. Zhu Chuanyu, *Songdai xinwen shi*, 166. See note 90 to Chapter 5 for a discussion of the dating of the memorial reporting this case of fraud.

92. Chia, *Printing for Profit*, 40–52. I discuss the strategies commercial printers deployed in the publication of the encyclopedias in more detail in "The Encyclopedia as Textbook."

93. This format is used in many of the late Southern Song Imperial College essays anthologized in *Standards for the Study of the Policy Response*. I discuss this anthology in Chapter 7.

94. In the fully extant Ming edition, the text is also annotated with critical marks. However, a reproduction of the first pages of a Song edition shows that that Song edition did not have critical marks. The corresponding pages in the Ming edition have jots; no such marks appear in the Song edition. See the Appendix.

explanatory subdivisions in *A Net* functioned as mnemonic devices. By memorizing these lines, students could more easily remember the arguments and the evidence set out at greater length in the matching prose passages (see Fig. 4).

In thirteenth-century examination manuals, Learning of the Way concepts and texts were thus incorporated into a larger framework derived from the curricula of examination teachers. Examination teachers explained and promoted Learning of the Way values through methods developed by the "Yongjia" teachers. Printers sold them to students in formats adjusted to the perceived needs of the reading public. The thirteenth-century encyclopedias and the Learning of the Way manuals discussed in the previous section suggest that Learning of the Way ideology infiltrated the examination field in two ways. The leaders of the Learning of the Way, Zhu Xi and his disciples, presented Learning of the Way pedagogy as an alternative to current practices in the examination field and adopted an adversarial attitude toward those using methods of instruction focusing on examination preparation. And in the thirteenth century, examination teachers and printers disseminated knowledge about the Learning of the Way through the repertoire of methods they had been using to train students for the examinations.

These teachers and printers are not known to have been personally involved in the transmission of the Learning of the Way; they were simply responding to the higher profile of the Learning of the Way in elite culture and at court. The shift of the larger community of teachers and printers toward the Learning of the Way occurred between the 1220s and the 1240s. Lin Jiong's case suggests that he did not make the shift until the 1220s. The manuals he wrote in the 1210s did not accord space to Learning of the Way topics. Subsequently Lin Jiong probably responded to local as well as court support for the Learning of the Way by incorporating its doctrines into his pedagogy. He taught in Zhangzhou prefecture, the hometown of Chen Chun. In the 1210s Chen Chun taught at the prefectural school of Zhangzhou. Chen's local reputation was cited as the main reason for his being awarded, by imperial fiat, the honorary rank of gentleman of meritorious achievement (*digonglang* 迪功郎) and an appointment as prefectural magistrate in 1222. Although Lin Jiong made no reference to Chen Chun in *Ultimate Essays*, he and

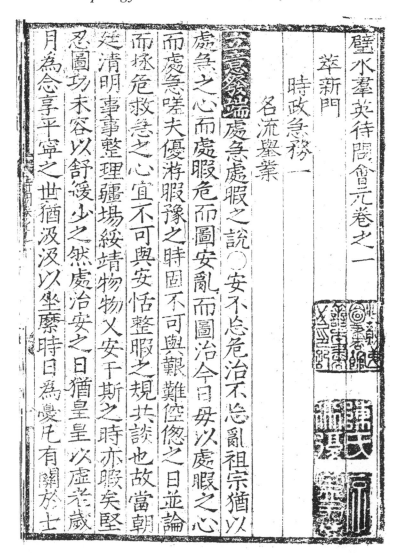

Fig. 4 First page of a typeset Song edition (1241–52) of *The Epitome of Eminent Men Responding at the Imperial College*. This page contains an excerpt from a successful examination essay and illustrates the high level of categorization in thirteenth-century commercial examination encyclopedias. The passage was marked as a model for "essay introductions" (highlighted block of text), in the subcategory of "urgent current affairs" under the main category "reform." The block of text below the highlighted heading and above the round separation sign marked the question or administrative principle addressed in the passage, in this case "theories of governing in a state of emergency and in peace time" (reproduced in Jiangsu guoxue tushuguan, Boshan shuying [1929], 1: 58; Harvard-Yenching Library).

Huang Lüweng frequently referred to imperial edicts conferring honors on the masters of the Learning of the Way in the later installments. Court support for the masters and their written legacy was a clear indicator for teachers and printers that knowledge of Learning of the Way tradition had become essential for examination candidates.

The combination of Learning of the Way concepts and "Yongjia" methods of examination preparation effected changes in the representation of both intellectual traditions. The boundaries of the Learning of the Way expanded. As it assumed a position of authority in literati culture, other traditions were subsumed under it. Conversely, "Yongjia" scholarship appeared as a shadow of the tradition represented in the works of Chen Fuliang and Ye Shi discussed in Chapter 3. The message of institutional reform that inspired the twelfth-century collections of "Yongjia" policy essays was replaced by the celebration of political unity under the aegis of a court appropriating the message of Learning of the Way moral reform.

## The Politics of Conciliation
## in the Examination Field

In thirteenth-century examination encyclopedias, the Learning of the Way shed the image of the combative opposition, an image so strikingly captured in Chen Chun's view of the examination field as a battle site. Examination teachers interpreted the Learning of the Way as the one tradition that encompassed other prominent types of learning and unified literati culture. Zhu Xi and other leaders of the Learning of the Way distinguished between various kinds of learning, separating the Learning of the Way from morally inferior or harmful types of learning, such as the Learning of the Su family, the Learning of Wang Anshi, and the Learning of "Yongjia." By the thirteenth century, their comments had become part of examination curricula. Students read criticisms of rote learning and were trained in composition and institutional history through Learning of the Way texts and anthologies. Such comments reshaped examination preparation, yet their interpretation was mediated through changing political needs. The authors and compilers of thirteenth-century encyclopedias denied any disagreement between the Learning of the Way and most other types of learning by relocating other types of learning within Learning of the Way ideology. The revaluation of Ancient Prose (especially the legacy of Su Shi) and Zhu Xi's main rivals in the interpretation of Cheng Learning (especially Lu Jiuyuan) illustrate the strategy of conciliation employed by thirteenth-century teachers.

Zhu Xi considered the appeal of the masters of Ancient Prose among examination candidates the greatest threat to students' ability to become true transmitters of the Way (see above). In contrast, in *Ultimate Essays*, *The Epitome*, and *A Net*, the legacy of the Northern Song masters of Ancient Prose was deemed fully compatible with the Learning of the Way. The author of a sample essay in *The Epitome* argued that the Ancient Prose of Ouyang Xiu and Su Shi was based on the cultivation of the self. They practiced writing in accordance with the Way, and their essays were distinct from the cut-and-paste work of vulgar examination writing. Ancient Prose could, in this author's view, be adopted in tandem with the Learning of the Way to transform current examination conventions:

The defects of literati culture today originate in the neglect of the rectification of the mind. We have to transform students' minds. The problems are also based in the lack of engagement with learning. We have to transform students' learning. . . . Transforming their minds is of the greatest importance. Transforming their learning is next. . . . We can transform their minds by relying on the means Yiluo [the Cheng brothers] used to bring forth writing of great sages and wise men. We can transform their learning by relying on the means the Su family used to bring forth writing of great masters and leading intellectuals.[95]

近日斯文之弊, 原于心之不正. 不可以不變其心也. 基於學之不講. 不可以不變其學也. . . . 變其心者上也. 變其學者次也. . . . 以伊洛之所以倡大聖大賢之文者, 而變其心. 以三蘇之所以倡宗工鉅儒之文者, 而變其學.

This passage manifests the ambivalence in the revaluation of Ancient Prose in Learning of the Way discourse. The author paid tribute to the learning of the Su family and recognized their contribution to literati culture. On the other hand, he clarified that such contributions did not match those made by the founders of the Learning of the Way. Mind-cultivation took priority over other forms of learning, and only the learning of the Cheng brothers cultivated sages rather than masters.

---

95. *Bishui*, 44.6b–7a.

This passage suggests a second dynamic in the new relationship between the Learning of the Way and Ancient Prose. The author's evaluation of Ancient Prose paralleled the view of other thirteenth-century examination teachers that Ancient Prose composition was a legitimate subject because it could be harnessed to express understanding of the Learning of the Way. When subordinated to the priority of mind-cultivation, Ancient Prose became a legitimate part of the Learning of the Way.

This perspective on the relationship between the Learning of the Way and Ancient Prose was widely shared among thirteenth-century examination teachers. Since they had been teaching composition courses featuring the works of the Su family and other leading authors of Ancient Prose, their efforts to legitimate Ancient Prose and reconcile it with the Learning of the Way were motivated by a concern about radical transformations of examination curricula. Lin Jiong also rose to the defense of the Ancient Prose masters. He recognized that many of his contemporaries mistrusted Su Shi for his preoccupation with literary and artistic values and the low regard he had occasionally shown toward Cheng Yi.[96] Lin objected that Cheng Yi and Su Shi believed in the same values: "Cheng and Su equally honored the sages of Antiquity, Yao and Shun, they equally imitated Confucius and Mencius, they equally took guidance from the Six Classics, and they equally esteemed humanity and rightness."[97]

Thirteenth-century encyclopedists endorsed Zhu Xi's narrow definition of the transmission but co-opted, unequivocally, Northern Song philosophers, such as Su Shi, whom Zhu Xi had treated with suspicion. They further extended this strategy of conciliation into Southern Song times. Zhu Xi had tailored his genealogy of the Way so that his interpretation of the Learning of the Way appeared as the only legitimate next stage in the transmission. Genealogy served as a means to exclude competing interpretations of morality. The official recognition of the Learning of the Way in the early thirteenth century ushered in a trend of reconciliation with the teachings of some of Zhu Xi's twelfth-century competitors. In 1209 Zhu Xi became the first of the Learning

---

96. *Yuanliu zhilun*, qianji, 4.20a–24a, 32a–34b.
97. Ibid., 4.32b.

of the Way teachers to receive posthumous honors, but similar awards were soon given Zhang Shi and Lü Zuqian, in 1215 and 1216, respectively. By 1261 all three were granted accessory sacrifices in the Confucian temple.[98] Zhang Shi and Lü Zuqian could easily be regarded as supporters of Zhu Xi's project.[99] The rapprochement between Zhu Xi and Zhang Shi had climaxed in Zhu Xi's redaction of Zhang Shi's collected works.[100] Lü Zuqian had collaborated with him on *A Record for Reflection* and had shown more deference to Zhu Xi than did his more outspoken intellectual competitors.

Lu Jiuyuan presented an alternative approach to Zhu Xi's definition of the path of moral self-cultivation. Zhu Xi labeled Lu's emphasis on the recovery of the "original mind" and his concentration on sudden enlightenment Chanist.[101] In the entry on Lu Jiuyuan in *A Net*, the editor collected negative evaluations of Lu's learning from Zhu Xi's recorded conversations. However, in the sections with sample arguments in both *A Net* and *The Epitome*, the authors unanimously defended Lu against charges of Chan Buddhist interests and contended that the learning of Lu Jiuyuan presented the same truth as Zhu Xi's Learning of the Way.[102]

The ecumenism in thirteenth-century examination encyclopedias was based on the centrality of Zhu Xi's legacy. The incorporation of selected Ancient Prose masters and Southern Song philosophers into Learning of the Way ideology was justified by their imputed agreement with it. The centripetal force of the conciliatory examination curricula was most evident in the encyclopedists' discussions of the appropriate approach to the study of the Classics. The authors proclaimed their confidence in a Learning of the Way curriculum, consisting of the Four Books,[103] the Classics, and Learning of the Way commentaries and monographs. Zhu Xi's legacy was acknowledged as the permanent core of the Learning of the Way:

---

98. Neskar, "The Cult of Worthies," chap. 6; Tillman, *Confucian Discourse and Chu Hsi's Ascendancy*, 232–33.

99. *Jie jang wang*, 33.36a; Tillman, *Confucian Discourse and Chu Hsi's Ascendancy*, 131.

100. Tillman, *Confucian Discourse and Chu Hsi's Ascendancy*, 46–47.

101. Ibid., chap. 8.

102. *Jie jiang wang*, 32.1a–16a; *Bishui*, 53.15a–17b.

103. *Yuanliu zhilun*, qianji, 1.23b–30b, 2:1a–5b, bieji, 1.11a–19b; *Bishui*, 47.1a–13b.

The classical texts of the sages of the period of the Three Dynasties and before were given a fixed shape when a great sage appeared. *The explanations of all scholars since the Three Dynasties were given their permanent place when a great scholar appeared.* . . .

Heaven gave birth to Zhongni [Confucius] in the late Zhou period, so that he could establish the merit of providing explanations after successive sages. Heaven also gave birth to Master Zhu [Zhu Xi] during our reigning Song dynasty, so that he could establish the merit of adding philological notes and corrections to the work of previous commentators. However, it was not the case that Zhongni had the ability to give explanations on the basis of his personal views. Master Zhu also did not have the ability to make corrections based on his personal views. All history goes back to principle, and all principles are comprehended in one mind. As for Zhongni's explanations of the Classics, he first matched their meaning with his mind, measured it against the standard of principle, and then gave explanations. Therefore, his explanations can never be altered. As for Zhu Xi's annotations to the Classics, he also first matched their meaning with his mind, measured it against the standard of principle, and then made corrections. Can't we expect that people will believe in Master Zhu the same way they believe in Confucius![104]

三代而上, 累聖之經, 至大聖而定. 三代而下, 諸儒之說, 至大儒而定. . . .

　是天生仲尼於晚周, 蓋爲累聖而闡述作之功也. 天又生朱子於我宋, 蓋爲諸儒加考訂之功也. 然仲尼非能以己見而述作. 朱子亦非能以己見而考訂. 古今一理; 万理一心尔. 仲氏之述作群經, 契之於心, 揆之一理, 而述作之也. 故萬世卒不可易. 朱子之考訂群經, 一契之於心, 揆之一理, 而考訂之也. 千載而下, 安知其不以信中尼者信朱子乎!

According to this examination candidate, Zhu Xi's interpretation of the Classics brought an end to the diversity of opinion concerning each of the Classics. Just as Confucius had ended the instability of the classical texts through his editorial work, Zhu Xi stopped the questioning of the authenticity of the Classics and the open-ended interpretations of his contemporaries. His commentaries provided the standards by which all other commentaries were to be judged. His commentaries were given canonical status in this reading because they expressed the unchanging principles of moral order embedded in the human mind and not just in the minds of their creators. The candidate's claims closely matched those Zhu Xi made for his commentaries, but their appearance in a

---

104. *Bishui*, 45.7a–b.

thirteenth-century encyclopedia indicated the broader acceptance of these claims in the decades following Zhu Xi's death.

Their broader acceptance transformed the stance toward former competitors of Zhu Xi in Learning of the Way discourse. In classical commentary, there was no longer a need to contrast the pros of Learning of the Way commentaries to the cons of competing commentaries. In the late twelfth century, when the Learning of the Way was the ideology of an embattled minority, Zhu Xi proposed to have examination candidates evaluate the pros and cons of a small range of influential commentaries, based on Learning of the Way criteria (see above). In the early decades of the thirteenth century, Zhu Xi's commentaries obtained a hegemonic position in examination preparation. As his commentaries began to dominate curricula in government schools and were adopted by private examination teachers, their truth went uncontested. The voices of dissent were either ignored or, as in the case of the Ancient Prose Masters, made to conform to Zhu Xi's interpretations. The official canonization of Zhu Xi's legacy, ironically, led to the blurring of the boundaries between the Learning of the Way and its erstwhile competitors. Intellectual controversies were relegated to a dark past, and those who had been emphatically excluded from Zhu Xi's genealogy of the Way were firmly located within its bounds.

The shift toward conciliation in intellectual culture paralleled changes in Song political culture. Thirteenth-century examination teachers and students perceived the Learning of the Way as a force for the unification of the body politic. Although it had been at the center of factional disputes throughout the last decades of the twelfth century, the Learning of the Way was portrayed as the solution to political divisions after Zhu Xi's death. The teachers' interpretation of the new role of the Learning of the Way in political culture accorded with efforts to claim the Learning of the Way at court, especially under and following the reign of Emperor Lizong.

In the early decades of the thirteenth century, the beleaguered Song government realized the potential of the Learning of the Way for reunifying a divided officialdom. After the rehabilitation of Learning of the Way advocates persecuted under the regime of Han Tuozhou, the court called leading proponents of Zhu Xi's legacy to court. These men, most notably Zhen Dexiu, who held court positions intermittently from the 1210s through the 1230s, called for the moral reform of the court and the bureaucracy at large.[105] Zhen Dexiu's reading of *The Great Learning* lay at the core of the early thirteenth-century campaign for moral reform. As court lecturer in the 1220s, Zhen advised Emperor Lizong that the model set by morally superior men would effect the moral regeneration of society at large.[106] The moral regeneration of the polity and society was, in this view, a prerequisite for the reinvigoration of the Song state in the face of the threat posed by the Mongols.

The moral reformation of society began with the self-cultivation of the emperor and required him to appoint the "morally superior" to office. The prioritization of the personal morality of politicians in government conflicted with earlier reform efforts, such as those of the "Yongjia" masters, who had advocated institutional reform as a solution to the Song state's security problems. Some critics recognized the contradiction in the moral reform program of Learning of the Way advocates at court. They objected that although the project of moral reform promised the unification and reinvigoration of a factionalized bureaucracy, the reliance on such criteria would most certainly have the opposite effect (Chapter 7). The history of the factional struggles in the late twelfth century, when Learning of the Way advocates campaigned against "morally inferior" officeholders, substantiated that view.

Despite this criticism, the court's support for the Learning of the Way program for moral reform decisively altered literati attitudes toward it. Examination teachers like Lin Jiong and Huang Lüweng and the compilers of thirteenth-century encyclopedias accepted the reformers' claims. Their essays expressed the belief that support for the Song house was inextricably linked to support for the Learning of the Way. The interpretation of institutional history and government, which remained core subjects for examination teachers, changed in accordance with the new alliance forged between the court and the Learning of the Way.

In their essays on institutional history, Lin Jiong and Huang Lüweng argued consistently for the superiority of Song institutions and attributed their superiority to the Song founders' adherence to Learning of the Way values. In their assessment, Song institutions and regulations

---

105. Zhen is also discussed in the next section and in Chapter 7.
106. De Bary, *Neo-Confucian Orthodoxy and the Learning of the Mind-and-Heart*, 86, 95–96.

surpassed those of the Han and Tang dynasties in a variety of areas ranging from accounting, the granting of titles, military organization and military discipline, the management of clerks, and the distribution of land.[107] Their conclusions were often completely opposite to those reached by the authors of *Detailed Explanations* and *To the Point in All Cases*. The "Yongjia" writers had pointed out the problems in Southern Song military organization and personnel administration and suggested that political reasoning and the historical study of these subjects would help in finding answers. In a lecture on the military system, for example, Lü Zuqian argued that the rise of military governors after the An Lushan Rebellion in the mid-eighth century marked a low point in the history of military institutions from which subsequent ruling houses had never fully recovered (Chapter 4).

By contrast, Huang Lüweng concluded an essay on military garrisons: "The founding fathers of our dynasty got rid of the problem that had lasted for several hundred years at once. That is humanity (*ren* 仁)!"[108] Huang's assessment radiated confidence in the Song military by attributing the Song founders' military and political success to their adherence to standards defined as core Learning of the Way values. In Huang's argument, Song emperors centralized military power because their cultivation of the virtue of humanity inspired them to do so; they did not arrive at this decision after a critical evaluation of institutional history or a utilitarian calculation of the pros and cons of this and other measures. The Learning of the Way was thus used to instill belief in the superiority of Song institutions, and the new relationship between the Learning of the Way and institutional history was used to bolster confidence in the Song project.

The principles of administrative thinking developed by the "Yongjia" masters were adopted in thirteenth-century encyclopedias, but their compilers reinterpreted them and made them part of a new bureaucratic ethic that was the result of the application of Learning of the Way

---

107. *Yuanliu zhilun*, qianji, 7.11b; houji, 3.27a; bieji, 9.4a–b, 11b, 10.9a, 15a, and *passim*. Deng Xiaonan's research on the shifting image of the Song founders also underscores the influence of the Learning of the Way on Southern Song representations of the first Song emperors; see her "Guanyu 'daoli zui da.'"

108. *Yuanliu zhilun*, bieji, 10.6b.

moral reasoning to government and history. Lin Jiong wrote several essays on principles of government, such as "employing experienced ministers," "recommending wise men," "showing respect for ministers," "guaranteeing meritorious ministers," and "combating the rat race for success."[109] Such principles were discussed in earlier encyclopedias and examination manuals, most notably in *To the Point in All Cases*.

In *To the Point in All Cases*, "combating the rat race for success" was one area of administration to which axioms of "Yongjia" political theory were applied.[110] The application of differing axioms to this problem suggested that, from the "Yongjia" perspective, there were multiple causes and that various management techniques could be used to reduce the ill effects of competition in the bureaucracy. In one of the essays addressing this problem, "Lower Administrative Levels' Laxness Toward Imperial Orders Derives from the Court's Own Laxness," the "Yongjia" Master argued that competition for official positions existed only when the process of recruitment and advancement was unclear. He criticized the Song court for its inconsistent policies in personnel administration, citing as an example the cancellation and reinstitution of the policy to appoint extra local officials and nominal officials without administrative functions.[111] Such policy shifts caused confusion about whether and how long such extra positions would be available and thus increased competition for them. Given the current confusion, the "Yongjia" Master argued, the problem of the scramble for office could be solved only if the upper stratum of the administration promulgated clear regulations.

Lin Jiong agreed that the emergence and the solution of the problem depended on central policy,[112] but he disagreed about the kind of policy necessary to counter competitiveness. For Lin Jiong, the solution lay in the moral example of the court and those at the top of the bureaucratic hierarchy. He wrote that the emperors of the reigning dynasty had

---

109. For these and other examples, see ibid., qianji, 8.29b–33b, 9.1a–19b; houji, 6.1a–9a, 8.10a–16b.

110. *BMF*, 10.72–73, 13.98, 1.6.

111. *BMF*, 10.72–73. For a history of the implementation and cancellation of extra local officials (*tianchai* 添差), see Umehara, *Sōdai kanryō seido*, 211–12.

112. *Yuanliu zhilun*, houji, 8.10b–13b.

taken the right measures by impressing the virtues of propriety, up-rightness, honesty, and a sense of shame (the four virtues that were, according to Mencius and Zhu Xi, latent in human nature) on their ser-vants. He invoked couplets from regulated poems composed by the Northern Song emperors Taizong 太宗 (r. 976–97) and Zhenzong that illustrated their censure of self-advancement and their appreciation of reserve. Such actions created an atmosphere of moral emulation and imposed restraint on those exposed to it. Lin Jiong substantiated this argument with a long list of examples of Song men who did not push their immediate career interests. In *Ultimate Essays,* "combating the rat race for success" became part of a bureaucratic ethic aimed at the re-unification of the Song body politic. This ethic was based on a core principle of Learning of the Way political philosophy, namely, the belief that the distinction between the morally superior and the morally infe-rior was central to the well-being of the body politic.

Lin Jiong adopted the same historical models as the "Yongjia" mas-ters to demonstrate the validity of the administrative principles ad-vanced in *Ultimate Essays.* The reinterpretation of these models illus-trates the emphasis on the distinction between the morally superior and the morally inferior in the bureaucratic ethic promoted in thirteenth-century examination encyclopedias. Scholar-officials active during the Qingli (1041–48) and Yuanyou (1086–94) reign periods became the fa-vorite models of historians and politicians during the Southern Song period. Their willingness to advocate reform while avoiding the ex-tremes of forced reform associated with Wang Anshi was widely ad-mired by Southern Song literati. The reasons cited for their greatness, however, varied.

In *To the Point in All Cases,* the "Yongjia" Master praised the collec-tive body of ministers under Renzong for their participation in serious political debates:

I have learned that during the reign of Emperor Renzong—when men like Du Yan [978–1057], Duke of Qi; Fan Zhongyan, with the title of Cultured and Just; Han Qi [1008–75], Duke of Wei; Fu Bi [1004–83], Duke of Zheng; Ouyang Xiu, Duke; Yu Jing [1000–1064] and Cai Xiang [1012–67] appeared successively in the highest official ranks—whenever there was a major issue at court, debates proliferated. For days no solution could be reached. Sima Junshi [Guang] and Fan Jingren [Zhen 范鎮 (1008–89)] were known to be best friends. Yet, in the affair of the musical pitches, they discussed dozens of problems without bad feelings.[113] So, at the time, nobody ever objected that even though "the debates went on for days in a row, but no solution was found; several days went by, but there was no resentment," they confused people and harmed the conduct of affairs![114]

聞仁宗朝, 杜祁公衍, 范文正公仲淹, 韓魏公琦, 富鄭公弼, 歐陽公脩, 余靖, 蔡襄之徒, 相繼在列, 每朝廷有大事, 議論紛然. 累日而不決. 司馬君實與范景仁號爲至相得者. 鍾律一事, 亦論難數日, 而不厭. 夫其所謂累日而不決, 數日而不厭者,當時亦曷嘗病其惑人, 而敗事也哉!

The "Yongjia" Master turned the politicians of the Qingli and Yuan-you reigns into exemplars of "Yongjia" administrative reasoning. The "Yongjia" scholars considered skill in argumentation an essential virtue of successful politicians; *To the Point in All Cases* and other manuals provided training in this skill for those preparing for a career in government. Moral stature did not figure in the evaluations of politicians of the past. The concept of "the morally inferior" (*xiaoren* 小人) did not appear in *To the Point in All Cases*.

With Zhu Xi began a revaluation of the merits of eminent Northern Song officials. *Record of the Words and Deeds of Famous Ministers*, compiled under his direction, redirected the interest of students from the voluminous compilations of memorials and the collected writings of these men toward their moral achievements.[115] Even though the anthology was tailored to focus student attention on moral achievement, Zhu Xi remained ambivalent about the legacy of Northern Song officials. He warned his students in personal conversation that these ministers had not gained insight into the foundations of morality.[116] Their accomplishments were therefore minor in comparison with those of the transmitters of the Learning of the Way, many of whom, including Cheng Yi, had not had the opportunity to serve in a ministerial position.

In *Ultimate Essays*, Lin and Huang continued the reassessment of Northern Song politicians based on the criterion of personal morality. They contrasted the high moral stature of the ministers of the Qingli

---

113. For their discussions on music, see *SB*, 307.
114. *BMF*, 6.42.
115. On Zhu Xi's involvement in the project, see Qian Mu, *Zhuzi xin xuean*, 5: 148–49.
116. *ZZYL*, 129.3090.

and Yuanyou governments to the moral indifference of the regimes taking over from them.[117] In an essay on keeping out the morally inferior, Lin Jiong presented the Qingli and Yuanyou reigns as the culmination of good government by upright ministers.[118] Key figures associated with the Qingli and Yuanyou spirit, like Ouyang Xiu, Shi Jie, Su Shi, and Cheng Yi, were credited for their roles in the struggle to restore the transmission of the Confucian tradition and drive out the heterodoxy of Buddhism.[119]

The reassessment of Northern Song political icons figured into examination teachers' efforts to co-opt eminent Song figures. The Qingli and Yuanyou politicians, whose membership partially overlapped with the masters of Ancient Prose, were portrayed as defenders of the values associated with the Learning of the Way. Through the portrayal of their cultivation of moral character and their opposition to the enemies of the Learning of the Way, they were seen as the embodiment of Learning of the Way political principles. The importance of the distinction between the morally superior and the morally inferior was in this view borne out by the history of subsequent reign periods. Lin Jiong argued that it was the intrusion of the morally inferior that led to the failure of reform during the Xining 熙寧 (1068–77) and Shaosheng 紹聖 (1094–98) periods.[120]

Thirteenth-century examination teachers' reinterpretation of institutional history and administrative principles reveals that they conceived of a symbiotic relationship between the Song court and the Learning of the Way. Learning of the Way values were used to define the institutions, policies, and eminent leaders of Song government and thus legitimated trust in the Song house. The teachers also clarified that the fate of the Learning of the Way depended on the support of the emperor or high ministers:

---

117. See the two essays on the main figures of the Qingli and Yuanyou reign periods in *Yuanliu zhilun*, qianji, 3.12b–21a. Another essay (ibid., 8.6a–11a) argues that, whereas Wang's government was characterized by the reliance on skills, before Wang Anshi's rise to power moral stature had been the standard for politicians.

118. Ibid., houji, 4.25a–28a.

119. Ibid., houji, 8.13a–16b.

120. Ibid., houji, 4.22b–29b.

During the period from the Yuanyou reign to the present, the Learning of the Way has been through several ups and downs, but the distinction between the human mind and natural principle has not faded away. In the end, the vitality and energy of the state depend on the power of preservation. *The Learning of the Way cannot turn its back on the state, but the state also cannot turn its back on the Learning of the Way.* Looking back now, we can see that when, in the last years of the Shaosheng period, known as the period of factional disaster, eminent intellectuals and experienced officials were exiled to remote corners of the empire, our Way was cut off. Showered by the generosity of Emperor Gaozong, good men came forward. Guishan [Yang Shi], an eminent intellectual, stood out. The Yang family, father and son, were employed in succession. As a result, the tradition from Yiluo flourished for the first time. This was thanks to Gaozong's recruitment of wise men. . . .

. . . As soon as Lord Sima [Guang] became councilor, he headed the alliance of our Way. Only then did the correct learning of Yiluo become known. After Zhao Zhongjian [Ding] asked for leave, the position of just men became isolated, and the disciples of Yiluo were again dispersed.[121]

自元祐至今日, 其間或興或廢不知其幾, 而人心天理不泯於頹靡之中. 國脉元氣終有賴於維持之力. 道學固無負於國家, 而國家亦無負於道學也. 自今觀之, 紹聖末年, 唱爲黨禍, 名儒故老海隅嶺表而吾道之派絕矣. 我高宗雨露洗濯善類萌蘗. 龜山大儒靈光獨立. 胡氏父子相繼擢用. 而伊洛之傳始振. 此高宗錄賢之功也. . . .

. . . 司馬公一相, 主盟吾道, 而伊洛之正學始明. 趙忠簡求去, 正人勢孤, 而伊洛之門人復散.

In this brief historical review of the transmission of the Learning of the Way in Song times, Lin Jiong mentioned two moments of restoration and one of reversal. Sima Guang receives credit for supporting the views of Cheng Yi when he became councilor after Wang Anshi's reforms. Cheng had been critical of Wang's views and was ousted from political power in 1087. Gaozong was credited with the restoration of the transmission four decades later, after the court had moved south of the Yangzi River. Gaozong's invitation to Yang Shi and his appointment as vice minister of works and academician-in-waiting in 1127 restored the visibility of Cheng Learning at court.[122] The persecution of

---

121. *Yuanliu zhilun*, qianji, 9.2b–4a. Later Li Xinchuan made the same argument in his *Record of the Way and Its Fate* (Chaffee, "The Historian as Critic," 320).

122. *SB*, 1228.

Learning of the Way followers after the exile of Councilor Zhao Ding in 1140 was a painful reminder of the importance of imperial or ministerial support. This more recent setback in the history of the Learning of the Way underscored the value of the construction of a symbiotic relationship between the history of the Song dynasty and the Learning of the Way, a project that inspired the thirteenth-century examination teachers and student authors discussed in this section. One examination candidate expressed his optimism about the guiding and unifying role of Learning of the Way ideology in the post–False Learning period as follows:

In the thirty or forty years since the restoration (*genghua*)[123] there have been no heterodox movements among the people, there have been no heterodox teachings among commentators, and there has been no heterodox learning among the scholar-officials.[124]

更化以來三四十年, 人無異端, 家無異説, 士無異學.

## Reading and Writing:
### The Syncretization of Canons

The Ancient Prose canon, the most influential body of prose literature in Chinese history in the second millennium CE, was fashioned and transmitted in twelfth-century private instruction in examination preparation. Ancient Prose models had a much wider application than examination writing, but they were first marketed and used as introductions to such writing. The "Yongjia" teachers' courses in composition and their anthologies and manuals played a crucial role in the formation of this canon. By the end of the twelfth century, its main characteristics were a core group of Tang and Song essayists and texts and the use of this canon as a tool to teach language, reading, and composition with an eye to the development of argumentative writing skills for use in the examinations (Chapter 4). This body of texts retained its position as the

---

123. There was a series of renovations announced by Ningzong and Lizong in the thirteenth century. For an overview of those in Lizong's reign, see Yang Yuxun, "Nan Song Lizong zhong- wanchao de zhengzheng," 7; for an example from Ningzong's reign, see Huang Kuanchong, "Wan Song chaochen dui guoshi de zhengyi," 26–27*n*29. Here the term refers to the cancellation of anti–Learning of the Way policies. For a similar use of the term, see *SS*, 394.12026.

124. *Jie jiang wang*, 33.36a.

dominant literary canon in the thirteenth century, but it went through significant changes. During the thirteenth century, the goals, the selection, and the methods of instruction were modified. The restructuring of the Ancient Prose canon reflected the tensions between an established literary canon and the emerging doctrinal canon of the Learning of the Way.

The continuing appeal of the Ancient Prose canon among examination teachers and students is evident from the growing number and expanding variety of anthologies available from thirteenth-century bookshops. The early anthologies were collections of prose pieces organized by author or by genre. By the early thirteenth century, different types of anthologies had appeared on the market. The variety of formats reveals the manipulation of the canonized masters for the purpose of instruction and the eagerness of commercial printers to market new tools for examination preparation.

Ancient Prose texts were packaged as introductions to composition. Xie Fangde became instructor at the Jianning Prefectural School shortly after he obtained the *jinshi* degree in 1256. Apart from this official appointment, Xie demonstrated his interest in education as a promoter of the construction of academies[125] and as the compiler of an Ancient Prose anthology that, like Lü Zuqian's, remains in print today. The selection of texts in his anthology, *Standards for Writing* (*Wenzhang guifan* 文章軌範; 1260s or 1270s), bore a close resemblance to those of twelfth-century anthologies. The vast majority were culled from the work of the core masters (Han Yu, Su Shi, Liu Zongyuan, Ouyang Xiu, and Su Xun). Xie's exposure to the competitive market in examination manuals in Jianning prefecture may, however, have inspired him to present the Ancient Prose canon in a new way.

He developed a course in Ancient Prose composition based on a pedagogical belief in the gradual acquisition of skills. He organized his selection of texts in a graded program, moving from texts he considered suitable for elementary learners to those requiring stronger reading skills and thus better suited for the advanced student. The first two chapters covered the stage of "developing courage" (*fangdan* 放膽). These chapters were to be read before students started writing.

---

125. Walton, *Academies and Society*, 79–80, 169.

Xie's anthology was specifically intended as a gradual introduction to examination writing. Its seven chapters were not numbered, but each was labeled by one of the seven characters constituting the line "Dukes, kings, generals, and ministers, how could it be that one is born to such classes?" (*Hou wang jiang xiang you zhong hu* 侯王將相有種乎). The line was unmistakably an encouragement for struggling examination candidates. In most of the chapters, Xie highlighted the specific benefits that would accrue to students studying the texts in that chapter. The proper way to organize an essay, the subject of the texts in the third chapter, was in Xie's view essential training for those eager to become expert at the composition of the examination exposition; the clarity of expression exemplified in the fourth chapter's pieces was preparation for the writing of essays on the meaning of the Classics and policy response essays; and the selections in the fifth chapter featured conclusions that stood out as models for candidates who found themselves under time pressure when attempting to finish the expositions and policy response essays in the second and third sessions of the examinations.

Other thirteenth-century teachers and printers expressed similar interest in the pedagogical pros and cons of different ways of presenting the material and repackaged the same prose texts in new formats. Fang Yisun, a student at the Imperial College in the early 1240s, compiled an anthology targeted at examination candidates and organized by categories representing compositional techniques. The reference to the Imperial College in the full title, *The Imperial College's Newly Compiled One Hundred Pieces of Brocade of Exquisite Pattern* (*Taixue xinbian Fuzao wenzhang baiduan jin* 太學新編黼藻文章百段錦; 1242), not only advertised Fang's association with the school but also nodded to this school's reputation as the best prep school for examination candidates.

In the preface, Fang Yisun explained that he found the traditional anthologies of Lü Zuqian and others too disorganized and confusing. Twelfth- and thirteenth-century examination teachers shared the goal of teaching compositional techniques through the analysis of Ancient Prose texts, but their anthologies featured the full texts of the prose pieces and discussed techniques in a hit-and-miss fashion only as they appeared in the texts. Fang's anthology was intended to provide a systematic overview of compositional techniques. He tried to convince the buyer that a focus on techniques and the use of short excerpts from various Ancient Prose texts would better illustrate the use of a particular technique in different contexts. The "pieces of brocade" were passages from texts mainly by eleventh- and twelfth-century masters, especially Su Shi and Lü Zuqian. The excerpts were organized by categories of compositional techniques, ranging from text formats and sentence construction to modes of writing such as explanation, description, and disputation to rhetorical devices such as inference, comparison, citation, and the use of historical evidence. Each of these categories was further subdivided to show different uses of the relevant technique. Fang's anthology carried the focus on instruction in compositional techniques in examination teachers' courses on Ancient Prose to its logical conclusion.

The texts of Ancient Prose masters were reformatted for students in yet another way. Thematic anthologies arranged excerpts of Ancient Prose texts according to a topical scheme similar to the examination encyclopedias discussed in the previous section and in Chapter 4. Two collections of annotated excerpts from the prose of Yang Wanli were published as *Excerpts from Yang Wanli, Annotated and Classified* (*Pidian fenlei Chengzhai xiansheng wenkuai* 批點分類誠齋先生文膾) around 1259. This anthology offered students quick access to arguments that could be used in papers on topics ranging from the attributes of the ruler and the minister to financial and military affairs to the textual tradition to historical figures throughout Chinese history. The proliferation of different types of Ancient Prose anthologies was a sign of the commercial revolution in the textbook business.

All the manuals mentioned above and others listed in the Appendix appeared in print, most of them in low-quality and therefore lower-cost editions. The increasing number of examination candidates, which reached several hundreds of thousands in the thirteenth century,[126] multiplied the opportunities for profit among commercial publishers. Such opportunities spurred a growth in the number of commercial publishers in the twelfth and the thirteenth centuries and heightened competition. Although the majority of such titles have not survived, doubtless because of their ephemeral value, the diversification of Ancient Prose anthologies, as well as the continuation and elaboration of

---

126. Chaffee, *The Thorny Gates*, 35.

the methods of annotation employed by twelfth-century examination teachers and printers, indicates that Ancient Prose anthologies claimed a substantial share of the thirteenth-century book market.[127]

The formation of the Ancient Prose canon in examination preparation met with resistance. As discussed above, Zhu Xi perceived the authoritative position of Ancient Prose in examination preparation as the major threat to students' adherence to the Learning of the Way and criticized the "Yongjia" teachers for their role in the formation of this canon. This section first investigates the Learning of the Way alternative to the proliferation of commercially printed Ancient Prose anthologies. Learning of the Way teachers compiled anthologies canonizing the work of the Song Learning of the Way masters and challenging the authority of contemporary Ancient Prose masters. They also devised methods of annotation to steer students away from the preoccupation with composition in reading. The tension between Learning of the Way curricula and the instruction of Ancient Prose eased when the Learning of the Way gradually gained wider acceptance among scholars and at court. As in the instruction of history and government discussed in the previous section, teachers reconciled Learning of the Way values with conventional training in composition in examination curricula from the 1230s on. The second part of this section discusses the transformation of the Ancient Prose canon as a result of this process of conciliation. It analyzes the restructuring of the literary canon that accompanied the growing dominance of the Learning of the Way doctrinal canon in the examination field and gauges the impact of the new relationship between the Learning of the Way and Ancient Prose on the definition of thirteenth-century Learning of the Way ideology.

## THE FORMATION OF THE
## LEARNING OF THE WAY CANON

Your warnings that among all the bad influences upon students none is bigger than current examination writing (*shiwen*) are corrective words. However, when we discuss the extremes thereof, it appears that Ancient Prose (*guwen*), because

---

127. For a discussion of the growth in commercial publishing during the Southern Song period, see Chia, *Printing for Profit*, 66; and for ephemeral literature, ibid., 143–46.

it causes students to abandon what is fundamental in order to pursue what is secondary, causes the same harm as current examination writing.[128]

所喻學者之害莫大於時文, 此亦救弊之言. 然論其極, 則古文之與時文, 其使學者棄本逐末爲害等爾.

After witnessing his contemporaries' fixation with Ancient Prose, Zhu Xi found the relationship between the Learning of the Way and Ancient Prose a vexing question. On one hand, he recognized the role of Han Yu and Northern Song Ancient Prose paragons like Liu Kai, Shi Jie, and Sun Fu in promoting the way of the ancient sages over the religious and literary alternatives of their times. On the other hand, he realized that, from the perspective of the Learning of the Way, Ancient Prose was a double-edged sword.[129]

The accounts of the proper transmission of culture by Tang and Northern Song Ancient Prose thinkers left an ambivalent legacy. He Jipeng distinguishes two kinds of accounts, depending on the standards for inclusion.[130] The genealogies constructed by Liu Kai, Sun Fu, Shi Jie, and Zu Wuze were based on a consideration of the transmission of the Way, a holistic cultural concept, and included Confucius, Mencius, Xunzi, Jia Yi, Dong Zhongshu, Yang Xiong, Wang Tong, and Han Yu. Starting with Ouyang Xiu, Ancient Prose advocates propagated genealogies of Ancient Prose that included authors not listed in the genealogies of the Way of the sages. In the estimation of Ouyang Xiu and Su Xun, the list should include Sima Qian and Ban Gu for their "potent and firm" writing style and the Warring States Period military strategists Sun Wu and Wu Qi for the "concise and cutting" qualities of their work (Chapter 4).[131] These choices reflect the wider intellectual and political interests of Ouyang Xiu and Su Xun; they also signify a transformation in the Northern Song Ancient Prose movement. After Ouyang and the Su family, Ancient Prose composition was subjected to concen-

---

128. Zhu Xi, *Zhu Xi ji*, 56.2824; cited in Guo Shaoyu, *Zhongguo wenxue piping shi*, 444.

129. For examples of Zhu Xi's mixed feelings about the contributions of Han Yu and the early Northern Song Ancient Prose masters, see *ZZYL*, 137.3276, 129.3089–91, 83.2174.

130. He Jipeng, *Tang Song guwen xin tan*, "Tang, Song guwen yundong zhong de wentong guan," 264–86.

131. Ibid., 273, 276.

trated analysis. Twelfth-century literati increasingly perceived Ancient Prose as a set of techniques and qualities of good writing best illustrated in selections of texts from Tang and Song authors.

The rulers and philosophers ranked in the Tang and Song Ancient Prose genealogies of the Way (including the ancient sage-kings, Confucius, and Mencius) also appeared in the Learning of the Way genealogies of Cheng Yi and Zhu Xi. Figures like Yang Xiong, Wang Tong, and Han Yu were occasionally included in genealogies of the Learning of the Way for their contributions to one aspect or another of the transmission of the Way, such as opposition to heterodox teachings or transmission through teaching.[132] Zhu Xi's frequent evaluations of these intellectuals and other writers promoted in Ancient Prose discourse indicate the influence of Ancient Prose curricula and bear witness to his attempts to keep the lines between Learning of the Way and Ancient Prose clearly drawn.[133] Even though he demonstrated a keener appreciation for writing than had Cheng Yi and accorded writing a (subordinate) place in his theoretical framework of the Learning of the Way, Zhu Xi felt compelled to continue the antagonism between Learning of the Way and Ancient Prose, set off by the legendary quarrels between Cheng Yi and Su Shi.[134]

As suggested in the quotation, Zhu Xi was especially concerned with his contemporaries' preoccupation with mastering the techniques used in the essays of the Tang and Northern Song champions of Ancient Prose and their twelfth-century followers. Ancient Prose, he claimed, was causing the same harm as the examination papers circulating among examination candidates. Zhu Xi merely stated the similarity; he could not discard the Ancient Prose legacy as a whole. Nevertheless, he decried the transformation it was undergoing as the Ancient Prose

---

132. De Weerdt, "Aspects of Song Intellectual Life," 11. For Cheng Yi's evaluations of the Han and Tang Ancient Prose masters, see Wing-tsit Chan, trans., *Reflections on Things at Hand*, 293–98. The commentary also cites evaluations from Zhu Xi's *Classified Recorded Conversations*. Cf. Wilson, *Genealogy of the Way*, 159.

133. For Zhu Xi's discussions of Xunzi, Dong Zhongshu, Yang Xiong, Wang Tong, and Han Yu, see *ZZYL*, 137.3255–76.

134. He Jipeng, "Zhuzi de wenlun," 1215–16; *Yuanliu zhilun*, qianji, 4.33a; He Jipeng, *Bei Song de guwen yundong*, 268–70.

redefined by Ouyang Xiu, Su Shi, and their successors in East Zhe increasingly set the standards for examination writing.

Zhu Xi and his immediate disciples challenged the dominance of the Ancient Prose canon by developing Learning of the Way alternatives to Ancient Prose anthologies. They compiled and published anthologies featuring the work of the four Northern Song Learning of the Way masters (Zhou Dunyi, Zhang Zai, Cheng Hao, and Cheng Yi), and they devised reading and annotation methods aimed at eliciting an approach to texts different from that found in the annotated anthologies of the work of the Tang and Song Ancient Prose masters.

*A Record for Reflection* played a crucial role in the history of Learning of the Way anthologizing. Earlier compilations containing sections on a range of thinkers associated with the Learning of Cheng Yi presented different voices.[135] *A Record for Reflection* offered a thematic introduction to the Learning of the Way and canonized the work of Zhu Xi's chosen predecessors, Zhou Dunyi, the Cheng brothers, and Zhang Zai. The arrangement in categories resembled the anthologies and encyclopedias produced for examination candidates. *A Record for Reflection* appeared in commercial editions and set a precedent for subsequent Learning of the Way anthologies.

Close readings of the collection became a central part in instruction in the Learning of the Way. As we have seen, Chen Chun had been introduced to the Learning of the Way by reading *A Record for Reflection*, and he promoted the collection as a basic source in his curriculum. One of his students wrote a paragraph-by-paragraph commentary on *A Record for Reflection*. This edition by Xiong Gangda 熊剛大 (*js.* 1214, fl. 1216–50) paraphrased the original text line by line.[136] Students also compiled supplements. Another student of Zhu Xi, Cai Mu 蔡模 (fl. 1220), compiled a sequel containing Zhu Xi's sayings.[137]

In the course of the thirteenth century, the growing share of new editions and compilations of Learning of the Way sources in the gamut of examination preparation manuals in commercial printing centers

---

135. For a short discussion of the earlier anthology *Zhuru mingdao ji* 諸儒鳴道集 in relation to *A Record for Reflection*, see Wilson, *Genealogy of the Way*, 152.

136. Wing-tsit Chan, trans., *Reflections on Things at Hand*, 338–39.

137. Postface to *All the Works on Nature and Coherence*, in ibid., xxxviin13.

attests to the Learning of the Way's shift from the periphery to the center of the examination field. As mentioned above, the Zeng family from Jian'an printed a supplement to *A Record for Reflection* and a special edition of the original, *A Record for Reflection, Classified for Use in the Examinations*. Students of Zhu Xi based in Jian'an prepared another commercial edition of the work of the Learning of the Way masters, entitled *All the Works on Nature and Coherence, with Line-by-Line Explanations* (*Xingli qunshu jujie* 性理群書句解). A re-edition of this work, titled *All the Works on Nature and Coherence, with Line-by-Line Explanations and the Correct Pronunciations—A New Edition* (*Xinbian yindian xingli qunshu jujie* 新編音點性理群書句解), is extant.[138]

In *All the Works on Nature and Coherence*, Xiong Jie 熊節 (*js.* 1199), a disciple of Zhu Xi, gathered core texts of the Learning of the Way. The first part was an anthology of these texts, arranged by genre. The peculiar configuration of the genres gives a sense of the diversity of the textual legacy of the Learning of the Way tradition; it also conveys the specificity of the message. The list begins with portraits and eulogies and then moves on to various kinds of instructions (e.g., concerning study and comportment, general exhortations, and school regulations), inscriptions, poems, prefaces, records, expositions, charts; it ends with biographies. Just as students in some late Southern Song Learning of the Way academies began their day by bowing in front of the images and portraits of the sages and wise men, this anthology first confronted the reader with Learning of the Way models for engaged learning. The exhortations on correct behavior and the instructions on moral education carried forward the manual's strong ritualistic flavor.

*All the Works on Nature and Coherence* exemplifies the formation of a Learning of the Way canon in thirteenth-century schools and the book market. The title derived from the inclusion of the major "books" (*shu* 書) on coherence, Zhang Zai's *Correcting Youthful Ignorance* (*Zheng meng shu* 正蒙書), Shao Yong's 邵雍 (1011–77) *Supreme Principles Governing the World* (*Huangji jingshi shu* 皇極經世書), Zhou Dunyi's *Penetrating the*

---

138. This work is preserved in a Yuan edition from Jianyang at the National Library in Taibei. Xiong Jie collected the texts, and Xiong Gangda prepared the commentary. Xiong Gangda had studied with Cai Mu and, like Cai Mu and Xiong Jie, hailed from Jianyang.

*Changes* (*Tongshu* 通書) and, in the second installment, *A Record for Reflection* and Cai Mu's two supplements, *A Record for Reflection, Continued* (*Jinsi xulu* 近思續錄) and *Another Record for Reflection* (*Jinsi bielu* 近思別錄). Besides the books, the anthology included other influential texts in the tradition, such as Zhu Xi's "Instructions for the White Deer Grotto Academy," Zhang Zai's "Eastern Inscription" ("Dongming" 東銘) and *Western Inscription*, and Cheng Yi's "Exposition on What Kind of Learning Did Yanzi Like" ("Yanzi suo hao he xue lun" 顏子所好何學論).

The difference between *A Record for Reflection* and Xiong Jie's early thirteenth-century anthology demonstrates the effect of canonization on Learning of the Way ideology. Zhu Xi presented his anthology as capturing the spirit behind the work of the Four Masters of the Learning of the Way. He pointed out that the anthology was not a substitute for their original writings; students should view *A Record for Reflection* as an introduction and move on to a careful reading of the masters' collected writings after finishing it.[139] Zhu Xi did not incorporate the works of the Northern Song masters as a separate item in his curriculum. In conversations with students or in written proposals, Zhu Xi typically listed the Classics, the Four Books and their commentaries (including those by Cheng Yi), and pre-Song historical and philosophical sources as required reading.[140] His followers gave priority to Learning of the Way sources in their teaching. Like Chen Chun, many of them recommended that the work of the Song transmitters of the Way be the first sources to be covered in Learning of the Way curricula. According to Zhou Mi 周密 (1232–98), as soon as Learning of the Way affiliates arrived at a new post, they printed the works of Zhou Dunyi, the Chengs, and Zhang Zai. He added that, in the course of the first half of the thirteenth century, the works of these masters had become the main source of instruction at local schools.[141]

The canon of Song Learning of the Way texts did more than prescribe what to read. Instruction in the canonical texts promoted ways of reading that differed from the methods used to teach Ancient Prose

---

139. "Preface by Chu Hsi," in Wing-tsit Chan, trans., *Reflections on Things at Hand*, 2.
140. Gardner, *Learning to Be a Sage*, 39–40; Zhu Xi, *Zhu Xi ji*, 69.3636–39.
141. Zhou Mi, *Guixin zashi*, 169; quoted in translation in Bol, "Neo-Confucianism and Chinese History: Position, Identity, and Movement," 28–29.

composition in examination preparation courses. The comments in *All the Works on Nature and Coherence, with Line-by-Line Explanations and the Correct Pronunciations—A New Edition* were of three kinds. There were phrase-by-phrase translations of all texts, phonetic glosses, and summary explanations of the meaning of each genre and each entry.

In contrast to commentarial practice in Ancient Prose anthologies, which was aimed at familiarizing students with the techniques of a classical written language, the annotation of Learning of the Way manuals was designed to neutralize the specifically written nature of the texts. By providing the right pronunciations, marking pauses and full stops, and by paraphrasing the text line by line, commentators brought the text as close to speech as possible. Philological emendations and introductory summaries contributed to a reorientation on the meaning of texts in their entirety.

The annotations in *All the Works on Nature and Coherence* derived from methods applied in Learning of the Way readings of the Four Books. Huang Gan 黃榦 (1152–1221), one of Zhu Xi's most trusted students, developed a set of rules to aid students in their reading of the Four Books:

A. Rules for (the punctuation of) sentences and phrases:

*sentence: lays out the main ideas, the meaning of the text is interrupted.

*short break: the particles *zhe* and *ye* correspond, the meaning of the text is not yet complete; the preceding text is reiterated, the argument is stated negatively first and then in the affirmative; the preceding text calls for the words to follow, the subsequent text follows up on the preceding words.

B. Rules for dots and lines:

Red lines in the middle for the general ideas, lateral lines in red for striking expressions and important passages, red dots for the central characters, black lines for research and corrections on institutions, black dots where omissions are made up for.

C. Rules for pronunciation:

We check both Xu Shuzhong [Xu Shen 許慎 (30–124)]'s *Explanations on Texts* (*Shuowen*) and Zheng Jiaji [Qiao 鄭樵 (1104–1162)]'s "Treatise on the Six Types of Characters" ("Liushu lue").[142] For each character that has two

---

142. The first work refers to Xu Shen's *Shuowen jiezi*, which appeared around 100 CE. "The Treatise of the Six Types of Characters" is one of the twenty treatises in Zheng Qiao's *Tongzhi*, published in 1161 (*SB*, 146–56).

pronunciations, we first rely on Jiaji's corrections of Shuzhong's mistakes; for the rest, we rely on Shuzhong. . . . For the pronunciation and the meaning, we consult Mr. Lu [Deming 陸德明 (ca. 560–630)]'s *Explanations for Canonical Texts* (*Jingdian shiwen*) and Mr. Jia [Changchao 賈昌朝 (998–1065)]'s *Distinguishing the Sounds of the Classics* (*Qunjing yin bian*);[143] overall, we mainly rely on Master Zhu [Zhu Xi].[144]

A. 句讀例
　句: 舉其綱文意斷
　讀: 者也相應文意未斷覆舉上文　上反言而下正上有呼下字下有承上
字

B. 點抹例
　紅中抹: 綱凡例
　紅旁抹: 警語要語
　紅點: 字義字眼
　黑抹: 考訂制度
　黑點: 補不足

C. 發音例
　並考許叔重説文, 及鄭夾漈六書略. 每字有兩音者, 先依夾漈所正叔
重之誤者. 餘方依叔重 . . . . 參陸氏經典釋文, 賈氏羣經音辨. 大抵依朱子
爲主.

The critical apparatus used in Learning of the Way instructional materials emphasized punctuation, that is, the division of the text into segments of meaning. Critical marks were used to highlight the central concepts of the text or philological problems. These same marks were applied in Ancient Prose anthologies to familiarize students with rhetorical and stylistic techniques, such as the composition of introductions and conclusions or the alternations between different modes of argument. By marking pauses and indicating the correct pronunciations, Huang Gan and those applying his reading instructions readied a text for continuous recitation and meditation. The application of this model of annotation, first developed for the reading of ancient classical

---

143. For the role of Lu Deming and his philological scholarship during the early years of the Tang dynasty, see McMullen, *State and Scholars*, 72, 77. On Jia Changchao, see *SB*, 197–200.

144. Cheng Duanli, *Dushu fennian richeng*, 2.10b–14a; cited in Ren Yuan, "Songdai jingdu zhi chuxin yu biduan," 59.

texts,[145] to the oeuvre of the Song Learning of the Way masters in early thirteenth-century anthologies such as *All the Works on Nature and Coherence* indicates that these texts had become unofficially canonized as classical texts.

Like the Ancient Prose canon, the formation of a Learning of the Way canon took place in educational institutions (academies, private and official schools). The unofficial canonization of the Learning of the Way masters fed into the official recognition of Learning of the Way ideology. In 1241 and again in 1270, Emperors Lizong and Duzong urged scholar-officials to study the work of the masters of the Learning of the Way (Chapter 5). The next section investigates the effect of official recognition on Learning of the Way ideology.

## THE TRANSFORMATION OF THE ANCIENT PROSE CANON

The recognition of the Learning of the Way among literati and at court effected a rapprochement between the Learning of the Way and Ancient Prose. This process occurred in two stages. In the 1220s and 1230s, Learning of the Way advocates integrated the study of the Learning of the Way canon and the instruction of Ancient Prose models of composition in their teaching practice, imposing Learning of the Way criteria on the selection and interpretation of Ancient Prose texts. After the official endorsement of the Learning of the Way canon in 1241, its supremacy in the examination field resulted in a redefinition of the scope and purpose of Ancient Prose anthologies. Examination teachers and commercial printers inserted large numbers of Learning of the Way texts into Ancient Prose anthologies and combined methods of instruction in reading and writing from both traditions. They justified the integration by claiming a common purpose for instruction in the Learning of the Way and Ancient Prose.

Between the 1210s and 1230s, Zhen Dexiu emerged as the most outstanding figure in the second generation of the Learning of the Way

---

145. For punctuation in classical texts, see Guan Xihua, *Zhongguo gudai biaodian fuhao fazhan shi*; and Zhang Bowei, "Pingdian si lun."

after Zhu Xi's death in 1200.[146] He mustered official support for the Learning of the Way at the capital while teaching at the Imperial College or serving in other capacities at court. He promoted Zhu Xi's teachings in the provinces through the implementation of social programs such as granary reforms. Zhen Dexiu became well known among examination candidates for advocating a curriculum in which the Learning of the Way canon was taught alongside more conventional courses in composition.

Building on a comment by Cheng Yi that it was permissible to spend a third of each month on examination writing if the rest of the time was spent on "real learning," Zhen Dexiu exhorted students to spend the first twenty days of the month reading and testing their knowledge of Learning of the Way texts.[147] Students should start by immersing themselves in the philosophical works of Zhu Xi and Zhang Shi and persevere until they had fully understood them. Next, they should carefully examine the foundations of Zhu's and Zhang's teachings in the writings of Zhou Dunyi and the Cheng brothers. Afterward, they were to look at the general histories of Sima Guang and Zhu Xi. The last ten days of each month were reserved for the acquisition of composition skills. Zhen's incorporation of authoritative texts of the Learning of the Way in a schedule of tests marked a further move in the establishment of the Learning of the Way as a major force in the examination field.

In contrast with Zhu Xi, who was preoccupied with defining the Learning of the Way against existing alternatives, and his first-generation disciple Chen Chun, who advocated the displacement of current examination writing through the allegedly unmediated language of the Learning of the Way, Zhen Dexiu made current examination writing (*shiwen*) and Ancient Prose (which had long dominated current examination writing) part of his Learning of the Way program. In 1232, he printed his Ancient Prose anthology, *The True Forefathers in Composi-*

---

146. Tillman, *Confucian Discourse and Chu Hsi's Ascendancy*, 241–45; *SB*, 88–90; Chu Ron-Guey, "Chen Te-hsiu and the 'Classic on Governance.'"

147. Zhen Duxiu, *Zhen Xishan xiansheng ji*, 7.106–7; cited in Chen Wenyi, *You guanxue dao shuyuan*, 222. For Cheng Yi's original statement, see *Er Cheng quan shu, Henan Cheng shi waishu*, 11.5a. It was also cited in *Jinsi lu*; cf. Wing-tsit Chan, trans., *Reflections on Things at Hand*, 199. Cf. Chen Wenyi, ibid., 217–18; and Gardner, *Learning to Be a Sage*, 19.

*tion* (*Wenzhang zhengzong* 文章正宗). Zhen presented his work not as "the key," "the secret," or "the model" of successful writing, as had Lü Zuqian and Lou Fang before him and Xie Fangde did after him, but as a compendium confronting students with "the true origins" of writing. The concern with true models indicated the priority assigned to Learning of the Way criteria in Zhen's anthology. Nevertheless, the model of previous Ancient Prose anthologies significantly impacted the selection and the methods of annotation.

Zhen developed an organizational scheme very different from the arrangement by author or by genre used by his immediate predecessors in compilation or by the editors of the standard literary anthologies, *Selection of Texts* and *The Best Tang Texts* 唐文粹 (*Tang wen cui*; 1011). He divided all writing into four modes. According to his analysis, the imperative mode (*ciming* 辭命) took shape in the orders of the sage-kings, first encountered in *The Book of Documents* and developed in the decrees of Han emperors. The argumentative mode (*yilun* 議論), which he found difficult to define, operated in the exchanges between ruler and minister and between friends and went back to examples in *The Analects* and *Mencius*. The narrative mode (*xushi* 敘事) was characterized by the chronological overview of a period, an affair, or a person; the true origins for historiography were *The Book of Documents*, *The Spring and Autumn Annals*, and the histories of Sima Qian and Ban Gu. The poetic mode (*shifu* 詩賦) of lyrics, rhyme-prose, and related genres had sprung from ancient songs, Confucius' redaction of *The Book of Songs*, and Qu Yuan's 屈原 *Songs of Chu* (*Chuci* 楚辭).[148]

Late Southern Song prose anthologies for examination candidates featured "argumentative writing" (*yilun*). Indeed, Lü Zuqian's *The Key* and the other Ancient Prose anthologies discussed above were entirely devoted to argumentative writing. Zhen Dexiu's anthology put argumentative writing back into the context of writing in general. In the preface, Zhen Dexiu pointed out that writing was but one part of learning. Learning was the process whereby "scholars investigate coherence and bring about practical results." Learning, and writing as part of the process of learning, was for Zhen Dexiu, as for Zhu Xi, self-cultivation and the social and political transformation brought about by

---

148. Zhen Duxiu, *Wenzhang zhengzong*, introduction, 1b–5a.

the moral self. Since writing should serve these and only these goals, he decided on the following selection criteria for *The True Forefathers in Composition*:

In what I have collected today, I went by the standards of the clarification of coherence and pertinence to sociopolitical use. I picked only those texts whose form goes back to Antiquity and whose meaning is close to the Classics. If texts do not conform to these standards, I do not include them, even if the diction is well crafted.[149]

今所輯, 以明義理, 切世用, 爲主. 其體本乎古, 其指近乎經者, 然後取焉. 否則辭雖工亦不錄.

*All* writing was to be judged by the same criteria. "Usefulness" was defined by the primary criterion of accordance with the theory of the moral self and its embedding in the human relationships (familial, social, and political) central to the Learning of the Way.

Liu Kezhuang, a well-known literary critic and contemporary of Zhen's, criticized Zhen Dexiu's rigid standards in the poetry section.[150] Zhen threw out many poems Liu would have included. Its more rigid standards notwithstanding, Zhen Dexiu's collection shared many of the characteristics of previous Ancient Prose anthologies. Despite his claims that argumentative writing derived from the Classics, *The Analects*, and *Mencius*, Zhen Dexiu began this section with texts from *Conversations from the States* (*Guoyu* 國語, 5th–4th c. BCE), *Zuo's Commentary*, and *Stratagems from the Warring States*, the models for the argumentative writing of Liu Zongyuan and Su Shi; the Classics were not excerpted. The chapters on argumentative writing further contained the same standard texts by Han Yu and Liu Zongyuan included in the Ancient Prose anthologies compiled by Lü Zuqian and Lou Fang.

In his comments Zhen combined the priority assigned to the evaluation of the meaning of the text in Learning of the Way anthologies with the attention paid to rhetoric and style in annotated Ancient Prose anthologies. Lü and Lou had used critical marks and notes to comment on the structural and stylistic highlights of the texts; problems of factual and moral truth, when raised, had been secondary considerations. In

---

149. Zhen Duxiu, *Wenzhang zhengzong*, 1a.
150. *SKTY*, 38.4154.

Zhen's comments, the emphasis was reversed. Their different approaches to Liu Zongyuan's "On Feudalism" can serve as an example.

Lü Zuqian's overall evaluation of "On Feudalism" and the interlinear commentary and annotation exclusively addressed the rhetorical characteristics and strengths of the piece (Chapter 4). By contrast, Zhen's final and only comment on it read:

> The staging of this piece is grand and powerful, the argumentation is vigorous and impressive; it can truly be considered a model for composition. However, as far as the truth is concerned, it is not like this. Therefore, Hu Zhitang [Hu Yin] says: "In the case of feudalism, the benefits are shared with the whole empire; this represents the fairness of the Way of Heaven. In the case of administrative units, the empire is given to one person; this caters to the selfishness of human desires. However, some contemporary scholars regard Liu Zongyuan's argument as incontestable. How could it be so?"

按: 此篇間架宏闊, 辨論雄俊; 真可爲作文之法. 然其理則有未然者. 故致堂胡氏曰: 封建與天下共其利天道之公也. 郡縣以天下奉一人人欲之私也. 而世儒乃有以柳宗元之論爲不可易者. 豈其然乎!

The note continues with an elaborate, eight-page refutation in small type of Liu's argument.[151] Zhen's appreciation for Liu's rhetorical skill was not backed up by interlinear comments or critical signposting. In extant Song and Yuan editions, only punctuation marks were added to the text; this practice illustrates the legacy of Learning of the Way critical methods in Zhen Dexiu's reading program.[152]

Composition had become a legitimate component in Learning of the Way education by the 1230s. In *The True Forefathers in Composition*, Zhen

---

151. Zhen Duxiu, *Wenzhang zhengzong*, 13.15a–19b.

152. This applies to an edition held at the National Library in Taibei, dated to the late Song or early Yuan, and a Yuan edition preserved at the Library of the Palace Museum, Taibei. In a short article on the extant *juan* of a Song edition preserved at the Library of Taiwan National University, Li Xuezhi ("Taida zang Songban *Xishan xiansheng Zhen Wenzhong gong Wenzhang zhengzong*," 79) notes that there are punctuation marks in the right-hand margin throughout the text. Takatsu Takashi ("Sō Gen hyōten kō," 135) writes that circles and jots "are visible" in the edition preserved at the National Library in Taibei; it is not clear, however, that these stylistic marks were, like the punctuation marks, part of the original edition. In the sixteenth century, Xu Shizeng (*Wenti ming bian xushuo*, 96–97) attributed an elaborate system of punctuation and critical marks to Zhen Dexiu; see Guan Xihua, *Zhongguo gudai biaodian fuhao fazhan shi*, 178–79.

Dexiu used Learning of the Way criteria to redefine the Ancient Prose canon and to adjust the tools used to teach it in contemporary examination preparation manuals. In Zhen's scheme, instruction in Learning of the Way texts and instruction in composition skills were kept separate. After the official endorsement of the Learning of the Way in 1241, the Learning of the Way canon occupied center-stage in examination preparation. The process of reconciliation between Learning of the Way and Ancient Prose, initiated by Learning of the Way advocates in the 1220s and 1230s, resulted in the co-optation of Ancient Prose in an expanded Learning of the Way ideology. Examination teachers and commercial printers issued literary anthologies featuring the prose of the Ancient Prose masters alongside the writings of the Learning of the Way masters. Although examination teachers continued to emphasize the importance of Ancient Prose in obtaining success, they now justified the practice of composition by claiming the practice of moral cultivation as the common purpose behind the Learning of the Way and Ancient Prose.

Around 1273, Liu Zhensun 劉震孫 and Liao Qishan 廖起山 published *The Legitimate Seal of Compositions from the Past and Present—A New Collection with Several Scholars' Criticism* (*Xinbian zhuru pidian gujin wenzhang zheng yin* 新編諸儒批點古今文章正印).[153] The collection was published in four installments, a common technique in the commercial printing of encyclopedias and anthologies in the thirteenth century.[154] Liu Zhensun, notary of the administrative assistant to the mili-

---

153. A wonderful Song copy of this work is preserved in the Library of the Palace Museum in Taibei (see the Appendix).

154. Encyclopedias like *Ultimate Essays, Inquiries into a Multitude of Books* (*Qunshu kaosuo* 群書考索), and *Classified Collection of Affairs and Compositions of the Past and the Present* (*Gujin shiwen leiju* 古今事文類聚) appeared in four or more installments. Southern Song anthologies were also published in installments: e.g., *The Best of Chen Liang and Ye Shi—With Stress Marks* (*Quandian Longchuan Shuixin xiansheng wen cui* 圈點龍川水心先生文粹) in two parts, *Awesome Expositions from Ten Masters, with Notes* (*Shi xiansheng aolun zhu* 十先生奧論註) in three parts, and *Awesome Expositions by Various Masters* (*Zhuru aolun* 諸儒奧論) in four parts. *The Legitimate Seal* offers convincing evidence that the model in four series (*qianji, houji, xuji, bieji*) was already in use during the late Southern Song. Most of the other encyclopedias and anthologies in four installments survive only in Yuan or Ming editions. I discuss the effects of serialization of encyclopedias in De Weerdt, "The Encyclopedia as Textbook."

tary commissioner of Wu'an 武安 in Tanzhou 潭州 (Jinghu South circuit 荆湖南路), collected and organized the texts; Liao Qishan, professor at the prefectural school of Raozhou 饒州 (Jiangnan East circuit 江南東路), did the collation.

In the preface, Liu guaranteed readers of *The Legitimate Seal* three things. First, he emphasized that the editor of *The Legitimate Seal* endorsed the transmission of learning from Antiquity as it had been delineated in Zhu Xi's Learning of the Way. In *The Legitimate Seal*, Zhu Xi's genealogy of the transmission of learning replaced eleventh- and twelfth-century genealogies of the transmission of Ancient Prose. Liu Zhensun clarified that he did not use "seal" in the title because of its association with skillful carving and patterning. In applying the metaphor of the seal to compositions, he claimed to be concerned mainly with "legitimacy." He deemed his work in line with the tradition of mind-cultivation and good government extending from the sage-kings Yao and Shun to Zhu Xi. Liu Zhensun paid tribute to the Learning of the Way canon by testifying that Zhu Xi and Shao Yong, whom Zhu Xi included as one of his intellectual predecessors in *A Record for Reflection*, had inherited the seal of the sages in their work. Zhu Xi had transmitted and clarified the significance of the Sixteen-Character Dictum attributed to the sage-kings of Antiquity in his commentary on *The Doctrine of the Mean*; Shao Yong had worked out Jizi's 箕子 (ca. 12–11th c. BCE) ideas about the "great ultimate" (*huangji* 皇極) in *Supreme Principles Governing the World*.[155] Jizi was the sage adviser to the last king of Shang and allegedly wrote "The Great Plan" ("Hong fan" 洪範), in which the concept *huangji* first appeared.

Second, Liu explained to potential book buyers that the application of Learning of the Way criteria distinguished his selection from those found in other anthologies. He advertised his anthology as a collection of Ancient Prose true to the Learning of the Way by noting sarcastically

---

155. I discuss the Sixteen-Character Dictum in more detail in Chapter 7. Jizi remonstrated against the government of Zhou, the last Shang king. He ended up in prison because of his remonstrations but was set free by King Wu of the Zhou for whom he allegedly wrote "The Great Plan" ("Hong fan"). Jizi's ideas are recorded in *The Book of Documents*, in the "Books of Zhou" (Legge, *The Chinese Classics*, vol. 3, *The Shoo King*, 332). For an interpretation of the cultural relevance of "The Great Plan," see Nylan, *The Five "Confucian Classics,"* 136–67, esp. 139–41.

that not all the texts collected in *Layers of Foamy Waves*[156] were great, not all of those in *Awesome Expositions*[157] were deep, and that not all of those in *Upholding Antiquity*[158] measured up to the legacy of Antiquity. He assured the reader that all Ancient Prose compositions in his selection were imprints of the legitimate seal—that is, they were products of the transmission of mind-cultivation.

References to the transmission of the Way in titles or to Zhu Xi in prefaces were more than just a sales gimmick. The supremacy of Learning of the Way in the examination field resulted in a redefinition of Ancient Prose. Examination teachers' and printers' adoption of Learning of the Way selection criteria altered the constellation of the Ancient Prose canon. Whereas the central figures of Zhu Xi's Learning of the Way had played a negligible role in earlier Ancient Prose anthologies, prefaces by Cheng Yi and Yang Shi, Zhang Zai's inscriptions, and Shao Yong's charts figured prominently in *The Legitimate Seal*. Prefaces, explanations, inscriptions, admonitions, eulogies, disputations, letters, and charts by twelfth-century advocates of the Learning of the Way, including Zhu Xi, Zhang Shi, Huang Gan, Zhen Dexiu, and Wei Liaoweng, also took up a sizable part. Zhu Xi and Zhang Shi were the most frequently cited authors in *The Seal*. With sixty-nine and thirty entries respectively, they outstripped the Ancient Prose masters Han Yu (twenty-five entries), Su Shi (twenty-two), Chen Fuliang (seventeen), and Liu Zongyuan (seventeen). The preponderance of Learning of the Way texts was partly due to the large selection of pieces in genres more representative of Learning of the Way writing, such as charts (*tu* 圖) and inscriptions (*ming* 銘)—fourteen of Zhang Shi's pieces, and thirteen of Zhu Xi's were short inscriptions. Zhu Xi's work was also well represented in the sections for favorite Ancient Prose genres such as the record (*ji* 記) and the preface/farewell (*xu* 序). In the latter section, Zhu Xi's prefaces on the Four Books and other Learning of the Way texts,

---

156. The text reads *Ceng lan* 層瀾. I know of no collection by that name. The author is probably referring to Lin Zhiqi's *Observing Foamy Waves* (*Guan lan*), discussed in Chapter 4.

157. See the description of *Awesome Exposition from Ten Masters, with Notes* in the Appendix.

158. This refers to Lou Fang's anthology, which also appeared under the title *The Key to Upholding Ancient Prose—With Notes by Master Lou Fang*. See Chapter 4.

eight in total, figured more prominently than farewell pieces and prefaces by Ouyang Xiu (four) or Su Shi (two).

Third, Liu vouched that official seals would accrue to those studying *The Legitimate Seal*. His anthology was inspired by the conviction that the acquisition of Ancient Prose composition skills remained crucial to examination success and a career in government. Because of this conviction, Liu included the same authors and texts as had previous Ancient Prose anthologies; he also preserved their methods of annotation. He presented his work as the most comprehensive collection of famous scholars' writings and previous critics' comments.

Students using Liu Zhensun's anthology learned creative writing by studying the annotated compositions of the masters of Ancient Prose. He borrowed from the general evaluations and interlinear comments and marks in the Ancient Prose anthologies of Lü Zuqian, Lou Fang, and, to a lesser extent, Zhen Dexiu and added notes from literary miscellanies by the Yongjia scholar Dai Xi and unidentifiable authors whose works are no longer extant.[159] The works of the Tang and Northern Song masters were fully annotated, and some pieces by Southern Song Ancient Prose authorities like Lü Zuqian and Chen Fuliang also came with interlinear notes on structure and style. Teaching the art of composition remained a crucial factor in examination preparation. In Liu's collection, "Yongjia" teachers such as Lou Fang and Lü Zuqian were considered better guides in this respect than Zhen Dexiu. In *The Legitimate Seal*, the text of Liu Zongyuan's "On Feudalism," for example, combined the comments and marks of Lou and Lü; Zhen Dexiu's observations were omitted.[160]

The preservation of methods of instruction that had been criticized and rejected by leading proponents of the Learning of the Way conflicted with Liu's ostensible goal of compiling an anthology based on

---

159. The Siku editors (*SKTY*, 38.4163) noted that such works as *Academician Dai Xi's Opinions* (*Xueshi Dai Xi biyi* 學士戴溪筆議) and *Leisurely Conversations at the Eastern Academy* (*Dongshu yantan* 東塾燕談) cited in *The Seal* were no longer available. The critics referred to as Huaicheng, Songzhai, Jiaozhai lang were also unknown to them. I have found no further information on these works and these critics. On Dai Xi, see Zhou Mengjiang, *Ye Shi yu Yongjia xuepai*, 291–92; and Hervouet, *A Sung Bibliography*, 25–26, 236.

160. Liu Zhensun, *Gujin wenzhang zheng yin*, xuji, 1.3b–7b.

Learning of the Way criteria. The contradiction was, in Liu's view, only apparent. It fit in with the general trend toward a conciliation between the Learning of the Way and Ancient Prose. According to examination teachers and editors adopting this approach, different methods could be used in the instruction of reading and writing, as long as they directed students toward the same goal—the cultivation of the moral mind. This corresponded with the core message of the Learning of the Way and conferred the seal of legitimacy on Liu's compilation.

Alongside the methods of instruction used in twelfth-century Ancient Prose manuals, *The Legitimate Seal* included the alternative of reading and writing instruction developed by Learning of the Way teachers. The writings of Learning of the Way masters such as Zhu Xi and Zhang Shi were not marked, and Liu discussed only the contents of the piece.[161] Unlike Zhen Dexiu, examination teachers and printers not affiliated with the Learning of the Way did not extend Learning of the Way methods of annotation to all Ancient Prose texts. They legitimated the instruction of compositional techniques by subsuming it under the process of moral self-cultivation.

In the preface to *The Legitimate Seal*, Liu promised that by working through the texts and following the criticisms, students were mentally preparing themselves for the art of writing. By the late Southern Song, Liu Zhensun and other compilers of literary anthologies and literary encyclopedias[162] denied that their work had literary value only. They justified involvement in the literary tradition by claiming it as a necessary part of the process of moral self-cultivation. The editors of late Southern Song examination encyclopedias proposed that Ancient Prose be used in conjunction with the Learning of the Way to transform literati learning. They argued that studying Ancient Prose engaged the individual in a continuous process of inner cultivation and was, therefore, very different from the kind of examination preparation advocates of the Learning of

---

161. For examples of annotated pieces by Lü Zuqian and Chen Fuliang, see Wang Tingzhen, *Guwen jicheng*, 6.12a–b; and for comments on a piece by Zhu Xi, ibid., 13.1a–3b. The commentators could not be identified—the Siku editors also included them in their list of unknown critics in their review on this work. The same comments appear in *The Legitimate Seal*.

162. Zhu Mu and Fu Dayong, *Gujin shiwen leiju*; cf. De Weerdt, "Aspects of Song Intellectual Life," 16.

the Way deemed detrimental to moral cultivation. Liu's work performed well in the competitive market of examination manuals. Shortly after the original compilation, a re-edition appeared. Wang Tingzhen's 王霆震 *A Complete Compilation of Ancient Prose—A Newly Printed Edition with Famous Scholars' Criticisms* (*Xinkan zhuru pingdian guwen jicheng* 心刊諸儒評點古文 集成) was a lower-cost edition of the same work and attests to the market value of anthologies like *The Legitimate Seal* in late Southern Song examination preparation (see Appendix A).

Before Liu Zhensun, Fang Yisun had justified his commitment to Ancient Prose on the same grounds and offered a more elaborate rationale for the conciliation between the Learning of the Way and Ancient Prose. Fang Yisun finished his guide to Ancient Prose composition, *One Hundred Pieces of Brocade of Exquisite Pattern*, around 1242, shortly after the official endorsement of the Learning of the Way in 1241.[163] In the prefaces, both Fang Yisun and Chen Yue 陳嶽 described the study of composition as a gradual, organized, and continuous process of self-edification:

Now, as for our engagement in writing, *it is a path*. Therefore, we have to walk slowly and learn to say "yes" first. Then from walking we move on to running, from saying "yes" we move on to argument, and via the path we reach the inner secrets. Who knows—we may turn out like Qing Ji,[164] Su Qin[165] (?–317), Han [Yu], Liu [Zongyuan], Ouyang [Xiu], or Su [Shi] some day! *We only have to keep working on it.*[166]

Even though quite a few compilations of Ancient Prose have appeared all over the book market, because those intended to be brief are not properly organized and because those intended to be detailed get lost in specifics, *they are of no help in discussing self-cultivation (wei ji)*. . . .

Crafting on the basis of genre conventions is definitely merely the position of those who have not received direct instruction by a great master or of students of later generations. If they are familiar with beautiful pieces of brocade

---

163. The National Library in Taibei possesses an early Ming edition of this work.

164. Qing Ji is the name of a water spirit, mentioned in *Guanzi*, 39.76. "Qing Ji looks like a human being. It is four *cun* tall, wears a yellow robe and a yellow hat, and carries a yellow shield. It rides a small horse that is pretty fast. If one calls its name, one can have it go 1,000 *li* away and return in one day."

165. Su Qin lived during the Warring States period and traveled from one kingdom to another, offering his political advice to their rulers. His biography is recorded in Sima Qian, *Shi ji*, 69.2241–77.

166. Author's postface (1242), in Fang Yisun, *Baiduan jin* (National Library, Taibei).

and learn how to compose through them, if they follow the structure and look for the intent, if they recite them with their mouths and ponder them in their minds, *then the work of previous wise men is not just something on paper. If one establishes oneself and comports oneself on this basis, and if one serves one's ruler and benefits the people with this, there is nothing one cannot accomplish. Surely this is not just a means to attain a higher examination ranking![167]*

今吾之爲文章, 蹊徑也. 是必先徐行, 學唯也. 由行而奔. 由唯而辯. 由蹊徑而堂奧. 異日之不慶忌, 不蘇秦, 不韓柳歐蘇爾! 其勉哉!

　古文之編書, 市前後凡幾出矣. 務簡者本末不倫. 求詳者枝葉愈蔓. 駁乎無以議爲已.

　至於以體式繡梓者, 特不過爲私淑後學地. 如果知其美錦而學製焉, 因其體而求其意, 誦諸口而惟諸心, 則前哲所作不在紙上矣. 以此立身行己, 以此致君澤民, 當無施而不可. 豈特梯級一第而已哉!

According to Chen Yue, who wrote a preface to the compilation in 1249, Fang Yisun had spent most of his time studying the works of the masters of Tang and Northern Song Ancient Prose. His lifelong immersion in Ancient Prose texts had resulted in great success at the Imperial College.[168] From a policy essay Fang Yisun wrote at the Imperial College around 1241, it appears that his success there was closely related to his support for the Learning of the Way. In an essay anthologized in *Standards for the Study of the Policy Response,* Fang Yisun defended the coherence of the Learning of the Way masters' views on the process of learning in response to skeptical questions about contradictory statements Cheng Yi and Zhu Xi had made on the subject.[169]

Like Zhu Xi in his denunciations of Lü Zuqian's concern with "the framework" of a text or in his critique of the rush for fame and riches

---

167. Chen Yue's preface (1249), in Fang Yisun, *Baiduan jin,* 1a, 2a–b.

168. Chen Yue (ibid., 2a) mentioned that Fang did very well at the Imperial College. It seems, however, that he did not obtain the *jinshi* degree. In the Ming edition, the author's name was prefaced with "a scholar from the Song dynasty"; *jinshi* degree–holders were commonly referred to as such.

169. (*Jingxuan*) *Huang Song Cexue shengchi,* 5, essay no. 1. The essay was written in response to a question about discrepancies between Zhu Xi's instructions for the White Deer Grotto Academy and statements about learning from Cheng Yi. Fang Yisun argued that Zhu Xi's instructions were designed to have students practice the very values they were reading about. The centrality of moral practice in Zhu Xi's curriculum made him, according to Fang Yisun, the legitimate successor to the legacy of the Classics and Cheng Yi.

among "Yongjia" scholars and examination candidates in general, Chen Yue berated formalism in learning and writing and a utilitarian concern with examination success. In his curriculum, Zhu Xi focused on the careful reading and pondering of texts as a major component in the development of moral character. Chen Yue responded to the suspicion against literary studies by arguing that by "following the path," by discovering the structure and the meaning of Ancient Prose texts, and by reciting and pondering them, students could build moral character and thus serve their ruler and society.

The Learning of the Way canon had originally been defined against the preoccupation with argumentative writing and the canonization of Ancient Prose in the examination field. When the Learning of the Way began to dominate examination preparation, the purpose of Ancient Prose composition was redefined in accordance with Learning of the Way ideals. The difference between Zhu Xi's and Fang Yisun's positions on Ancient Prose reflects a transformation in Learning of the Way ideology. Thirteenth-century teachers and printers responded to the official endorsements of the Learning of the Way. They interpreted the critical and antagonistic stance of Zhu Xi and his disciples in the late twelfth and early thirteenth centuries as an unfortunate byproduct of the tense political atmosphere of the late twelfth century and broadened the scope of the Learning of the Way by subsuming diverse currents in examination preparation and literati culture under it.

# 7

# The Learning of the Way
## Transformation of Examination Standards
## (ca. 1200–1274)

In the second half of the twelfth century, "Yongjia" teachers instructed hundreds of students in the exegesis of the Classics and histories and the discussion of contemporary policy and scholarship. The commercial success of anthologies of expositions and policy response essays by Chen Fuliang and Ye Shi throughout the last decades of the twelfth century suggests that "Yongjia" scholarship reached an even broader audience. In response, Learning of the Way teachers challenged the authority of "Yongjia" teachers in the examination field and taught students alternative methods for interpreting the Classics and histories and new ways of discussing current affairs.

As we have seen, over the course of the thirteenth century, Learning of the Way teachers increasingly defined the standards for successful examination writing. This chapter traces the gradual ascendancy of Learning of the Way ideology in the examination field through an analysis of examination expositions and policy response essays written from the 1180s to the 1270s. As in Chapter 3, the analysis is divided by genre to clarify the impact of the Learning of the Way on both the exegesis of classical and historical texts (in expositions) and the discussion of issues in administrative policy and scholarship (in policy response essays).

The expansion of the Learning of the Way into these areas of scholarly endeavor followed the same trajectory. This trajectory can be divided into three stages. The first stage corresponds to the last two decades in Zhu Xi's life. During this stage, examination candidates affiliated with the Learning of the Way adopted the confrontational language of philosophical debate associated with the recorded conversations of the Learning of the Way masters. After the official endorsement of Zhu Xi's commentaries in the 1210s and 1220s, the earlier antagonistic stance gave way in the 1220s and 1230s to a confident, lengthy elaboration of Learning of the Way beliefs in examination expositions. Learning of the Way political theory was acknowledged as a legitimate, albeit not necessarily satisfactory, competitor of other approaches in policy questions and response essays. Once the court fully endorsed its position of authority in the examination field in 1241, the adaptation of the Learning of the Way to the conventions of examination writing entered its final phase. In the last decades of the Song dynasty, the Learning of the Way canon, the legacy of Zhu Xi in particular, became the standard body of reference in exposition and policy questions and response papers. An analysis of examination essays from this period further demonstrates that the ideology espoused by examination candidates after the 1240s differed from that practiced by earlier Learning of the Way affiliates in several respects. Following the trend of examination teachers and commercial publishers (Chapter 6), examination candidates in the final decades of the Song period broadened the scope of the Learning of the Way by subsuming within it types of examination preparation opposed by its founding figures.

## *Standards for Expositions*

### THE ANTAGONISTIC VOICE OF THE LEARNING OF THE WAY IN THE TWELFTH CENTURY

In a report issued after the departmental examinations of 1196, Ye Zhu called for the destruction of recorded conversations (Chapter 5), which had become symbolic of the impact of the Learning of the Way on examination writing and on literati culture more generally. Daniel Gardner has demonstrated how the genre embodied a new mode of intellectual discourse.[1] Discussion of the meaning of the classical heritage,

---

1. Gardner, "Modes of Thinking and Modes of Discourse in the Sung."

unrestricted by the texts of the Classics and previous commentaries, invested authority in contemporary voices, those of Learning of the Way teachers and those of their students. The conversations recorded the question-and-answer sessions through which the teacher guided an individual student in understanding and practicing the cosmic imperative of moral cultivation. The notes students took during and after these sessions were transposed not into standard written forms of the Chinese language but into a form that partially incorporated the spoken vernacular into the written accounts.

The recorded conversations incorporated both colloquial expressions widely shared among Song dynasty speakers of Chinese and terms that shared the morphology of the colloquial (namely, the prevalent use of binominals) but that were part of a narrower technical vocabulary of moral philosophy.[2] The use of colloquial and technical language clearly set the Learning of the Way teachers' discussion of the Classics, histories, and contemporary issues apart from the scholarship of their contemporaries. The language of the recorded conversations reflected a broader antagonism toward contemporary literati trends and rival intellectuals. As both Daniel Gardner and Ichiki Tsuyuhiko have shown, the ideas of contemporaries received much attention in Zhu Xi's recorded conversations, most of it negative.[3]

A small number of examination expositions dating from the 1180s and 1190s validate the perception of examiners that examinees were adopting the discourse of recorded conversations in their essays. Late twelfth-century examination candidates learned about the Learning of the Way in lectures or through transcripts of lectures and question-and-answer sessions featuring Learning of the Way teachers. They applied the model of exegesis exemplified in those sessions to the passages of the Classics, histories, and philosophers raised in exposition topics.

Chen Deyu 陳德豫 and Feng Yi were among the earliest successful candidates to express their commitment to the Learning of the Way in

---

2. Because of the incorporation of colloquial language, linguists have used the recorded conversations as source material for the history of Chinese grammar; see, e.g., Zhu Minche, *"Zhuzi yulei" jufa yanjiu*; Huang Jinjun, *Er Cheng yulu yufa yanjiu.*

3. Gardner, "Modes of Thinking and Modes of Discourse in the Sung," 582; Ichiki, *Shu Ki monjin*, 483–509.

examination writing. Chen Deyu obtained the *jinshi* degree in 1187, the year in which examiners Chen Jia and Hong Mai submitted a report complaining about the prevalence of Learning of the Way terminology in examination papers (Chapter 5). Feng Yi obtained the *jinshi* degree in 1193. The commitment of both Chen and Feng to the Learning of the Way is well attested outside the examination halls. Chen Deyu protected *The Surviving Works of the Cheng Brothers* during the ban on False Learning in the 1190s.[4] Feng Yi was a disciple of Zhu Xi and published several works on famous titles in the Learning of the Way canon such as Zhou Dunyi's *Diagram of the Supreme Ultimate* and Zhang Zai's *Western Inscription*.[5]

The papers of Chen Deyu and Feng Yi focused on a central tenet in the Learning of the Way. They explained moral self-cultivation as a continuous and gradual process. The topic of Chen's essay, "Yao practiced the Way, and Shun performed acts of filial piety,"[6] was taken from the following passage in *The History of the Han Dynasty*:

Therefore, Yao attentively practiced the way every day, and Shun conscientiously performed acts of filial piety unfailingly. When their good acts accumulated, their names became known. When their virtue became manifest, their status became elevated.[7]

故堯兢兢日行其道, 而舜業業日致其孝. 善積而名顯. 德彰而身尊.

Chen Deyu agreed with Dong Zhongshu, the Han court adviser to whom this statement was attributed, that practicing the way and performing acts of filial piety were the defining characteristics of Yao and Shun as models of sagehood. He explained the consistency of the sages in performing their daily duties and cultivating the virtues appropriate to their social and political roles as evidence of their understanding of moral coherence:

In the world, there are moral principles that really do not allow for full realization. The sage has a mind that truly does not allow for fulfillment.

---

4. *SRZJZL*, 2640–41.

5. Ibid., 2745; Wing-tsit Chan, *Zhuzi menren*, 252–53.

6. *LXSC*, 6.19a–24a.

7. Ban Gu, *Han shu*, 56.2517.

Now, as for "fully realizing the Way of the sovereign" and "discharging one's offices as a son,"[8] only sages are up to it. However, that they still exert themselves and do not allow for fulfillment—*some say* that it is because the sages are modest. This is wrong. The principles of the world, in reference to human relationships, do not allow for full realization. Actually, it is not so much that they do not allow for full realization as that we cannot fully realize them.

天下有實不容盡之理. 聖人有誠不容已之心.

　夫盡君道共子職, 至於聖人足矣. 而猶拳拳不容已者, 或謂聖人之謙. 非也. 天下之理惟人倫爲不容盡. 非不容盡也, 不可得而盡也.

According to Chen Deyu, Dong Zhongshu's statement highlighted two aspects of sagehood. First, Yao's practicing the Way and Shun's acts of filial piety manifested their insight into the natural principles pertaining to human relationships. Yao devoted himself to the Way of the ruler, and Shun to his duties as a son. They thus demonstrated their understanding of the moral pattern that structured the human world into a hierarchical moral order and their ability to live in accord with it. Second, it was their unfailing moral efforts in their respective capacities that defined Yao and Shun as sages. Chen Deyu argued that, on a metaphysical level, the principles of the world, particularly the principles of human relationships, do not allow for full realization. Even if this were theoretically possible, human beings would be incapable of achieving it. Therefore, Yao and Shun were sages, not because they performed their duties perfectly or fully realized the virtues associated with their social roles but because they continued their efforts in spite of knowing that their goals were ultimately unachievable.

Feng Yi made a similar point in his essay on a passage from *The Analects*, "The benevolent and the wise give extensively and provide for the multitude."[9] Feng Yi's exposition focused on the words attributed to Confucius that followed this sentence. Confucius averred that Yao and Shun found it difficult to "give extensively and provide for the masses and thus reach sagehood." Feng Yi argued that Confucius' apparently puzzling statement about the ancient models of sagehood stemmed from his insight that sagehood was an ideal never to be fully reached but

---

8. *Mengzi* VA:1; cf. Legge, *The Chinese Classics*, vol. 2, *The Book of Mencius*, 343.
9. *LXSC*, 6.13a–18a; D. C. Lau, *The Analects*, 85.

always worth pursuing.[10] Yao and Shun were the best models, Feng argued, because their example taught that there were stages (*fenliang* 分量) in the process of self-cultivation. They approached sagehood through accumulated effort; even they would have been unable to jump to the stage of "providing for the multitude" without continuous exertions.

Chen Deyu admitted that his contemporaries did not necessarily share his interpretation of the sagehood of Yao and Shun. As the passage from his essay illustrates, he made explicit reference to dissenting arguments. The technique of bringing in dissenting voices only to refute them was frequently used in Zhu Xi's conversations with his students during the 1180s and 1190s. Zhu Xi took this scholastic technique one step further. As Ichiki Tsuyuhiko has demonstrated, Zhu Xi discussed his criticisms of some of his rivals with his students.[11] The discussions developed into metacriticisms in which the students learned to evaluate why the rival was wrong and why the teacher's criticism of his work was right. Such exercises fit into Zhu Xi's efforts to present his interpretation of Cheng Learning as its true transmission. The technique of metacriticism trained students to confirm Zhu Xi's reconstruction of the legacy of Cheng Learning.

Although Chen Deyu did not violate the implicit rule that contemporaries not be mentioned by name in examination writing, the dispute in his examination exposition was not a figment of his imagination. Chen Fuliang's collection of examination expositions includes two essays on topics similar to those addressed in Chen's and Feng's papers. In these essays, Chen Fuliang described at length the sages' modesty and Confucius' humility. In a piece titled "Confucius Never Went Beyond Reasonable Limits,"[12] Chen argued that the sage could not afford to set his expectations of others too high and distance himself from the average human being. In response to the topic "Shun and Yu held aloof

---

10. Earlier commentators did not interpret Confucius' discussion of benevolence and sagehood in this passage as a reference to particular stages of the path of self-cultivation. According to Xing Bing 邢昺 (932–1010; *Lunyu zhushu*, 6.16b–17a), the passage was about the Way of Benevolence, which he explained as the Confucian version of the golden rule. On Xing Bing's commentary, see *SB*, 415–17.

11. Ichiki, *Shu Ki monjin*, 501–7.

12. *LXSC*, 6.25a–30a; D. C. Lau, *Mencius*, 129.

from the empire even when they were in possession of it,"[13] he maintained that Shun and Yu did not consider the empire vast because they did not have a high opinion of themselves. Both expositions were examples of Chen Fuliang's dramatic situationalism: his analysis of statements from *The Analects* and other Classics focused on the psychology of the ancients when making those statements.

Chen Fuliang's interpretations were unacceptable to Chen Deyu. He rejected Chen's view that Yao's and Shun's never-ending efforts at self-cultivation were inspired by modesty. He defined the sage as someone who knows that he cannot fully realize his moral obligations and, therefore, never slacks off. "What kind of modesty is that!" he exclaimed. An understanding of the coherence of moral order, not modesty, underlay their actions and words.

Not only contemporary scholars but also previous commentators in the classical tradition were subject to critical treatment. Like Zhu Xi, Chen Deyu was critical of Han scholars like Dong Zhongshu. Although Dong Zhongshu fared better than other Han classical scholars in Zhu Xi's discussions about the value of their commentarial work, Zhu did not acknowledge him as a fully legitimate transmitter of the Learning of the Way (Chapter 6). In his exposition, Chen Deyu distinguished the Learning of the Way interpretation of sagehood from the comments of "Yongjia" scholars and Dong Zhongshu alike.

Somebody may say that this [Dong Zhongshu's statement cited in the topic translated above] was set up for Emperor Wu. Even if it was indeed set up for Emperor Wu, it is just not worth arguing along those lines.
    Students should reflect on this and take their lead from the sage.

或曰: "是爲武帝設也." 果爲武帝設, 則亦不足辨也已.
    學者姑反而求諸聖人.

Chen rejected a situationalist reading of Dong Zhongshu's statement, which had been uttered in an audience with Emperor Wu. Instead, he measured Dong Zhongshu's response by universal standards. In his view, Dong demonstrated an inadequate understanding of moral coherence. This judgment was based on the sentence following the topic sentence in *The History of the Han Dynasty* (see above). Chen censured

---

13. *LXSC*, 8.83a–86b; D. C. Lau, *The Analects*, 94.

Dong's point that fame could be achieved by moral discipline: "Fame and status are not things that the sage is aiming at; what the sage knows is fully realizing his mind. . . . Master Dong's words were still very close to arguments for calculating results."[14]

The examinees' criticism of well-known commentators and contemporary thinkers was part of a broader application of the Learning of the Way language of philosophical debate to examination writing. Chen Deyu and Feng Yi used colloquial and technical expressions listed in the 1187 report of Chen Jia and Hong Mai and attested in Zhu Xi's recorded conversations. Feng Yi's argument hinged on the definition of self-cultivation as a gradual process. The concept he used to define the stages in the process, "dividing one's efforts" (*fenliang*), was one of the terms on the 1187 list (Chapter 5).

The voices of critics were cited at length and refuted in a written language that mimicked the recorded interactions between Learning of the Way teachers and disciples. The freedom from the rhetorical patterns of classical prose and from classical language and the preference for colloquial expressions and a more relaxed style of philosophical debate[15] characterized Chen Deyu's essay:

> Somebody has said, "Mencius said, 'The compass and square represent the ultimate of roundness and squareness. The sage represents the utmost of human relationships.'[16] If Yao cannot fully realize the Way and Shun cannot fully actualize filial piety, then none in the world will ever be able to realize these things. If there are principles in the world that do not allow for full realization, we will see that because of this reason the great human relationships will dissolve, and nobody will care about those in need."
>
> I say: "Although his is a discriminating remark, his insight is not substantial. Does he really think that the principles of the world can be fully realized?! Talking is easy!"

---

14. *LXSC*, 6.24a.

15. The use of dialogue in prose writing was not an innovation of Learning of the Way expositions. Tang and Song Ancient Prose authors made frequent use of the technique. Among the examples preserved in *Standards for the Study of the Exposition*, however, the technique appeared rarely. The arguments in the dialogue most closely resemble those in dialogues and lectures on the Learning of the Way.

16. *Mengzi* IVA:2; cf. Legge, *The Chinese Classics*, vol. 2, *The Book of Mencius*, 292.

或曰: "孟子曰: '規矩方圓之至也. 聖人人倫之至也.' 使堯而不能盡道舜而不能盡孝, 則古今天下當無能盡之人矣. 使天下而有不容盡之理, 則人之大倫將坐視其湮斁, 而莫之省憂也."

　曰: "辨則辨矣, 而見則未實也. 彼果以為天下之理可得而盡乎! 談何容易也."

The staging of a debate in which the author quotes his own voice underscores the value attached to the personal articulation of the truth in Learning of the Way discourse. The model of exegesis Zhu Xi recommended to students in discussions with them called for consideration and evaluation of various commentaries on the Classics followed by the student's personal interpretation. Zhu Xi included this model of classical exegesis in "Private Opinion on Schools and Selection," his famous proposal for the reform of the examinations (Chapter 6). A slightly different model of exegesis was embedded in Zhu Xi's discussions with students and the recorded conversations that resulted from them. Zhu Xi did not explicitly recommend this model; rather, the discursive strategy underlying conversations with students set a pattern of exegesis that students adopted. This model of exegesis was based on a similar logic, but it was not limited to discussion of specific commentaries on the classical passage under consideration. The teacher articulated an authoritative interpretation of passages from the Classics by combating more general modes of thought deemed incompatible with the Learning of the Way.

The personal articulation of the truth involved the student in the struggle to gain recognition for the Learning of the Way. Stridency, the overt display of conviction, and an effort to convert others were the hallmarks of early attempts to fashion examination essays compatible with the contemporary spirit of the Learning of the Way movement. Chen Deyu underscored his confidence in the authority of the Learning of the Way through the emphatic and unconventional use of modifiers meaning "real" (*shi* 實), "true" (*cheng* 誠, *xin* 信), and "authentic" (*zhen* 真). Such terms rejected official portrayals of the Learning of the Way as False Learning and revealed the conviction of its followers that it provided ultimate answers to the larger questions underlying any examination question. The stridency of examinees resulted from the proselytizing thrust of the Learning of the Way movement in the late twelfth century. Their essays combined attacks on rival views with explicit calls on all literati to listen to the message of the Learning of the Way. In the conclusion of the paper translated above, Chen Deyu called on all students to take Yao and Shun as their models and engage in moral self-cultivation. The characteristics outlined above alarmed examiners and court officials and were read as signs of factionalism (Chapter 5).

## STANDARDIZATION, RECONCILIATION, AND THE DOMINANCE OF LEARNING OF THE WAY DISCOURSE IN THE THIRTEENTH CENTURY

In contrast to the small number of examination essays defending Learning of the Way ideas in the late twelfth century, dozens survive from the following decades. Although the vast majority of examination papers from the twelfth and the thirteenth century do not survive and thus make accurate estimates of the impact of the Learning of the Way on examination writing impossible, the number of surviving Learning of the Way examination papers from the thirteenth century unmistakably reflects its adoption among a growing number and wider variety of agents in the examination field.

Table 3 in Appendix B tabulates the datable expositions in the collection *Standards for the Study of the Exposition*. The table shows that until the 1220s only a minority of expositions made explicit use of Learning of the Way concepts and explanations. From the 1230s to the 1250s, more than half the expositions applied Learning of the Way thought and often more explicitly and elaborately than before. By the 1260s, three-quarters of the essays drew on Learning of the Way elements of style. Tables 2B and 2C show that Learning of the Way essays appeared at the departmental and Imperial College examinations as well as at local examinations.[17]

Zhou Mi, a famous writer and critic, observed that after Xu Lin 徐霖 (1215–62) gained first place in the departmental examinations of 1244,

---

17. Most of the pre-1260 essays referring to the Learning of the Way were written for the departmental examinations. Due to the large selection of Imperial College examinations from the 1260s, *Standards for the Study of the Exposition* included a higher number of Learning of the Way essays written for internal college tests in the 1260s.

people vied to imitate his work to achieve examination success. Zhou wrote that Xu's work "put high store on nature and coherence."[18] He added that, following Xu Lin's success, the Four Books, "The Eastern Inscription" and *The Western Inscription, The Diagram of the Supreme Ultimate, Penetrating the Changes*, and the recorded sayings were at the center of literati interest. This suggests that Xu Lin's familiarity with the Learning of the Way canon was the reason behind his success and that the Learning of the Way canon became standard in the examination field thereafter.

Authors of Learning of the Way essays had taken first place in the departmental examinations before Xu Lin in 1244. *Standards for the Study of the Exposition* contains an examination exposition by Ye Dayou 葉 大有, who ranked first in the departmental examinations of 1232.[19] This

---

18. I thank Chu Ping-tzu for first bringing this to my attention. Zhou Mi, *Guixin zashi*, houji, 65. This passage is also discussed by Ishida, "Shū Mi to Dōgaku," 41; and Zhu Ruixi, "Song Yuan de shiwen," 35. For more information on Zhou Mi, see *SB*, 261–68.

An examination exposition by Xu Lin has been preserved in *Standards for the Study of the Exposition*. The topic for Xu's essay was a passage in which Tang Emperor Taizong declares: "With regard to the foundations of governing the people, nothing is more important than local officials" (*LXSC*, 1.69a–75a; Ouyang Xiu and Song Qi, eds., *Xin Tang shu*, 197.5616). In the first and middle parts of his essay, Xu Lin alluded to Liu Zongyuan's argument that the emergence of the feudal system of Antiquity was the outcome of impersonal historical developments. Cf. Chen Jo-shui, *Liu Tsung-Yüan and Intellectual Change*, 96. In "On Feudalism," Liu Zongyuan argued that people naturally formed associations and set up the most capable among themselves as their leaders. Given the expanse of the empire, the rulers of Antiquity were naturally inclined to divide the territory and let others oversee the localities. According to Xu Lin, Taizong understood the beneficial effect of the emperor's sharing his power and therefore charged local officials with the benevolent government of the people. However, he modified Liu's position on historical necessity by arguing that feudalism worked during the Three Dynasties because the sage-kings "cultivated their minds with pure humanity" and disciplined themselves day and night in order to "cut off *human desires* and solely preserve *natural coherence*." Taizong's understanding of the benevolent government of the sage-kings did not reach its true foundations: "what he called the root [selecting local officials] was really just an auxiliary aspect." Zhu Xi's criticism of Liu Zongyuan's interpretation of "sociopolitical reality" (*shi* 勢) was very similar. Cf. De Weerdt, "Canon Formation," 114.

19. *LXSC*, 4.1a–6a. Brief comments of the examination official(s) precede the text of the exposition. This fact and the positive tone of the comments enhance the likelihood that this was the exposition he wrote for the departmental examination of

paper used Learning of the Way ideas in the exegesis of historical passages in the same way as the winning Learning of the Way expositions from the 1240s and later. Ye explained the successes and failures of historical actors by reducing them to moral values and underscored the importance of mental self-cultivation. Unlike the early Learning of the Way examination papers discussed in the previous section, his explanations borrowed from Zhu Xi's work. As Zhou Mi's comment suggests, however, Ye's success did not mark a decisive turn in the impact of the Learning of the Way on examination writing. The expositions by the top departmental examination candidates in the examinations of 1235, 1238, and 1241 showed no particular concern with the Learning of the Way.[20] The success of Learning of the Way ideas mounted in the examinations after 1244.[21] During the 1250s, the expositions of three of the top graduates advocated Learning of the Way ideology.[22]

This evidence suggests that Emperor Lizong's 1241 endorsement of Zhu Xi's legacy and of the work of the founders of the Learning of the Way solidified the authority of the Learning of the Way in the examination field. The analysis of thirteenth-century expositions below shows that the unofficial canonization of the work of the Song masters in twelfth- and thirteenth-century academies and schools laid the foundation for this. It also shows that it was not until the subsequent official

---

1232. The names of the examiners are unknown. Cf. Mo Yanshi and Huang Ming, *Zhongguo zhuangyuan pu*, 148.

20. See the expositions by Yang Maozi 楊茂子, who won first place in the departmental examinations of 1235; Miu Lie 繆烈, first in 1238; and Liu Zi 劉自, first in 1241 (*LXSC*, 1.52a–57b, 1.35a–41a, 4.13a–18b). Cf. Mo Yanshi and Huang Ming, *Zhongguo zhuangyuan pu*, 149–50.

21. Whether this has anything to do with Xu Lin himself is unclear. Zhou Mi (*Guixin zashi*, bieji, 290–92) wrote elsewhere that Xu Lin showed his low estimation of Shi Songzhi 史嵩之, who had become grand councilor on the right in 1239, right after the examinations—Shi Songzhi had to resign under pressure of Imperial College students soon afterward (*SB*, 878). However, I have found no evidence that he was directly responsible for the change in fortune of Learning of the Way essays.

22. Chen Yinglei 陳應畾 won top place in 1250 (*LXSC*, 3.27a–30b); Ding Yingkui 丁應奎 in 1253 (ibid., 3.7a–13b); Li Leifen 李雷奮 in 1259 (ibid., 3.14a–20a). Little biographical information exists for Chen and Li; about Ding we know only that he went on to have a successful career at court (*SRZJZL*, 17; *SRZJZL-BB*, 1182, 424). The names of the examiners were not mentioned. Cf. Mo Yanshi and Huang Ming, *Zhongguo zhuangyuan pu*, 151–55.

recognition of their work that the Learning of the Way canon, and Zhu Xi's work in particular, became a prerequisite for examination success. My discussion is based mainly on essays on topics picked from the histories. Expositions on *The Analects* and *Mencius* attest to the influence of Learning of the Way more graphically, but the application of Learning of the Way interpretations to questions on two of the Four Books can be seen as natural. The use of Learning of the Way explanations in historical essays, however, is better proof of the growing power of Learning of the Way ideology.

The official recognition of Learning of the Way exegesis resulted in changes in the formulation of examination questions. In contrast to policy questions, which contained lengthy directions, exposition questions provided few indicators of the examination officials' tastes and inclinations. The examiners used direct quotation or recast the major elements of a passage into a question of no more than ten characters. My analysis of the provenance of the topics did not reveal changes in the emphasis on particular classical, historical, or philosophical sources. Neither did I find significant trends in the citation of specific chapters or sections from these sources.[23] However, changes in the formulation and the focus of the questions demonstrate the mounting interest of thirteenth-century examiners in Learning of the Way ideology.

In the 1250s and 1260s, almost all exposition topics were questions rather than direct quotes from the Classics, the histories, or the philosophers. In questions about the Classics, students were asked to correlate major concepts in Confucian ethics drawn from a sequence of sentences from the original source. For example, in the departmental examinations in 1250, students were asked the following question: "How about 'to gain insight is to be active, to be humane is to be at rest' and 'finding joy' and 'enjoying a full lifespan'?" (*The Analects,*

---

23. Some questions addressed subjects that received much attention in Learning of the Way moral metaphysics. Mencius' four germs, for example, appear in two questions (*LXSC*, 5.29a, 35a). The source of the topic is the same for both: *Mengzi* IIA:6 (Legge, *The Chinese Classics*, vol. 2, *The Book of Mencius*, 203). The relation between the Yellow River Chart and the Luo River Diagram is the subject of a question drawn from *The History of the Han Dynasty* (*LXSC*, 10.1a). The quotation is based on a passage from *Han shu*, "Wuxing zhi," 7.1316. For Zhu Xi's views on the subject, see Adler, "Chu Hsi and Divination," 175–76.

VI:21)[24] Another exposition topic in the same examination read: "How about 'fully realizing the mind,' 'understanding one's nature' and 'preserving and cultivating'?" (*Mencius*, VIIA:1)[25] Mid-thirteenth-century topics on historical sources exhibited a similar pattern. Examiners typically asked for an evaluation of a historical figure on the basis of specific categories such as humanity and rightness: "How about the Way and virtue, humanity and rightness, in the case of Emperor Wen?" or "How about the Way and virtue, humanity and rightness, in the case of the sages of Antiquity?"[26]

Topics from the twelfth century, in contrast, almost never included the interrogative "how about?" (*ruhe* 如何). In my count, it was used in only one topic, and then only to inquire about the meaning of the passage referred to.[27] By excerpting a full sentence as the topic and by retaining the declarative voice, twelfth-century examiners promoted traditional exegetical approaches, such as the line-by-line commentaries of the Han exegetes. "Yongjia" exegesis was equally suited for this type of question since it focused on the conditions under which the topic sentence was uttered.

By questioning a passage or, more often, the components deemed most significant in a given passage, thirteenth-century examiners were encouraging candidates to move beyond a mere contextual interpretation of a quotation. The reformulation of the passages and the questioning attitude toward the sources accorded with Learning of the Way reading methods.

Zhu Xi's exhortations for thorough study of the Four Books and the Five Classics were aimed at encouraging an experiential and integrative

---

24. *LXSC*, 3.27a. The author, Chen Yinglei, won first place in the departmental exams of 1250. Cf. Legge, *The Chinese Classics*, vol. 1, *Confucian Analects*, 192.

25. *LXSC*, 3.31a. Legge, *The Chinese Classics*, vol. 2, *The Book of Mencius*, 448.

26. *LXSC*, 3.1a–6a, 14a–20a (Ban Gu, *Han shu*, 65.2858, 57.2586).

27. For this question, see *LXSC*, 6.1a, 8.87a. The essay was by Lü Zuqian. Three more questions ended with "*ruhe*," but in each case, the question was part of the original quote. For example, "The sovereigns of Antiquity only cared about *how* (the Way of the true king) was practiced," and "In government care about *how* to practice government energetically" (*LXSC*, 4.88a, 95a, 7.55a). It was more common to ask for an evaluation of historical figures. The pattern "who was best" (*shu you* 孰優) occurred twice in the twelfth-century questions collected in *Standards for the Study of the Exposition* (ibid., 5.15a, 6.74a).

reading of the text rather than a contextual explanation of a quotation. To a thirteenth-century audience, Zhu Xi's readings of the Four Books and the Five Classics demonstrated an ability to "explain everything." Zhu Xi acquired the ability to explain the classical tradition as a whole by applying a vertical reading to all texts.[28] Even though Zhu Xi stressed the importance of reading a text in full and pondering the connections between sentences and sections, his reading method disengaged the words in a sentence from one another and linked them vertically to an analytical supertext. This supertext derived from the moral philosophy of the founders of the Learning of the Way as Zhu Xi read and synthesized it.

The systematic application of an analytical supertext became a distinguishing feature of Learning of the Way expositions before the change in the formulation of examiner questions in the 1250s. In the first half of the thirteenth century, examinees interpreted passages from the Classics and the histories as evidence validating the Learning of the Way's theory of the mind and clarified the Learning of the Way interpretation of the mind in detail. The lengthy and systematic discussion of Learning of the Way moral philosophy marked a transition from its representation in late twelfth-century examination writing. The twelfth-century candidates discussed in the previous section defended the Learning of the Way call for moral self-cultivation in polemical writing that focused on the refutation of rival exegetical approaches. Like the recorded conversations to which these essays were compared by their critics, their papers read like excerpts from conversations and did not expound the philosophical arguments on which the call for moral self-cultivation was based.

By the thirteenth century, students preparing for the examinations studied the Learning of the Way through a systematic reading of the Learning of the Way canon rather than through personal conversations with teachers claiming a position in the Learning of the Way genealogy. Zhu Xi's disciples and examination teachers not affiliated with the Learning of the Way incorporated the canon into their curricula. Printed editions of the new canon and of Learning of the Way anthologies appeared in the printing center of Jianyang, which thrived on the

---

28. *Yuanliu zhilun*, qianji, 5.2b. Cf. De Weerdt, "Aspects of Song Intellectual Life," 20.

appetite of examination students (Chapter 6). The greater availability of the main sources of Learning of the Way exegesis and their gradual adoption in school and academy curricula promoted an exegetical approach that appeared both less partisan and more grounded in a systematic set of ideas on the operations of the mind and its impact on government and society.

Lin Xiyi's 1235 exposition on a passage from *The History of the Han Dynasty* illustrates the interpretation of the Learning of the Way theory of the mind in thirteenth-century examination expositions. Lin was asked to write on the topic "Emperor Xuan 宣 [r. 73–49 BCE] sharpened his spiritual force to govern" (*lijing weizhi* 力精爲治).[29] Emperor Xuan was favorably portrayed in *The History of the Han Dynasty* as engaged in government affairs but without going to the extremes of expansionist policy making for which his predecessor, Emperor Wu, had become infamous.[30] In his essay, Lin Xiyi chose to focus on the workings of the emperor's mind, more appropriately, the workings of the mind in general, and showed little interest in the specifics of his government. In his own words, "A discussion of Emperor Xuan's rulership is insufficient to explain this; the whole meaning behind the reinvigoration of the government can be understood only from the application of this one mind. I got something from this."

Lin Xiyi divided the topic up into two components and read "sharpening one's spiritual force" (*lijing*) and "to govern" (*weizhi*) as instantiations of the distinction drawn in Learning of the Way philosophy between "the state of the mind when it is not yet manifest" and "the mind in its already manifest state" (*weifa* 未發 / *yifa* 已發). He thus uncovered two periods in Xuan's reign, one of quiet cultivation (before the Dijie era, 69–66 BCE) and one in which the beneficial effects of mental cultivation were manifest in good government (after 69 BCE). Xuan's cultivation of the mind resulted in a temporary reformation of Han government; the reason for the limited effects of the reformation derived,

---

29. *LXSC*, 1.58a–63b. The larger passage from *Han shu*, preface to the "Biographies of Officials," reads: "Emperor Xuan emerged from among the people. He knew the hardships of the affairs of the people. He personally attended to all situations and sharpened his spiritual force to govern" (Ban Gu, *Han shu*, 89.3624).

30. Twitchett and Loewe, eds., *The Cambridge History of China*, 1: 190.

in Lin's view, from the emperor's inability to fully comprehend the workings of the mind:

The order during the intermediate revival was called "sharpening the spiritual force." That it is still in people's eyes and ears until today, how can this derive from anything else besides the emperor's mind! How did the emperor in the end achieve this?

When a human being's one heart is moved, then it is confused; when it is still, then it is pure. During the period of obscurity, there is the benefit of being still and stable. After a while, his vision becomes pure, and, with patience, his thoughts become unadulterated. What the emperor achieved, I know it comes from this!

However, as for the heart-and-mind, only after coherence (*li*) and psycho-physical energy (*qi*) are combined is there this name.[31] If coherence can control the psycho-physical stuff, then its operation will be pure. If the psycho-physical stuff predominates over coherence, then its operation will be impure. The governments of Tang, Yu, and the Three Dynasties were pure and unmixed, sublime and unadulterated, because these rulers kept the moral principle in their mind pure. From the Qin and Han on, illustrious rulers and righteous emperors have occasionally made great accomplishments in their time, but, on the whole, their minds were dominated by psycho-physical stuff.

中興之治號爲屬精, 至今在人耳目, 是豈出於帝心之外乎! 帝果何以得此哉?

人之一心動則汩, 静則精. 當其韜晦之時, 蓋有静定之益. 閲歷之久, 則其見精, 容忍之積, 則其慮精. 帝之所得, 愚知其出於是矣.

雖然心也者, 合理與氣而後有是名也. 理足以御氣, 則其用也純. 氣得以勝理, 則其用也駁. 唐虞三代之治粹而不雜, 精而無間. 純乎心之理也. 秦漢而下, 英君誼辟時獲有爲於斯世, 而大抵皆以氣主之.

Lin Xiyi explained the ups and downs of Emperor Xuan's reign on the basis of the concepts of moral principle/coherence (*li*) and psycho-physical stuff (*qi*), which encapsulated Learning of the Way metaphysics. These concepts explained the formation and operation of the cosmos and, when used to explain the workings of the human mind, grounded moral philosophy in cosmological law. Lin argued that at the

---

31. The commentator indicates that this is a quote from Zhang Zai's *Correcting Youthful Ignorance*. The closest statement I have found in *Zheng meng* is "In the unity of nature and consciousness, there is the mind" (Wang Fuzhi, *Zhangzi Zheng meng zhu*, 17). I have adopted Wing-tsit Chan's translation in *Sourcebook*, 504n9.

beginning of his reign Emperor Xuan spent time and effort on meditation and was able to rein in the desires produced by psycho-physical stuff, which becloud the moral principle inherent in human nature. The effect of the emperor's efforts was only temporary, however, since he proved unable to halt the tendency of the mind to lose track of the moral coherence inherent in it.

Besides clarifying the metaphysical underpinnings of the Learning of the Way theory of the mind, Lin's exposition also explained its theses on historical development. The dominance of metaphysical stuff in the mind was not due solely to human psychology; it was also interpreted as a historical trend. History was essentially the history of moral degeneration; the theory of the interaction between the cosmic forces explained why. Learning of the Way ideology was characterized by a dehistoricizing thrust in which post-Antiquity history was reduced to the antithesis of the ideal order of Antiquity.[32] The message of the mind was presented as the recovery of the insight of the sages in the moral order of the world. The sage-kings of the Three Dynasties provided the ultimate and only appropriate standards of mental attentiveness and moral practice. Accordingly, in Learning of the Way expositions, the emperors of subsequent dynasties were measured against the models of the Three Dynasties. In Lin Xiyi's evaluation, Emperor Xuan's efforts remained subject to the trend of degeneration:

Thanks to the emperor's concentration and sharpness and the temporary reinvigoration, there were definitely abundant results, but he was not able to fulfill the principle of coherence of this heart so as to advance into the realm of transmitting the heart and "purifying and unifying" [this is a quote from the

---

32. I am borrowing the term "dehistoricizing thrust" from Terry Eagleton (*Ideology*, 59). I find it particularly appropriate as a characterization of Learning of the Way ideology since I take dehistoricizing to mean the *reduction* of historical factors. History was relegated to the background, reduced in Learning of the Way ideology, but not obliterated as was the case in the ideology imposed by Wang Anshi. Insofar as Zhu Xi recognized that moral and sociopolitical perfection, as it purportedly existed in Antiquity, cannot be reached by grand centralized planning, his approach can be seen as "a compromise with history" (Hartwell, "Historical Analogism," 690–93). However, for Zhu Xi historical events and actors had to be consistently measured against the highest moral standards, which the Learning of the Way teachers had derived from their reading of the Classics.

Sixteen-Character Dictum]. He let the Han just be the Han and stopped at this; this is regrettable!

以帝之精銳，一時之振厲固有餘用，而不能充此心之理以進於傳心精一之地. 使漢之爲漢僅止於斯. 是可慨嘆也已!

The Sixteen-Character Dictum that Lin Xiyi cited as the highest form of mental cultivation derives from *The Book of Documents*. Zhu Xi selected it as the core message of the Learning of the Way because it outlines succinctly the difference between the moral mind that pre-serves moral coherence and the human mind that is prone to human desire, and because it points out the path to the restoration of the for-mer. In the preface to his commentary on *The Doctrine of the Mean* (*Zhongyong zhangju*), he described it as the core of the Learning of the Way, a tradition that had been transmitted in these terms among the sage-kings of Antiquity.[33] After the endorsement of Zhu Xi's commen-taries on the Four Books, the phrase became a confession of faith in Learning of the Way expositions. Li Guan 李瓘 (*js.* 1268) defined the Way as follows:

This Way, which Way is it? Since the incipient forces of the human mind and the moral mind have been distinguished in the Sixteen Characters, Yu trans-mitted it to Tang, Tang transmitted it to King Wen, King Wu, and the Duke of Zhou. All of them followed this Way.[34]

斯道也何道也? 自人心道心之幾一決於十六字之間，而禹傳之湯湯傳之文武周公. 率是道也.

All Song dynasty thinkers addressed the significance of the Three Dynasties and the sage-kings. The sage-kings of Antiquity figured prominently in the Classics, which constituted the foundation of literati education. Interpretations of their significance in history differed widely, however, as the sage-kings were imagined as the embodiment of each intellectual group's central beliefs. Affiliates of the Learning of the Way portrayed them as the source of universal moral values that transcended historical time. As Lin Xiyi suggested in the passage translated above,

---

33. "The human mind is insecure; the moral mind is barely perceptible. Have utmost refinement and singleness of mind" (Tillman, *Confucian Discourse and Chu Hsi's Ascendancy*, 123; cf. Legge, *The Chinese Classics*, vol. 1, *The Doctrine of the Mean*, 61).

34. *LXSC*, 10.23b.

the goal for later emperors was to become like the sage-kings of Antiquity and relinquish their identity as Han or Tang emperors.

Chen Fuliang and the "Yongjia" teachers saw the sage-kings as wise administrators who ruled according to historicist principles. In an exposition entitled "What Are the Methods of the Kings Like?" Chen Fuliang chastised Emperor Cheng and Du Qin of the Han dynasty for passing over the recent past in favor of the rulers of Antiquity. Emperor Cheng was singled out for neglecting the legacy of his Han dynasty predecessors (Chapter 3). Chen Fuliang defended his approach toward history by projecting it as the established practice of the sage-rulers and thus the appropriate model for later generations.

Thirteenth-century expositions advocated the Learning of the Way theory of the mind by anchoring it in metaphysics and remote Antiquity. Examinees simultaneously acknowledged that the Learning of the Way was a fundamentally new movement. They relied on the eleventh- and twelfth-century treatises and commentaries of the founders of the movement to interpret the classical legacy. Expositions from the 1250s and 1260s also quoted from the work of the masters of the Learning of the Way. Because of the unofficial canonization of Zhu Xi's legacy in school curricula and the subsequent imperial approval in the first half of the twelfth century, Learning of the Way commentaries and treatises began to replace Han and Tang commentaries as the standard commentaries for examinees.

Li Guan quoted and paraphrased passages from Zhou Dunyi's *Penetrating the Changes* and *Diagram of the Supreme Ultimate* for his exposition on a passage from *The History of the Han Dynasty*. His selection from Zhou's work indicates that he understood it in accordance with Zhu Xi's reading of it in *A Record for Reflection*. The topic for Li's exposition was "How about 'the Beginnings of the Way of the (True) King,'"[35] a line from a policy essay attributed to Dong Zhongshu. Li Guan constructed his argument around the two main concepts in the title: "the beginnings" and "the way of the true king." He interpreted the beginnings (*duan* 端) as the subtle, incipient, activating forces giving rise to

---

35. Ibid., 10.21a–25a. The topic referred to a passage from one of Dong Zhongshu's policy response essays in *The History of the Han Dynasty* (Ban Gu, *Han shu*, 56.2501–2).

good and evil (*ji* 幾), a concept Zhou Dunyi had introduced in *Penetrating the Changes*.[36] "The Way of the true king," Li Guan associated with two pairs of concepts, kingship/hegemony and moral principles/desires. These oppositions were not new in the history of Chinese philosophy or unique to the Learning of the Way in the Song dynasty; the linkage between kingship and moral principle, on one hand, and hegemony and human desire, on the other, was, however, a distinctive feature of Learning of the Way discourse.[37]

Li Guan adopted Zhou Dunyi's argument that even though human beings' original mind was purely good and impartial, the state of incipient activation, which was the spring of activity, might give rise to either good or evil. In order to act morally, human beings needed to return to the original mental state of tranquility and nondifferentiation; when acting on the incipient forces, they had to remain true to the original moral nature. On the question of true kingship, Li Guan represented Zhou's argument as follows: "If our ruler goes for the good side of the incipient forces and puts that in practice, he will realize kingship in a pure fashion, and he will not mix in practices of hegemony; he will realize moral coherence in a pure way and he will not mix in desires." Li Guan's association of "the beginnings of the Way of the king" with the incipient forces may have resulted from the prominence given this concept in *A Record for Reflection*. Zhu Xi had excerpted Zhou Dunyi's ideas about tranquility and incipient activation in its first pages.[38]

Thirteenth-century examination candidates combined quotations from the authoritative sources of Learning of the Way with an explicit personal confirmation of the truth of the masters' words. Several thirteenth-century expositions ended with a direct quotation[39] from one of the newly canonized Learning of the Way thinkers and a personal

---

36. Wing-tsit Chan, comp. and trans., *Sourcebook*, 466–67.

37. On Zhu Xi's views on *ba/wang* 霸/王, see Tillman, *Utilitarian Confucianism*, 143–45. For debates on *li/yu* 理/慾, see Wing-tsit Chan, "The Principle of Heaven Vs. Human Desires."

38. *LXSC*, 10.20a–b, 22a–b; cf. Wing-tsit Chan, trans., *Reflections on Things at Hand*, 5–8.

39. In contrast to policy response essays, citations in examination expositions were usually unidentified.

confirmation of the truth of the statement.[40] Chen Yinglei 陳應靁, who won first place in the departmental examinations of 1250, concluded his exposition:

> Therefore [*Master Cheng*] *said*, "There is no starting point in the cycle of action and rest; there is no beginning to the interplay of *yin* and *yang*." *I say* that the same is the case for "to be humane" and "to gain insight."[41]
>
> 故曰:"動靜無端, 陰陽無始." 愚於仁知亦云.

The quotation from an essay by Cheng Yi, followed by the examinee's endorsement, illustrates the different status of the Learning of the Way in the thirteenth-century examination field. In the late twelfth century, Learning of the Way affiliates like Chen Deyu and Feng Yi quoted the positions of opponents and rivals of the Learning of the Way and refuted them. By the mid-thirteenth century, the antagonistic defense of Learning of the Way moral philosophy had given way to the confident display of the examinee's mastery of the Learning of the Way canon.

As the Learning of the Way became an official ideology, its stance toward other modes of scholarship in the examination field became more accommodating. The most influential thinkers of the Learning of the Way, Cheng Yi and Zhu Xi, were suspicious of the teaching of composition as a major component of examination learning. Cheng Yi had little regard for the major literary model of his time, Su Shi. Zhu Xi, too, warned his disciples about the dangers of reading Su Shi uncritically. In Zhu Xi's estimation, literary writing was valuable only if it conveyed moral self-cultivation. The study of model essays for purely stylistic reasons, irrespective of their moral content, was deemed self-destructive for those engaged in the Learning of the Way (Chapter 6).

---

40. The citation from a Learning of the Way thinker, followed by the phrase "*yu gu* . . ." 愚故 ("I therefore . . . ," "on this basis I . . ." made this argument) is a recurring pattern in thirteenth-century expositions. See, e.g., *LXSC*, 2.76a, 3.64a, 6.97b.

41. *LXSC*, 3.30a–b. For the quote from Cheng Yi, see Wing-tsit Chan, comp. and trans., *Sourcebook*, 571; Cheng Hao and Cheng Yi, *Er Cheng quanshu*, 1.1b–2a. Chen was responding to the question, "How about 'to gain insight is to be active, to be humane is to be at rest' and 'finding joy' and 'enjoying a full lifespan'?" (*Analects* VI:21; Legge, *The Chinese Classics*, vol. 1, *The Confucian Analects*, 192). Chen explained "to gain insight" (*zhi* 知), "to be humane" (*ren* 仁), "finding joy" (*le* 樂), and "enjoying a full lifespan" (*shou* 壽) with terms commonly used in Learning of the Way philosophy: "action" (*dong* 動), "rest" (*jing* 靜), "essence" (*ti* 體), and "effect" (*xiao* 效).

For many of Zhu Xi's contemporaries with an interest in the Learning of the Way and for Learning of the Way affiliates attempting to institutionalize Learning of the Way thought in examination preparation, the Tang and Northern Song paragons of Ancient Prose and Southern Song literary talents like Chen Fuliang retained their appeal. Given the proper ideological framework, elements of the prose of Han Yu, Liu Zongyuan, and Su Shi and of other established literary models taught in twelfth-century anthologies could be used to express Learning of the Way truths persuasively.[42]

The same belief underlay the compilation of *Standards for the Study of the Exposition*, the anthology on which much of the research for this section is based. The collection was published in the last decade of Song rule. Through it, students preparing for the last examinations held during the Song dynasty learned about the authoritative position of Zhu Xi's commentaries on the Four Books and the philosophy of the Song masters of the Learning of the Way in the explication of passages from the Classics, the philosophers, and the histories. The top expositions for the departmental examinations from the previous three decades and the most recent successful Imperial College expositions included in this anthology showed how the Learning of the Way canon had become the standard body of reference for textual exegesis in the examinations.

At the same time, *Standards for the Study of the Exposition* exemplified how teachers and editors combined the teachings of the Learning of the Way with the strengths of its erstwhile rivals in their curricula and textbooks. Whereas twelfth-century proponents of the Learning of the Way objected to the situational arguments furthered by the grand master of exposition writing, Chen Fuliang, Wei Tianying, the editor of *Standards for the Study of the Exposition*, included an inordinate number of his expositions. Wei incorporated Chen's model essays in an anthology otherwise devoted to actual examination expositions, because he believed that the analysis of Chen Fuliang's work remained a prerequisite for the mastery of the genre. He noted the influence of Chen's "writing method" on later essays, including those advocating Learning of the

---

42. For a discussion of the Ancient Prose of Han Yu, Liu Zongyuan, Ouyang Xiu, and Su Shi, see Yu-shih Chen, *Images and Ideas in Chinese Classical Prose*.

Way values. As we saw in the previous chapter, the official recognition of the Learning of the Way resulted in the integration of the study of the Learning of the Way canon with pre-existing types of examination preparation, particularly Ancient Prose composition, in examination curricula. Thirteenth-century examinees adopted the ideology of reconciliation advocated by examination teachers and encouraged by examiner standards.

## *Standards for Policy Response Essays*

### MORAL LEARNING, POLITICAL CRITICISM, AND GENEALOGICAL DISCOURSE

"It is a shame not to practice one's moral principles when at court."[43] In a postscript to an examination question posed in an internal Imperial College examination in 1181, Wei Liaoweng 魏了翁 paid tribute to Shen Huan 沈煥 (1139–91) for having his students ponder the implications of this passage from *Mencius*.[44] Shen Huan had won acclaim for instructing students by moral example, but a censor took offense at his question. He accused Shen Huan of spreading sectarian teachings and had him demoted to a professorship in Gaoyou 高郵, Huainan East circuit 淮南東路.[45] This incident was one of many in the conflict in the central government over the rise of "the learning of nature and coherence" starting in the 1170s. The concern with the moral stature of officials in policy questions characterized the growing influence of the moral philosophy of the Cheng brothers.

More so than other examination genres, such as the essay on the meaning of a passage from the Classics or the exposition, policy questions stimulated political and intellectual controversy. The format of the policy question engaged examiners and examinees in debates about a

---

43. Wei Liaoweng, *Heshan ji*, 62.14a–b. *Mengzi* VB:5; Legge, *The Chinese Classics*, vol. 2, *The Book of Mencius*, 384.

44. Shen Huan was one of the Four Masters of Mingzhou. For further information on the Four Masters, see Walton, "The Institutional Context of Neo-Confucianism," 468–72.

45. *SS*, 410.12338–39.

wide range of issues concerning government or scholarship. Like the memorial, it was a powerful and prestigious means to discuss policy alternatives to contemporary administrative issues or to vent political protest.

Others have rightfully pointed out the Song government's use of the examinations as a means to promote its political agenda and as a channel for partisan recruitment.[46] Imperial and central government decisions clearly affected the examinations during the campaign against False Learning or after the full-scale adoption of the Learning of the Way; yet, the history of the Learning of the Way equally shows that thirteenth-century decisions were molded by more complex developments in the examination field.

Designing policy questions, writing policy response essays, and circulating them factored into twelfth- and thirteenth-century intellectual and political debates. This section first discusses the dissemination of Learning of the Way criticisms of political ethics and examination preparation through examination questions. Next, it traces the debate over the representation of Cheng Learning in twelfth-century policy questions and response essays. Zhu Xi's challenge to other types of learning compelled his rivals to question his definition of true learning in their examination questions.

Before the impact of Cheng Learning became visible in higher-level examinations, its advocates had been appealing to students at local and private schools. During his first appointment, as assistant magistrate in Tongan 同安 (Quanzhou 泉州, Fujian) in the early 1150s,[47] Zhu Xi published several announcements in which he admonished local students against a preoccupation with examination skills and examination success. He notified them that the questions he would ask of them were intended to gauge their personal understanding. The questions were designed to redirect students' attention from the textual analysis of passages from the Classics and the histories to an awareness of their own progress in moral learning. He justified the announced changes in the tests at the local school on the grounds that his questions were in

---

46. Ning, "Songdai gongju dianshi ce yu zhengju."

47. For Zhu Xi's years in Tongan, see the account in Shu Jingnan, *Zhuzi da zhuan*, 116–40.

keeping with the spirit of Antiquity, when "responding to questions" had served as a guide to effective moral cultivation.[48]

The policy questions Zhu Xi designed for the students of the Tong-an District School marked a radical break with contemporary conventions in content and form. The vast majority of the thirty-three questions were about learning. In contrast, two-thirds of Lü Zuqian's policy questions and all but one of Chen Fuliang's surviving questions dealt with administrative issues, institutions, and history.[49] Zhu Xi's questions also differed from those of his intellectual rivals in terms of their approach to the subject matter. Their questions were long, listing specific cases for the students' consideration. As we saw in Chapter 3, a question from Lü Zuqian about *The Analects* focused on specific passages from the text and required students to solve paradoxical statements about humanity and sagehood. Lü's question was typical of the "points of doubt in the Classics" genre (*jingyi* 經疑), a standard question pattern analyzed in Zeng Jian's *Secret Tricks for Responding to Policy Questions*:

Questions about points of doubt in the Classics are of just two kinds: there are those that cite contradictory passages from the Six Classics to pose questions; and there are those that ask about agreements and disagreements in scholars' commentaries.[50]

經疑之問不過兩說. 有援引六經自相牴牾爲問. 有即諸儒註疏異同爲問者.

Zhu Xi asked the Tongan students the following question about *The Analects*:

Recently, you students have been devoting yourselves to the text of *The Analects*. You have pondered all that is said in its twenty chapters. Even though you have not as yet analyzed the meaning of the text as thoroughly as you have the text itself, still, one cannot say that you have not applied yourselves to this [investigating the meaning]. You have already firmly absorbed the abstract and the general parts in your minds, but can you also let the examiner know how

---

48. Zhu Xi, *Zhu Xi ji*, 74.3868–73. Cf. *Mengzi* VIIA:40.

49. My calculation is based on a survey of questions in Lü Zuqian's *Donglai ji* (thirty-four questions) and Chen Fuliang's *Zhizhai ji* (fourteen questions).

50. Zeng Jian and Liu Jinwen, *Dace mijue*, section 11.

you have applied this in practice? If not, you would be studying here not because you have your minds set on the path of morality but only because of self-interest; this is not what I expect from you.[51]

頃與二三子從事於論語之書. 凡二十篇之說者, 二三子盡觀之矣. 雖未能究其義如其文, 然不可謂未嘗用意於此也. 惟其遠者大者二三子固已得諸心而施諸身矣. 亦可以幸教有司者耶? 不然, 則二三子之相從於此, 非志於道, 利焉而已耳. 非所望於二三子也.

The question was strikingly short. Zhu Xi did not list specific passages. Rather, his pedagogy was aimed at the formation of moral subjects. He believed a systematic, careful, and engaged reading of texts was central to this project. The standards applied in the question foreshadowed Zhu Xi's guidelines for reading and learning as set out in later conversations with students. Students were asked to reflect on the text as a whole and connect its meaning to their own lives. The emphasis on personal development and moral cultivation conflicted with textual analysis as practiced in twelfth-century examination preparation. Zhu Xi attempted to integrate exegesis of the classical text and reflection on individual practice in examination preparation. He maintained that the student's application of moral reasoning to his own life contributed to what Zhu Xi saw as the ultimate goal of the civil service examinations, namely, the recruitment of morally superior men for office.

Zhu Xi's efforts to make students focus on the larger implications of what they were studying in examination preparation classes were most evident in the following question—this must be the shortest policy question asked in the Southern Song dynasty, if not the whole of the history of the Chinese civil service examinations:

When people are young, they learn things; after they have grown up, they want to put these things into practice. Can we hear from you gentlemen what you are learning today and what you will put into practice in the future?[52]

人幼而學之; 壯而欲行之. 諸君子今日之所學, 他日之所以行, 其可得聞歟?

Apart from the questions on textual exegesis, Zhu Xi's policy questions on institutions and current affairs also challenged standard exami-

---

51. Zhu Xi, *Zhu Xi ji*, 74.3880.
52. Ibid.

nation practice. Zhu Xi was adamant about the relevance of his teachings to the selection of officials. His questions included a radical critique of the civil service examinations. In the six questions dealing with education and selection, he drew a sharp contrast between his time and Antiquity, when moral training had formed the core of the curricula at schools of all levels and moral stature constituted a determining factor in the recruitment of officials.[53] The establishment of the *jinshi* track in the Song dynasty as the "sole" channel for entry in the bureaucracy and the exclusive focus on texts and writing fostered by this institution were, in his view, key problems, and he wanted students to come to grips with them, for, "in a scholar, writing is of secondary importance, it does not lead to *a firm distinction between the morally superior and the morally inferior.*"[54]

Zhu Xi's preoccupation with the definition of the profile of the morally superior carried decidedly political overtones. Examination regulations forbade the discussion of contemporary court politics. In his instructions for the policy response tests, Zhu Xi claimed to abide by those rules and urged students not to exceed their station by discussing court politics. He did, however, not hesitate to ask sensitive questions about the moral quality of contemporary officialdom:

The censors function as the eyes and ears of the Son of Heaven. They have license to discuss all matters in the subcelestial realm. In the last ten years or so, matters of recruitment have been decided by the private opinion of the grand councilor. He has filled the ranks of the censors with the dullest, the most opportunistic, and the most shameless men of our generation; they have formed a faction and made alliances in order to engage in their treacherous practices together.[55]

臺諫天子耳目之官. 於天下事無所不得言. 十餘年來用人出宰相私意. 盡取當世頑頇嗜利無恥之徒以充入之. 合黨締交. 共為姦慝.

The question was a direct criticism of the administration of Qin Gui, which ended in 1155, during Zhu Xi's tenure in Tongan. From the late 1130s to the early 1150s, Qin Gui and his allies launched a series of attacks on his political rival, Zhao Ding, and purged Zhao's associates.

---

53. Ibid., 74.3875, 3876, 3881, 3882 (questions 28 and 31 are about schools: questions 8, 10, 13, and 27 concern selection).

54. Ibid., 74.3877.

55. Ibid.

Zhao Ding had found a political common ground with advocates of Cheng Learning. As a result, Qin Gui combined the overhaul of the bureaucracy with a campaign against confined learning (see Chapter 5).[56] In another policy question, Zhu Xi asked about the rationale behind the recent denunciation of "the Learning of Specialized Schools." Against official allegations of arrogant Chengist claims to exclusive truth, Zhu Xi pitted his own argument that the teacher held a central position in the transmission of specialized arts and skills. He asked students to evaluate his claim that scholars should commit themselves to the learning that had been transmitted to them and, if need be, resist the government's reaction to this claim.[57]

After his tenure in Tongan, Zhu Xi continued his educational offensive (Chapters 5 and 6), but he never again relied on examination questions to remold students' patterns of thought and action. Zhu Xi may have felt that drawing up examination questions for the district school students in Tongan was part of his official responsibilities. He could rename the classrooms, substituting names like "setting one's mind on the way of morality" (*zhi dao* 志道) and "relying on humanity" (*yi ren* 依仁) for previous appellations like "the advancement of wise men" (*huizheng* 彙征),[58] which, in his view, promoted careerism.[59] He could lecture on *The Analects* in a way that perplexed all students.[60] Yet, he could not omit written examinations. After he established himself as a private teacher, he removed examination preparation from the curriculum and discussed it only in contradistinction to the Learning of the Way (Chapter 6).

The questions Zhu Xi designed in Tongan suggest that, despite his unwillingness to present his teaching as examination preparation and his unwillingness to instruct students in writing examination essays when not serving in office, he assumed that examination writing could be modified to serve the pedagogical and political goals of the advo-

---

56. Schirokauer, "Neo-Confucians Under Attack," 164–66.

57. Zhu Xi, *Zhu Xi ji*, 74.3877–78.

58. The expression is based on the line commentary for nine at the beginning under the hexagram "peace" (*tai* 泰) in *The Book of Changes*. Cf. Wilhelm, *The I Ching*, 49.

59. Zhu Xi, *Zhu Xi ji*, 74.3868.

60. Ibid., 39.1753; quoted in Shu Jingnan, *Zhuzi da zhuan*, 129.

cates of Cheng Learning. Examination writing was more than just a means of defending Cheng Learning against its political enemies; it also had a role to play in the increasingly acrimonious debate about the meaning of Cheng Learning among scholars sympathetic to it. Zhu Xi played a key role in this debate. His attempts to define his version of "the Learning of the Way" as the true transmission of the Way caused alarm among literati sympathetic to Cheng Learning as well as among the ruling factions at court. Those sympathetic to Cheng Learning questioned Zhu Xi's narrow definition of the Learning of the Way, and he responded in kind. This exchange is a clear example of my larger contention that examination preparation and writing provided a forum for literati debates on contemporary scholarship.

In the only policy question Zhu Xi designed after his term of office in Tongan, he asked students what tradition represented the true Learning of the Way.[61] The question was occasioned by the reconstruction of the White Deer Grotto Academy in 1180.[62] As the prefect of Nankang (Jiangnan East circuit), his second significant post after two decades of private study and teaching, he used official funds to restore this private academy on Mount Lu. The White Deer Grotto Academy had ranked among the top educational institutions early in the Song dynasty, but it had fallen into disuse following the central government's repeated efforts to organize education into a network of official schools. Zhu Xi repeatedly involved local governments in the restoration of private schools. This effort signaled a further attempt to imbue official education with the spirit of the Learning of the Way.[63]

In his question, Zhu Xi singled out three issues related to the transmission of the Way for the consideration of the academy's students. The first concerned the contested position of Mencius in the transmission of the Way. Zhu Xi prefaced the question with the assertion that after the death of Confucius and his disciples, the followers of Yangzi

---

61. Zhu Xi, *Zhu Xi ji*, 74.3884.

62. Tillman, *Confucian Discourse and Chu Hsi's Ascendancy*, 108–14; Chen Wenyi, *You guanxue dao shuyuan*, 29–42; Chaffee, "Chu Hsi and the Revival of the White Deer Grotto Academy."

63. For a history and analysis of the impact of official schools on academies and the influence of academy teaching on official schools, see Chen Wenyi, *You guanxue dao shu-yuan*. She discusses Zhu Xi's role in the "academy movement" in chap. 2.

and Mozi dominated the scene. Mencius clarified the learning of Confucius to rectify the situation, but the transmission of his teachings was cut short after his death. Having established the primacy of Mencius in the transmission of Confucius' learning, he asked students to discuss the reasons for the recent controversy over the position of Mencius.

Zhu Xi's question about Mencius' position in the transmission of the Way was the reverse of questions posed by some of his intellectual challengers. In a policy question on the connection between schools of learning and orderly government, Chen Fuliang dealt with the same subject matter.[64] What needed to be re-examined, in his mind, was the elevation of Mencius above other eminent intellectuals and the unanimously accepted argument that true learning was not transmitted after his death. Chen Fuliang questioned whether this was really so and entertained the possibility that the exaltation of Mencius since Han Yu had prevented the presentation of different arguments. He cautioned examinees not to rely on other people's arguments in answering this question.

Similarly, Lü Zuqian remarked in a question on the transmission of the Way that Han Yu's severance of the transmission after Mencius in "Finding the Origins of the Way" implied a rather narrow definition of the tradition, whereas Han Yu's claim to have succeeded to the legacy of Confucius, Mencius, and Yang Xiong seemed rather generous.[65] Lü Zuqian cited conflicting evidence from a variety of sources. In contrast with Zhu Xi, he asked his examinees to consider a variety of figures in their attempt to draw a genealogy of transmitters of the Way.

Second, Zhu Xi asked students to evaluate the various types of learning that had gained prominence since the eleventh century. Zhu Xi noted that the intellectual scene had never been so vibrant as during the Song dynasty when the learning of Ouyang Xiu, Wang Anshi, and Su Shi "gained influence at court" and when the teachings of Hu Yuan and Cheng Yi were "passed on to students." He indicated that there were differences among these teachings and noted that Cheng Yi found the thought of Wang Anshi and Su Shi particularly uncongenial. The central question in the definition of true learning was who among

---

64. Chen Fuliang, *Zhizhai ji*, 43.17b–19b.
65. Lü Zuqian, *Donglai ji*, waiji, 2.1b.

these men had succeeded to the Way of Confucius and who had lost sight of it.

Zhu Xi knew from his experience in Tongan that students were reading the works of all these men, especially those of Su Shi and Wang Anshi, in preparation for the examinations. From the beginning of his teaching career, he made students reconsider the authorities on which their education was based. In one of the questions designed for the Tongan students, he reminded them that classical authorities warned against prioritizing composition skills over morality in choosing models to emulate:

Mencius said, "Reciting their poetry and reading their writings without knowing their personality, is this permissible?"[66]

孟子曰: "頌其詩, 讀其書, 不知其人, 可乎?"

Zhu Xi listed the various masters and asked students to evaluate the moral worth of each in order to determine whether they figured in the genealogy of true learning.

In the question mentioned above, Chen Fuliang also wrote a review of the most prominent scholars of the eleventh century. He pointed out the strengths of their respective legacies: students of Ouyang Xiu excelled in literature and government; Sun Fu and Shi Jie were models of moral excellence and brilliant classical scholarship; Hu Yuan trained students who proved to be effective administrators; Zhou Dunyi, the Cheng brothers, and Zhang Zai clarified the essentials of the teachings of Confucius and Mencius. Chen Fuliang asked his examinees to consider the contributions of all these scholars. Like Zhu Xi, he explicitly limited the definition of "the Learning of the Way" to the Four Masters, but he was critical of those claiming it as an exclusive identity. He countered claims from some followers of the Learning of the Way that "ever since the Duke of Zhou and Confucius, the masters from Luo have been the only wise men."

Chen Fuliang concurred with some of the criticisms court officials leveled against the Learning of the Way. In his question, he chided

---

66. This is the opening sentence of a policy question on the transmission of true learning Zhu Xi designed for district school students in Tongan (Zhu Xi, *Zhu Xi ji*, 74.3874; cf. *Mengzi* VB:8; Legge, *The Chinese Classics*, vol. 2, *The Book of Mencius*, 392).

those followers of the Learning of the Way who "behave politely out-
wardly and pretend they agree," but, when they cited passages from *The
Great Learning*, such as "extending knowledge and investigating things,"
had "no facts to argue about and no evidence that can be examined."
He stressed that the moral philosophy of the Learning of the Way was
of great value but encouraged students "not to get fixed on one or two
ideas." Chen's question was representative of the ambivalent attitude of
the "Yongjia" teachers toward the legacy of the Cheng brothers. The
"Yongjia" teachers shared the Chengist view that the moral reform of
the emperor and high officialdom was essential to the reinvigoration of
the Song state. They also shared the view that the classical texts the
Cheng brothers had selected to propagate the message of moral cultivation contributed to this end. They were, on the other hand, averse to
exclusivist claims for the truth of Cheng learning. For the "Yongjia"
teachers, Cheng learning was but one of many schools in the intellectual legacy of the past that provided answers to contemporary issues.[67]

The third issue that factored into the discussion of true learning was
the impact of Daoism and Buddhism. Although there was more consensus among Zhu Xi's peers about the negative side-effects of Daoist
and Buddhist learning, he proved most strident. Unlike his peers, he included a discussion of Daoism and Buddhism in a more general question about the transmission of the Confucian Way to underscore the
need for doctrinal purity in Chengist circles. Lu Jiuyuan, another con-
tender in the debate to define the scope and meaning of Cheng Learning, proposed a conciliatory approach toward Buddhism in examining
students.[68]

Zhu Xi's 1180 question fit in with his antagonistic campaign for the
Learning of the Way, which occupied him during the last two decades
of his life. He strongly encouraged students to explain the validity of a
genealogy of transmitters of the Way that linked Mencius to Cheng Yi.

---

67. De Weerdt, "The Ways of the Teacher." Zhu Shangshu (*Songdai keju yu wenxue
kaolun*, 436) notes that Chen Fuliang's expositions were influenced by "Lixue" but does
not explain how he interprets this influence and how it manifested itself.

68. Lü Zuqian (*Donglai ji,* waiji, 1.11a–12b) asked one question on the differences between deviant teachings and the Way. For a translation and discussion of a policy question by Lu Jiuyuan on "the discussion of deviant teachings," see Foster, "Seeking a
Tradition," 9–14, 17–19; and Lu Jiuyuan, *Lu Jiuyuan ji,* 24.288–89.

This was the genealogy that Zhu Xi defined as the transmission of true Confucian learning in scholarly exchanges and in his published work. This genealogy and the idea of a genealogy that excluded other types of learning more generally ran counter to the inclinations of his powerful rivals in literati culture, including those who shared his interest in Cheng Learning. The questions of Chen Fuliang and Lü Zuqian encouraged students to be critical of narrow definitions of "the Learning of the Way." They asked students to discuss the pros and cons of different branches in the history of Confucian learning and to incorporate Cheng Learning in a broad and nonexclusive tradition of learning.

## The Relevance of the Learning of the Way in Thirteenth-Century Policy Debates

The stridency of Zhu Xi and his followers fueled arguments about the infiltration of Learning of the Way discourse in examination papers and motivated the campaign against the Learning of the Way under Han Tuozhou (Chapter 5). Court debates about the impact of the Learning of the Way continued into the thirteenth century. Whereas the political debates of the twelfth century were concerned with the sheer appearance of the Learning of the Way language in examination essays, debates over the next few decades focused on the relevance of Learning of the Way tenets to the administrative problems of the Song government. Learning of the Way political theory was acknowledged as a legitimate, albeit not necessarily satisfactory, competitor with other approaches in policy questions and response essays.

After the rehabilitation of those persecuted in the early 1200s, Learning of the Way advocates felt confident enough to expound the political message of the Learning of the Way in their policy questions and response essays at the departmental and palace examinations. Zhen Dexiu stood at the forefront of those candidates and examiners who inquired into the political theory of the Learning of the Way and maintained that it provided the key to the rehabilitation of Song administration.[69]

---

69. For biographical information on Zhen Dexiu, see Chu Ron-Guey, "Chen Te-Hsiu and the 'Classic on Governance,'" esp. the introduction and chap. 1; de Bary, *Neo-*

In 1199, Zhen Dexiu obtained the *jinshi* degree at the very young age of twenty-two. Barely six years later, he passed the polymath (*boxue hongci* 博學宏詞) examinations, a success that qualified him for secretarial posts in the capital.[70] Zhen spent most of the rest of his life in office. His political career was interrupted for a seven-year period of retirement (1225–32) due to his protest against the government of Councilor Shi Miyuan 史彌遠 (1164–1233). Zhen's prolonged presence in the capital and his active political profile made him the most conspicuous of the thirteenth-century advocates of the Learning of the Way.

Zhen Dexiu passed the *jinshi* examinations while the ban on False Learning was in effect. In examination essays and questions written after the ban was lifted, Learning of the Way moral philosophy significantly impacted the discussion of contemporary governmental problems. Zhen's 1205 policy response for the polymath examination was an early example of the new, outspoken advocacy of a Learning of the Way program for political reform in examination writing. In the question, the examiner stressed the severity of current economic and military problems.[71] He noted that problems such as the slackening of military discipline, the unruliness of local militias, insufficient funds, and the impoverishment of the people had been around for a while but cited evidence that the situation was worsening. He entreated his examinee to demonstrate "practical learning" (*you yong zhi xue* 有用之學) by outlining detailed and practical proposals appropriate to the current situation. A final instruction emphatically encouraged the candidate to speak out frankly, without fear of the effect of his words on his future career.

Neither point was lost on Zhen Dexiu. He maintained that the cultivation of the mind, central to the Learning of the Way, was the only practical method to reunite the forces tearing Song society asunder. In his view, the moral malaise pervading all levels of society was the worst threat to the security of the Song dynasty. He cited healing the rifts between the outer court and the inner court and the disputes, especially

---

*Confucian Orthodoxy and the Learning of Mind-and-Heart*, esp. 83–87; *SB*, 88–90; and Tillman, *Confucian Discourse and Chu Hsi's Ascendancy*, 241–45.

70. Langley, "Wang Ying-Lin," chap. 2; De Weerdt, "Aspects of Song Intellectual Life," 6.

71. Zhen Dexiu, *Xishan wenji*, 32.3b–5b.

over diplomatic issues, between "petty men" and those rehabilitated after the attack on False Learning as the key factor on which the success of the efforts to implement change would depend. In Zhen's view, "useful learning" was not primarily about the discussion of administrative measures; solutions to these problems could bear results only if a disciplined body politic, starting with the emperor and the court, was in charge.

The source for Zhen Dexiu's program of political reform was *The Great Learning*. In court audiences and memorials, he upheld this Learning of the Way classic as a complete guide for imperial government. In 1234, following his reappointment to prestigious central government posts such as minister of revenue and Hanlin academician, Zhen Dexiu presented his programmatic reading of the text to Emperor Lizong. He titled his work *The Extended Meaning of "The Great Learning"* (*"Daxue" yanyi* 大學衍義). This work, first printed in 1229, was the crystallization of years of thought and lecturing on *The Great Learning*.[72] During his short tenure as Classics mat lecturer in 1225, Zhen delivered presentations on this text on at least sixteen occasions.[73]

In *The Extended Meaning*, Zhen discussed four stages in the eight-step program that linked moral self-cultivation with the realization of peace and order in the empire. He began with the internal aspects of moral cultivation. Among the related four steps in the original text of *The Great Learning*, Zhen focused on the first two ("the investigation of things" and "the extension of knowledge"); he left out "making the will sincere" and "rectifying the mind." He then covered the outward manifestation of self-cultivation associated with the last four steps in the original sequence. Again he focused on the first two steps ("the cultivation of the body" [in Zhen's interpretation], and "the regulation of the family"). Zhen omitted the last two steps in the sequence, "ordering the state" and "keeping the empire at peace." He did so intentionally because he believed that the emperor and the court needed to apply themselves to the first stages rather than focus on administrative questions. The omission of the last two topics in the sequence underscored

---

72. For a detailed discussion of Zhen Dexiu's work on *The Greater Learning*, see de Bary, *Neo-Confucian Orthodoxy and the Learning of the Mind-and-Heart*, 106–35; see also Hervouet, *A Sung Bibliography*, 215–16.

73. De Bary, *Neo-Confucian Orthodoxy and the Learning of the Mind-and-Heart*, 85–98.

Zhen's argument that the moral reformation of the court was the key to solving the Song dynasty's administrative issues, including its military and financial troubles.

As examiner for the departmental examinations of 1235, Zhen Dexiu asked examinees to discuss the model of rulership outlined in *The Great Learning* and to analyze contemporary history on the basis of this model.[74] The question clearly suggested the line of argument Zhen Dexiu expected his examinees to pursue. He hailed the renewed interest of Emperor Lizong in the records of the sage-kings and in their methods of the cultivation of inner virtue. He saw the emperor's interest in self-cultivation as evidence that progress was being made with regard to the first two steps in the program for good government outlined in *The Great Learning*, "the investigation of things" and "the extension of knowledge." He noted that the emperor appeared more solemn and exercised more self-restraint in the luxuries of "carriages and robes" and in entertainment than he had earlier in his reign. This was a sign that he was advancing on the path of "the cultivation of the body / outward appearance" (*xiu shen* 修身), the fifth step in the original program. Mind-cultivation also influenced his management of the imperial household: he no longer visited the women's quarters to indulge in selfish desires. This meant that he was making headway on "the regulation of the family" (*qi jia* 齊家), the sixth step.

In *The Great Learning* model, the emperor's efforts should have produced administrative results. Zhen admitted that, despite attempts to reinvigorate the polity and society, "we can't say that the state is well-ordered" (國治) or "that the empire is at peace" (*tianxia ping* 天下平). The last two steps in *The Great Learning* program for political reform appeared not to follow automatically on the earlier steps. Zhen asked examinees to explain why the sovereign's virtue had not produced a visible effect on the Song's financial and military problems, but he denied that the current situation exposed flaws in the political theory of *The Great Learning*.

He cited historical cases to help the examination candidates examine the reasons behind the contradiction. The cases of two Han and two Tang emperors demonstrated that, despite their intentions to clear their

74. Zhen Dexiu, *Xishan wenji*, 32.29a–31a.

minds of unprincipled thoughts, they failed to carry through their reso-
lutions. In the logic of the sequence from the investigation of things to
the ordering of the empire, the ruler's failure to fully manifest his virtue
would have a negative impact on society. Zhen Dexiu's question carefully guided students' analysis of the Learning of the Way theory of
moral cultivation and pre-emptively silenced the criticism of those who
focused on the results of his proposed program of moral reform.

Zhou Mi and Huang Zhen 黃震 (1213–80), writing after Zhen's death,
recounted that many had invested high hopes in Zhen Dexiu and his
project but grew impatient with his inability to implement change after
his successive appointments.[75] Zhen's political program was questioned
at court and in examination questions in the 1230s. Changes in the international arena, particularly the disruption in the north caused by the decline of the Jin empire and the rise of Mongol power, resulted in intense
debates about issues of war and peace, the formation of alliances, and
Song military organization. The strong opinions voiced at court rever-
berated through the examination halls and school compounds.

Wu Yong 吳泳 (*js.* 1208), who participated in these debates at court,
questioned the participants in decree examinations, Imperial College
examinations, and local examinations on military organization, regional
armies (the armies on the Yangzi and Huai rivers), local militias, and
strategic matters.[76] In 1234 he asked Imperial College students sitting
the lower-level examinations there to reflect on the apparent recent
success of the Song troops in recapturing the northern cities of Kaifeng
and Luoyang and confronted them with the policy options facing the
Song government at the time. Wary of the increasing influence of
Learning of the Way ideology on examination writing and on court debates on these matters, Wu Yong admonished candidates not to rely on
arguments relating defense to the cultivation of the mind.[77]

---

75. Yang Yuxun, "Nan Song Lizong zhong- wanchao de zhengzheng," 8. For a discussion of how Zhen Dexiu's commitment to the Learning of the Way shaped his
stance on foreign policy, namely, his insistence on self-strengthening, see Zhu Hong,
"Zhen Dexiu ji qi dui shizheng de renshi."

76. For Wu Yong's participation in court debates about war and peace, see Huang
Kuanchong, "Wan Song chaochen dui guoshi de zhengyi," 30, 41, 57, 67*n*12, 89, 124–25.
For his policy questions, see Wu Yong, *Haolin ji*, 33.9a–19a.

77. Wu Yong, *Haolin ji*, 33.19a.

Similarly, in a question written for a decree examination in 1235, the year in which Zhen wrote his departmental examination question, Zhao Rutan 趙汝談 (ca. 1160–1240) expressed a low regard for those advocates of the Learning of the Way who relied on *The Great Learning* to solve the Song court's military problems:

Yet, the scholar-officials are chanting the subtle words from *The Great Learning* about "extending knowledge" and "investigating things" haughtily and in unison, and they speak very highly of the great plans to recover the land. This, my mind cannot grasp.[78]

而搢紳先生方且雍雍然峩峩然交誦致知格物之微言, 深贊佳兵闢土之偉畫. 此愚心所竊怪而絕不喻者也.

Despite Wu Yong's and Zhao Rutan's skepticism about the relevance of Learning of the Way moral philosophy to the Song's military and financial problems, Zhen Dexiu's program of political reform and his promotion of Learning of the Way moral and political values in his teaching and administrative career inspired students to follow in his footsteps.[79] He provided support for those who dared to speak up in favor of the Learning of the Way in their examination essays. He funded the publication of the policy response essays of a former student whose work he thought deserved a better ranking.[80] He wrote a review on the essay of another palace examination graduate, praising his courage in "spitting out what he had learned in daily practice" and his ability to distinguish between "self-interest and rightness."[81]

Wang Mai, a student of Zhen's who participated in the decree examination of 1235, waged an outspoken defense for Learning of the Way values. He maintained that *The Great Learning* was of the greatest importance to the reinvigoration of the Song body politic. He reasoned, as had Zhu Xi and Zhen Dexiu before him, that the ousting of morally inferior men from officialdom was a priority. Xu Yuanjie 徐元杰 (?–1245?), another disciple of Zhen's, defended Zhen's program of political

---

78. Wang Mai, *Quxuan ji*, 1.24b.

79. For a discussion of Zhen's involvement in examination preparation manuals, see Chapter 6.

80. Poon, "Books and Printing in Sung China," 108.

81. Zhen Dexiu, *Xishan wenji*, 36.7b–8b. I have found no further information about the author of the essay, Huang Ruyi 黃汝宜.

reform in the 1232 palace examinations.[82] He argued that moral self-cultivation was the permanent foundation of good government.[83] Unity, based on the Learning of the Way vision of the moral rehabilitation of self, polity, and society, was the main item on the political agenda of early thirteenth-century advocates of the Learning of the Way. The implementation of the program of moral reform outlined in *The Great Learning* was the means for creating the united bureaucracy that was to restore Song power. Both Wang Mai and Xu Yuanjie succeeded in the examinations. Their success demonstrated that by the 1230s Learning of the Way arguments had become legitimate in examiners' eyes, even if they were skeptical about their applicability to current administrative questions.

Despite court acceptance of Learning of the Way political proposals, the focus on moral behavior continued to entwine Learning of the Way advocates in factional disputes. Examiners and examinees with Learning of the Way inclinations continued to view examination writing as an appropriate channel for voicing political protest. Following Zhu Xi's example in Tongan, they voiced criticism of court officials in local and school examinations.

In the 1230s administrators of school examinations had students offer opinions on court politics. An examiner at the Shaowu 邵武 Prefectural School (Fujian) told his students:

We who reside at the local school also have a responsibility to discuss government. Suppose the place is different and you find yourself in the company of wise men. I wonder, what should we learn from the words of Confucius and Mencius, and what should we pick up from the conduct of the upright men of our dynasty? Let's investigate this together. Don't just say that those who are not serving yet should not discuss the issue of presenting oneself and offering one's resignation.[84]

吾儕身游鄉校; 不無議政之責. 使易地而處諸賢之地, 不知於孔孟之言宜何所師, 我朝先正言宜何所法? 當相與索言之. 毋但曰未仕者不當議進退!

---

82. *SB*, 429.
83. Xu Yuanjie, *Meiye ji*, 5.1a–2ob.
84. (*Jingxuan*) *Huang Song Cexue shengchi*, 9, essay no. 1.

The question could be read as an attempt to make students reflect on political responsibility, but it was more than an innocent exercise in moral theory. According to the editor, the question was occasioned by the appointment of a new councilor. No names were mentioned in this case, but another essay from the Shaowu Prefectural School suggests that the criticism was part of the protest over the rise of Shi Songzhi 史嵩之 (1189–1257) in the late 1230s. Shi Songzhi's appearance at court, first as assistant councilor in 1238 and soon afterward as grand councilor of the right in 1239, resulted in massive discontent.[85] It had been only six years since Songzhi's uncle, Grand Councilor Shi Miyuan, had succumbed in his official residence in Lin'an. The specter of another autocratic councilor from the Shi family inspired the examiner's critical question.

The examiner awarded first place to the essay of a student who argued that, under the rule of an evil minister, officials should either hang on and speak out or, if they could not speak out, leave their post and make a statement by setting a moral example at home. The question and the essay expressed disapproval over the silence and indecision of high officials such as the grand councilors, supervising secretaries, and attendants to the emperor.

In another question from the Shaowu school examinations, the examiner asked students to reflect on the potential harm of having another member of the Shi family as councilor of state and to evaluate current prospects on the basis of a comparison with two episodes from the early Southern Song period. The first episode concerned the issue of imperial succession. The examiner juxtaposed the exemplary relationship between Emperors Gaozong and Xiaozong to the purge and subsequent murder of Zhao Hong 趙竑 (d. 1225) in 1224–25. Zhao Hong was first in line to succeed Ningzong, but court attendants orchestrated the enthronement of Zhao Yun 趙昀, better known to history as Lizong. Zhao Hong was sent away from court, accused of planning a rebellion, and summarily executed—many pointed to Grand Councilor Shi Miyuan as the mastermind behind the scheme. The cruel fate of the prince remained a sensitive issue; memorials were written throughout the following decades demanding his reinstatement and the

---

85. Davis, *Court and Family in Sung China*, 148.

adoption of an heir to his line.[86] The second episode centered on the threat posed by powerful councilors. The question recounted the negative impact Qin Gui had had on proper administrative procedure and asked students to draw lessons from this episode in view of the possibility that Shi Songzhi would succeed his uncle.

The winning essay exposed both problems as violations of the same principle: the sanctity of human relationships. The first two examples demonstrated disregard for the bonds between father and son and between brothers; the second problem derived from violations of the proper relationship between ruler and minister. The author insinuated that Shi Miyuan and Shi Songzhi disregarded the basic human relationships on which an orderly society was founded. Their moral unfitness was sufficient reason to call for Shi Songzhi's resignation. Shi Songzhi's career came to an end upon allegations of unfilial behavior at the death of his father. When his father passed away in 1244, Shi Songzhi decided not to resign and return home for the customary three-year mourning period. His continued occupancy of his post provoked student protests in all the capital schools. Over 141 students at the Imperial College signed a petition asking for his resignation.[87]

The protest in 1244 is often cited as an example of the political activities in which students at the metropolitan schools engaged. Political protests have been portrayed as side activities separate from students' curricular tasks; the examples from Shaowu suggest that political protest was part of the examination culture not only in the capital but also in the provinces. Learning of the Way values played an important role in these protests.

At the same time, literati exploited the space created by the conventions of the examination field to advertise their intellectual and political agendas. Examination writing opened up avenues for political opinion. Teachers like Chen Fuliang and Lü Zuqian paid singular attention to composition because they believed in the power of the written word in

---

86. On the affair and its aftermath, see ibid., 96–104; and idem, *Wind Against the Mountain*, 36, 86.

87. Davis, *Court and Family in Sung China*, 150–55; Wang Jianqiu, *Songdai Taixue yu Taixuesheng*, 307, 368–69; cf. Lee, *Government Education and Examinations in Sung China*, 193–94.

setting out projects for good government. The practice of "designing mock examination questions" and "writing test policy response essays" (*ni ce* 擬策) attests to the wider cultural significance of the examinations. The use of the genre of the policy response legitimated the expression of political criticism, since it was presented as loyal political advice to the emperor. Consequently, the adoption of this genre lent the work of people who were not participating in, and at times even not preparing for the examinations, strong political overtones.

Now, during years when it is decreed to select *jinshi*, mock policy response essays appear from all the publishing-houses in my native place. . . . As for the authors, who have appeared from underneath dilapidated walls in the deep valleys and attend one another like elder and younger brothers or teachers and students in difficult circumstances, they discuss the following questions among themselves: How can the mind of the emperor be rectified? How should the heir apparent be taught? How should the people's hardships be alleviated? How should the border troubles be solved? . . . Take heed! If in the present situation the avenues for speech for the scholars were opened widely, then [our words] would be transmitted upward and come to the emperor's ear; they can't be of no use. Since it cannot be like this, [our proposals] are printed and circulated, they are distributed to those scholars who are taking the examinations so that those who adopt our explanations have something to tell to the examination officials. If we consider the examinations this way, why would we turn our backs on the examinations? The circulation of these materials will also ensure that when responding to the emperor, we won't answer with old events and reproduce stereotyped words.[88]

然則詔舉進士之歲吾鄉諸齋擬策四出. . . . 而擬策者出於窮澗頹壁之下, 兄弟師友不朝夕溫飽之間相向輒言曰: 上心若何正? 東宮若何教? 民病若何甦? 邊憂若何解? . . . 嗚呼! 如使當世大開古者士傳言之路, 則轉而上聞未必無益. 既不能然, 則刊刻流布傳於同試場屋之士. 使得吾說者皆有以告有司. 如此而應科舉, 亦何負科舉哉! 使得對天子, 其不應故事襲腐語.

Like Chen Fuliang in *Awaiting Reception* and Ye Shi in *Presented Scrolls*, Liu Nansou 劉南叟, the author of the printed collection of essays for which Ouyang Shoudao wrote this preface, chose to take advantage of the opportunities offered by the examination system and secure at least

---

88. Ouyang Shoudao, *Xunzhai wenji*, 9.12b–13b.

the attention of examination candidates, if not an imperial audience. The workings of the civil service examinations as a cultural field shaped by the interactions and changing authority of different interest groups are most evident in the adaptation of the Learning of the Way discussed in the next section.

## The Learning of the Way as Official Ideology

The court permitted the expression of Learning of the Way views through its endorsement of Zhu Xi's commentaries on *The Analects* and *Mencius* in 1212 and of all Four Books in 1227. The court's acts of recognition from the 1210s through the 1230s represented a series of mild concessions to demands for the full installation of the Learning of the Way as state orthodoxy. Continued pressure finally bore a result in 1241 (Chapter 5).

Emperor Lizong's 1241 decree was very different from the earlier official pronouncements in support of the Learning of the Way. Edicts from 1209, 1212, 1215, 1216, 1220, 1227, and 1235 concerned the conferral of posthumous titles and other honors on individuals and the authorization of commentaries. The 1241 decree called for a full-fledged adoption of "the Learning of Coherence," as it was christened in the decree. As teachers and students began paying homage to the images of Zhou Dunyi, the Cheng brothers, and Zhu Xi at the Imperial College and official schools, they began to treat the work of these masters as canonical sources in their examination writing. This section discusses the characteristics of official Learning of the Way ideology as it was represented in policy questions and response essays from the 1240s to the 1270s. Questions and essays corroborate the transformation of Learning of the Way ideology traced in Chapter 6 and the section on exegesis in expositions in this chapter. In examination writing, official Learning of the Way ideology was characterized by the centrality of Zhu Xi's legacy, the adoption of genealogical discourse and Zhu Xi's version of the genealogy of the Way, and the confirmation of doctrinal consistency and unity among its major representatives.

Lizong's praise for Zhu Xi's achievements solidified the image of Zhu Xi as the great synthesizer of the Learning of the Way tradition. Students turned to his work for explications of Learning of the Way moral philosophy and commentary on the Classics. The centrality of

Zhu Xi in official Learning of the Way ideology is borne out by the numerical surveys of the expositions and policy response essays anthologized in *Standards for the Study of the Exposition* and *Standards for the Study of the Policy Response* (Tables 3, 2B, and 2C in Appendix B). According to Table B3, of the total number of expositions referring to the Learning of the Way in the 1250s and 1260s, 55 percent and 73 percent, respectively, relied on the authority of Zhu Xi's words. These numbers represented a marked increase over the preceding decades. Tables 2B and 2C suggest that Learning of the Way beliefs already had a considerable impact in departmental and Imperial College examinations in the 1220s and 1230s. Of the total number of expositions referring to the Learning of the Way during this period, less than 30 percent referred to Zhu Xi. To judge from the evidence, the 1240s marked not only the breakthrough of Learning of the Way ideology in the examinations but also the ascendancy of Zhu Xi's legacy.

By the 1250s and 1260s, successful examination candidates were intimately familiar with Zhu Xi's written work. They quoted from his commentaries on the Four Books, his memorials, his anthologies and editions of the work of his chosen predecessors in the Learning of the Way, recorded conversations, and other work. Explicit references to the names of contemporary authors and long quotations from their work were prohibited and unusual in examination writing before the 1240s. Only pre-Song authors, especially those associated with the Classics and their commentaries, were cited. In the last decades of the Song dynasty, examinees quoted Zhu Xi to substantiate arguments on all issues.

An Imperial College examination question, attributed to an otherwise unknown student named Cheng Shenzhi 程申之, illustrates the effect of the canonization of Zhu Xi's work. The examiner asked students to evaluate the truth of the argument that the mind of the sovereign was central to the well-being of the state. Cheng Shenzhi affirmed the truth of this political theorem and presented the argument that all administrative and social dysfunction derived from partiality in the sovereign's mind.[89] This argument had become increasingly popular during the campaigns of dynastic reinvigoration inspired by Learning of the Way advocates since the beginning of the thirteenth century. It had

---

89. (*Jingxuan*) *Huang Song Cexue shengchi*, 8, essay no. 2.

been most elaborately and forcefully advanced by Zhen Dexiu. As shown in the previous section, Zhen Dexiu and his associates defended this position as a conclusion drawn from *The Great Learning*. They relied on classical sources and did not refer to the Song masters of the Learning of the Way.

In contrast, Cheng Shenzhi started his essay with an explicit quotation from a sealed memorial Zhu Xi submitted to Emperor Xiaozong in 1188:

The emperor has been worrying about state affairs and hopes to bring about order; it is not so that his mind has not striven for this. How could it be that he does not aspire to strengthen the hold of fundamental principles and to ameliorate customs! The only problem is that he has not yet been able to clear the obstruction of selfish evil in his thought.[90]

夫以陛下之心憂勤願治, 不爲不至. 豈不欲夫綱維之振風俗之美哉! 但以一念之間未能去其私邪之蔽.

The memorial vocalized Zhu Xi's oppositional politics. He submitted this memorial after a court audience the same year but declined the court post he was offered after the emperor received it. This refusal to serve in office expressed Zhu Xi's opposition to the opponents of the Learning of the Way at court. In Cheng Shenzhi's reading, the memorial was proof of the legitimacy of Learning of the Way political theory.

In accord with the examiner's further instruction that students bring up any related arguments, Cheng Shenzhi wrote that the enforcement of the dynasty's fundamental principles and the state of the people's conduct depended not only on the emperor's self-cultivation but also on high officials' monitoring the emperor's mind (*ge xin* 格心). This, too, was a familiar thesis in Learning of the Way political theory. Zhen Dexiu had suggested the same line of argument to his examinees in 1235. In Cheng's essay, this argument also needed Zhu Xi's seal:

Duke Wen [Zhu Xi] also said that it will be impossible to strengthen the hold of a crumbled constitution and to renew people's conduct if the proper ministers are not in place. I felt free to draw on Duke Wen's words to encourage

---

90. Zhu Xi, *Zhu Xi ji*, 11.460–88; the passage quoted is on p. 473. Shu Jingnan, *Zhuzi da zhuan*, 635–38.

our ruler in the first part of my essay, and I used his words to encourage our ministers in the last part.

文公又謂宰相有不得人則何以振已頹之紀綱, 已壞之風俗. 愚敢前以文公之言勉吾君, 後以文公之言勉吾相.

According to the commentator on this essay, Cheng's piece was representative of a prevailing trend among examination candidates:

Our sagely dynasty upholds the true learning (*zhengxue*). The explanations of Master Huian [Zhu Xi] have found widespread approval in our time. Those who respond to policy questions regularly cite him to have a starting point for further discussion. In this essay, the concept of selfish desires in the ruler's mind functions as the thread throughout the whole piece. This argument and the concluding advice that high ministers monitor the [sovereign's] mind are both substantiated with evidence from Huian's work. *It is like officials deciding a case on the basis of the law.*

聖朝崇正學. 晦翁先生之說盛行于世. 對策者多引用爲話頭. 此策全篇以君心之私主張. 結尾以大臣格心獻策. 皆引晦翁之説爲証. 如官員坐法斷案相似.

This comment impressed on examination candidates buying the annotated anthology in which it appeared the fact that the Learning of the Way had become state orthodoxy. It clarified that its official recognition had resulted in the canonization of Zhu Xi's work. For examinees this implied that Zhu Xi's work had the status of a law code: students were encouraged to check his work to find answers to examination questions.

Zhu Xi's impact on literati culture was not limited to moral philosophy and political theory; it also extended into the discussion of socioeconomic policy. A policy question for a school examination held at the Shaowu Prefectural School in the mid-thirteenth century asked students to evaluate the rationale for and the effectiveness of community granaries.[91] The community granaries were an innovative approach to rural credit and famine relief promoted by Zhu Xi while serving as intendant for ever-normal granaries, tea, and salt for Liang Zhe East (Zhejiang) in the early 1180s. The institution of the granaries was in agreement with a new trend among Southern Song elites to replace the antagonistic rela-

---

91. (*Jingxuan*) *Huang Song Cexue shengchi*, 10, essay no. 1.

tionship between the rich and the poor with an awareness of mutual dependence and the active participation of elites in social welfare programs.[92] The granaries provided grain loans to peasants and were managed voluntarily by prominent men from the local community.

The student who obtained first place in the examination defended the establishment of community granaries. He argued that Zhu Xi had demonstrated that they were of great benefit to the local community when managed well. He attributed the reasons behind the complaints about the community granaries to malfeasance. The solution to the difficulties community granaries had faced was of a moral kind. The student recommended that care be taken to nominate men of benevolence as managers of the granaries. Men of benevolence were defined as those among the local elite who were driven by the same moral principles as Zhu Xi.

The centrality of Zhu Xi in Learning of the Way ideology was also evident in the adoption of his genealogy of the Way. During the examinations of the last three decades of the Song Dynasty, examiners no longer questioned the Learning of the Way genealogy. They endorsed Zhu Xi's lineage of transmitters and extended the lineage in the direction he had mapped out:

The Way was obscured for a long time. After Confucius, Zi Si and Mencius carried on his legacy. After Zi Si and Mencius passed away, Xun [Qing] and Yang [Xiong] dissented. Master Han [Yu]'s criticism that their selection was indiscriminate and their explanations superficial was to the point. However, he also had no way out. Very few over a thousand years! The masters of Guan and Luo [Zhang Zai and the Cheng brothers][93] came forward, and the master from Xin'an [Zhu Xi] carried on their legacy.[94]

道不明久矣. 夫子之後, 思軻繼之. 思軻既沒, 荀楊舛駁. 韓子所謂, "擇焉而不精, 語焉而不詳者," 得所譏矣. 然韓子亦未免焉. 寥寥千載. 關洛勃興. 新安繼續.

---

92. Oberst, "Chinese Economic Statecraft and Economic Ideas," 148–71.

93. Technically Guan and Luo refer only to these masters. It is clear from the rest of the question that the term refers to the four masters Zhu Xi saw as the founding fathers of the Learning of the Way. In another binomial, "Lian Luo" 濂洛, Zhou Dunyi is substituted for Zhang Zai. Both terms seem to refer to all four masters in this context.

94. (*Jingxuan*) *Huang Song Cexue shengchi*, 7, essay no. 1.

The examiner reiterated Zhu Xi's suspicions about those thinkers who claimed to transmit the true version of Confucius' Way after Mencius. He fully endorsed Zhu Xi's version of the core Song figures of the transmission (Zhou Dunyi, the Cheng brothers, and Zhang Zai) and added Zhu Xi, in line with the new official image of Zhu Xi as the grand master of the Learning of the Way.

The meaning of genealogical thinking shifted and was reduced in the reproduction of the Learning of the Way in examination writing. For Zhu Xi, intellectual genealogy, the act of tracing one's intellectual ancestors, was a defining characteristic of the teacher of the Learning of the Way. Through it, advocates of the Learning of the Way laid claim to a position of authority in its transmission. The genealogy underscored the significance of personal interaction with a teacher in the genealogy and of face-to-face transmission of the tradition of mind-cultivation. Zhu Xi presented the revival of the Way initiated by the Cheng brothers as a continuous chain of transmission. The Cheng brothers transmitted their understanding and practice of moral self-cultivation to their immediate disciples, who then transmitted it to their disciples, and so on. Disciples became teachers by successfully inserting themselves in a chain of transmission that led back to the Cheng brothers.[95]

Zhu Xi's genealogy of the Way was also a template of the coherence of the Learning of the Way legacy. Zhu Xi constructed, invalidly according to some of his contemporaries, a direct line of transmission among the Four Masters. The direct line of transmission supported his view that their teachings constituted a coherent philosophy. This meaning dominated the interpretation of the genealogy of the Way in thirteenth-century examination writing. The genealogy of the Four Masters, with the addition of Zhu Xi, became a mental picture for students, a picture that compelled them to imagine the unity of the Learning of the Way canon and its accord with the Way of the sages of Antiquity. The examiner who summarized the lineage of the Learning of the Way in the passage just quoted asked students to defend the rationale behind it. The question continued in the vein of the questions of the "points of doubt in the Classics"

---

95. De Weerdt, "The Ways of the Teacher."

type.[96] The examiner quoted contradictory statements from the body of the Learning of the Way classics, but added that these quotations provided examinees an opportunity to sort out the main principles scattered in the masters' writings and in the records about their lives.

The examinee whose essay was selected for inclusion in the anthology elaborated on some of the discrepancies but concluded that students needed to move beyond textual contradictions: "In reciting the books of the Four Masters [in his essay he opted to focus on the Cheng brothers, Zhu Xi, and Zhang Shi],[97] and in learning about their conduct, we absolutely have to set our minds on what they set their minds on and learn what they learned." Examinees confirmed doctrinal consistency as an article of faith; the lineage of the canonical masters supported that faith.

In the twelfth century, questions about the genealogy of the Learning of the Way were limited to those with an interest in Cheng Learning and were part of a debate over the membership in and appropriateness of an exclusive line of transmitters of the Way. In the thirteenth century, questions about Zhu Xi's genealogy of the Way became an identifiable type of policy question. This development was indicative of a larger trend. Starting in the 1240s, questions about Learning of the Way doctrinal matters became increasingly common. As in the case of the questions on the genealogy of the Way, questions on other doctrinal matters were formulated in such a way as to have students represent the Learning of the Way as a coherent and consistent ideology.

On a visit to the Imperial College in 1241, Emperor Lizong presented a copy of Zhu Xi's "Announcement of the White Deer Grotto Academy" in his own calligraphy to the school.[98] One of the teachers

---

96. An example of this type of question is Lü Zuqian's question on *The Analects* discussed in Chapter 2.

97. In the last decades of the Southern Song dynasty, the catchphrase "the Four Masters" applied either to Zhu Xi's grouping or, by extension, to the Chengs and their officially acknowledged successors, Zhu Xi and Zhang Shi. The latter use equally accords with Zhu Xi's genealogy of the Way. In contrast to his other rivals, he co-opted Zhang Shi in the lineage. Cf. Tillman, *Confucian Discourse and Chu Hsi's Ascendancy*, 81–82. For another use of the term, see ibid., 244.

98. Zhu Xi intentionally avoided using the term "school regulations" (*xuegui* 學規). He called his precepts for the students of the academy "*jieshi*" 揭示, a term used for

seized on the event to ask questions about the contents of the announcement:

Now, I, the examination official, don't feel at ease in generalizing about the [Learning of the Way] masters' learning of the nature in my question. I will focus on those words of the masters that relate to "the list of five precepts" and "the process of learning" [discussed in the announcement]. Can we try to analyze a couple and verify them against each other?[99]

今也有司不敢泛獵諸儒性學以問. 姑即諸子之言, 取其關於五教之目, 爲學之序者, 試繹一二相與質訂焉, 可乎?

Zhu Xi's announcement was a short list of quotations from the Classics. In Daniel Gardner's interpretation, the quotations were organized according to a predetermined logic. They defined step by step the content of learning, the method of learning, the aim of learning, and the effects of proper learning on the individual's dealings with others.[100] The examiner asked students to explain the meaning of and the relationship between items listed in the first two sections of the announcement: the virtues that should obtain in the five basic human relationships (the content of learning; from *Mencius*) and the method of learning (studying extensively, inquiring carefully, pondering thoroughly, sifting clearly, and practicing earnestly; from *The Doctrine of the Mean*).

Fang Yisun,[101] who performed well enough on this part of the examination to be included in the anthology, focused on the sequence of learning. He argued that the sequence of learning, moving from the investigation of things and mental preparation to moral practice, could be reduced to one core value of the Learning of the Way. "Mental atten-

---

public announcements. For a comparison between previous regulations and Zhu Xi's "Announcement," see Chen Wenyi, *You guanxue dao shuyuan*, 56–106. Cf. Chaffee, "Chu Hsi and the Revival of the White Deer Grotto Academy," esp. 54–62. On the presentation of the placard, see Yuan Zheng, *Songdai jiaoyu*, 74.

99. (*Jingxuan*) *Huang Song Cexue shengchi*, 5, essay no. 1. The "list of five precepts" and "the process of learning" constitute the first two sections of Zhu Xi's "Announcement." Cf. Gardner, *Learning To Be a Sage*, 29–30; and Chaffee, "Chu Hsi and the Revival of the White Deer Grotto Academy," 54, for translations.

100. Gardner, *Learning To Be a Sage*, 30.

101. Fang Yisun was engaged in examination teaching. His Ancient Prose manual is discussed in Chapter 6.

tiveness" (*jing* 敬) encompassed both the contemplative side of learning and its active side. "Mental attentiveness" encapsulated the different stages in the process of learning and the different interpretations they had been given. It was an aggregate value that, like other core values of the Learning of the Way such as humanity (*ren* 仁), could be used to reconcile conflicting and confusing canonical passages.

Fang also relied on genealogical thinking to underscore the unity and coherence of the Learning of the Way as state ideology. He argued that the emperor's endorsement of Zhu Xi's announcement was evidence of the court's affiliation with the true transmission of the Way. Zhu Xi's announcement had rightfully been adopted as the basis for instruction empire-wide because his work continued the legacy of the founders of the Learning of the Way and, through them, the teachings of the sage-kings and sage-teachers of a more remote past. He added that the classical quotations in Zhu Xi's announcement were proof *ipso facto* that his views on the process of learning bore the true imprint (*zheng yin* 正印) of the Way.

Around the same time, the side-effects of the canonization and official adoption of the Learning of the Way masters' writings became an issue. The transformation of Learning of the Way ideology had come full circle. The critique of examination learning had been a major impetus for Zhu Xi's Learning of the Way. By the mid-thirteenth century, it had become clear that examination learning remained a nagging question, even after the canonization of the Learning of the Way. Wei Liaoweng, a leading advocate of the Learning of the Way in the early thirteenth century, expressed concern over the new conformism that was spreading in the name of the Learning of the Way.[102] One examiner later wondered whether the work of Zhu Xi and Zhang Shi had not been deprived of its force after its commercialization as successful examination writing:

When we look in the examination halls for those with profound scholarship, sincere manners, and substantive writing, on the whole we can't find more than three or four in ten. In most cases, their writings all sound alike and are superficial and confused. That is the average. Could it be that the style of the

---

102. Chen Wenyi, *You guanxue dao shuyuan*, 225.

Qiandao [1165–73] and Chunxi [1174–89] periods has vanished without a trace? *Could it be that penetrating discussions in the style of Zhu [Xi] and Zhang [Shi] have disappeared in fact and exist in name only?*[103]

然場屋間求其學問之深醇, 氣象之渾厚, 辭章之典實者, 大地拔十之中未能三四. 往往雷同一律者, 多荒疎雜犯者, 眾是. 豈乾淳之風流敢斷而影絕! 朱張之講貫, 名存而實亡!

---

103. This excerpt comes from another question for a monthly examination; see (*Jingxuan*) *Huang Song Cexue shengchi*, 5, essay no. 2.

# Conclusion

## From Northern Song to Southern Song, Revisited

The transition from the Northern to the Southern Song coincided with a shift in the political dynamics of the imperial order from state activism to elite activism. The turn from central government intervention in, to local elite management of, the local economy, local security, and social welfare did not bring an end to state activism, but it did set out the basic parameters for elite activism throughout the rest of imperial history. During the past two decades, social historians of Song China have traced this transition in local histories of social, economic, and religious life.[1] This study of the imperial civil service examinations addresses the question of the political meaning of the transition from state to elite activism. What was the impact of the reorientation of Song elites from the center of imperial politics to local society on those institutions that had embodied the centralizing policies of the Song state since its establishment in 960 and that remained firmly in place after the relocation of the court to the south in 1127?

The tiered civil service examinations, culminating in the palace examination at the court, were but one of many institutions that the first Song emperors designed to centralize power and tie scholar-officials across the empire closer to the dynastic state. The growing numbers of

---

1. See, e.g., Hymes, *Statesmen and Gentlemen*; idem, *Way and Byway*; Hymes and Schirokauer, *Ordering the World*; Hansen, *Changing Gods in Medieval China*; Oka, "Nan Sō ki no chiiki shakai ni okeru 'yū'"; idem, "Nan Sō ki Onshū no chihō gyōsei o meguru jinteki ketsugō"; and von Glahn, *The Sinister Way*. Cf. De Weerdt, "Amerika no Sōdaishi kenkyū ni okeru kinnen no dōkō."

participants, numbering in the tens of thousands in the eleventh century and in the hundreds of thousands by the mid-thirteenth century, turned the examinations into a political communications network with wide reach. This study has demonstrated that the examinations were not only a site of the transition from state to elite activism but also a catalyst enabling it and hastening it.

The examinations played the role of a catalyst in the transformation of elite strategies because the meaning of examination participation and examination success changed. The sheer numbers of candidates and the even greater numbers of students preparing for the examinations are indicators of the change in elite perceptions of the value of examinations. Even though literati were aware of the low quotas of graduates and the declining odds of success at even the lowest level of the examinations, they continued to invest in examination preparation. Participation was no assurance for success, and graduation no guarantee of bureaucratic employment, but participation in and of itself yielded social capital that bought power and prestige locally.

My investigation of examination preparation and examination writing suggests a second reason why the civil service examinations accommodated and promoted elite activism during the Southern Song period. Throughout most of the Northern Song period, political elites focused on the court and the capital because information on court politics was concentrated there. The increasing numbers of students preparing for the examinations increased demand for information about court and capital politics. Students in the provinces needed such information in order to prepare for policy questions on current administrative issues and in order to keep abreast of changing trends in examination writing at the apex of the tiered examination system. The examination encyclopedias analyzed in Chapters 4 and 6 suggest that teachers, editors, and commercial printers provided such information locally in the twelfth and thirteenth centuries. Further research remains to be done on the exchange of information among the court, capital, and local literati, but the evidence gathered here suggests that the growing numbers of students and the associated development of commercial printing transformed the examinations into a site for the exchange of information on current affairs and scholarly trends. The reach of the examinations into the prefectural levels of the administration made such information available to elites locally and reduced the need to take up residence at the capital.

This book has further illustrated how the elite redefinition of the meaning of examination participation led to the restructuring of the examination field. Following the demise of the interventionist educational policies of the reformist regimes of the late eleventh and early twelfth centuries, the central government's impact on examination curricula declined. The Classics and some Tang and Song commentaries were required reading for students at government schools throughout the Song period,[2] but the Southern Song government did not prescribe a curriculum that met the larger needs of examination candidates. Official regulations defined the basic structure and the procedural details of the civil service examinations, but the court shied away from imposing lists of required commentaries and other titles as it had done during Wang Anshi's reforms. Emperors and high court officials in the first century of Southern Song rule presented themselves as the protectors of broad scholarship and intellectual diversity and refused to endorse curricula considered partisan.

As an institution, the examinations were, moreover, not linked to a comprehensive network of official schools. The eleventh-century effort to subordinate the examinations to a network of official schools, extending from district schools to the Imperial College in the capital,[3] was not repeated during the Southern Song period. Official schools trained only a small minority of the large numbers of students attempting the local-level examinations. The prefectural school in Fuzhou, the prefecture with the largest number of *jinshi* graduates in the Song period, served 300 students in 1165; that same year, 17,000 candidates took the triennial prefectural examination.[4] Official schools could not absorb the

---

2. Yuan Zheng's (*Songdai jiaoyu*, 7–76) overview of the changes in government school curricula throughout the Song period is almost exclusively concerned with the Classics and commentaries. Classical scholarship was, however, but one of the skills tested in the examinations.

3. Examinations were integrated into the operation of the schools (ibid., chap. 2).

4. Lee, *Government Education and Examinations in Sung China*, 176. Even if the district schools, which usually had a much smaller number of students, are included, the official schools fell far short of accommodating all students. For more information on the prefectural and district schools in Fuzhou, see Zhou Yuwen, *Songdai de zhou-xianxue*, 416–18.

ever-increasing number of candidates, and the majority of students relied on private tutors and various kinds of manuals to prepare for the
examinations.[5]

The early Southern Song government's commitment to a noninter-
ventionist policy in matters pertaining to examination curricula weakened its position in the market for examination materials. Until the end
of the eleventh century, editions of the Classics, the histories, and
rhyme books published by the Directorate of Education monopolized
the market. Starting from the last decades of the Northern Song period,
commercial printers inserted themselves in the market by publishing
cheaper editions of Directorate works, modified in various ways mostly
for the convenience of the student.[6] During the twelfth century, commercial printers took over the market for examination anthologies and
encyclopedias. They published manuals attributed to famous examination teachers and contracted literati to compile and annotate anthologies. Conflicts between the government and commercial printers en-
sued, caused not only by the government's general misgivings about
private profiteering but also by the ideological tensions between officials and teachers amplified by the press.

Examiners were a formidable force in the examination halls. The
court selected examiners for the higher-level examinations, and the

---

5. Lee (*Government Education and Examinations in Sung China*, 105), citing Chaffee's dissertation, writes that by the Southern Song there were at least 588 government-run local
schools, 72 of which were prefectural and 516 of which were county schools. Chaffee
(*The Thorny Gates*, 75, 136) gives a total of 750 schools throughout the Song dynasty, 516
county schools and 234 prefectural schools. Zhou Yuwen's (*Songdai de zhou-xianxue*, 259)
numbers are slightly higher than Chaffee's estimates. He has found 571 county schools
and 271 prefectural schools.

Lee (ibid., 175–76) writes that in the Southern Song county schools usually enrolled
only a few scores of students and prefectural schools a couple of hundred; this is
slightly less than during the Northern Song. Given Chaffee's (ibid., 35) estimates of the
number of examination candidates, 79,000 around 1100 for the whole of Song China,
climbing to about 400,000 by the mid-thirteenth century for the south, it is evident that
the number of students from official schools must have been a minority among the total number of examination candidates.

For a discussion of the development of the teaching business in the twelfth and thirteenth centuries, see Liang Gengyao, "Nan Song jiaoxue hangye xingsheng de beijing."

6. Poon, "Books and Printing in Sung China," 102–4, 122–27; Chia, "Printing for
Profit," 160–61; Liu Hsiang-kwang, "Yinshua yu kaoshi," 194*n*53.

regional intendancies assigned them for the lower-level examinations.[7] Despite mid- and high-level official supervision, examiners enjoyed a considerable degree of autonomy. The head examiners set their questions independently and were responsible for grading and ranking all papers. The power of individual examiners at the higher levels of the examinations was in most cases restricted to the confines of one examination setting.[8] The Song court instituted a variety of measures to curtail the power of individual examiners. Incumbents were assigned on an ad hoc basis and were swiftly and continuously rotated out of examiner positions. To candidates, examiners at higher-level examinations were more a source of anxiety than the guardians of examination standards.

The main authorities in the formation of examination standards in the twelfth century were teachers. As the government withdrew as the principal trendsetter in examination preparation, traditions of teachers came to occupy positions of authority in the examination field. The expanding pool of examination candidates and the proportionately high number of unemployed degree-holders gave teaching a greater appeal as a temporary or permanent career. The success of an examination teacher was measured by his examination success and his teaching credentials—experience in the Imperial College, the number of his students, and their examination results.

Famous examination teachers like Lü Zuqian and Chen Fuliang had passed the examinations with flying colors, gained a reputation at the Imperial College, and taught several hundreds of students, some of whom followed in their footsteps. Lü Zuqian admitted that his success as a teacher depended largely on examination preparation courses. His lectures on the Classics and history attracted large audiences. He compiled historical reference works and took especial care in the teaching of composition. Chen Fuliang was noted for his scholarship on *The Spring and Autumn Annals* and *The Book of Songs*, his mastery of institutional history and political reasoning, and his splendid Ancient Prose writing, which transformed current conventions for the examination exposition.

---

7. Araki Toshikazu, *Sōdai kakyo seido kenkyū*, 18–36, 182–88, 321–30.

8. For a different interpretation of the power of the examiners in the eighteenth century, see Man-Cheong, *The Class of 1761*, 169–75.

The teaching careers of both Lü Zuqian and Chen Fuliang grew out of a tradition of examination teaching in the East Zhe region. Generations of teachers in the prefectures of Wenzhou and Wuzhou established the reputation of East Zhe as a center for examination learning. Elsewhere in the empire, in the mountainous regions of Fujian, Zhu Xi and his disciples turned to teaching and developed curricula that increasingly integrated the moral philosophy of their movement, the Learning of the Way, with the requirements of twelfth-century examination preparation.

The impact of teachers on examination standards extended beyond the traditional channels of lecturing and examination preparation courses. The proliferation of commercial printshops in the twelfth century, a direct result of the expansion of examination participation, made teachers' course materials and annotated examination writings available in relatively affordable editions. According to one official estimate, the production costs for a printed text with a relatively high circulation were one-tenth of the cost of manuscript copies.[9] Even though the price of one set of collected writings or a commentary on the Classics was equivalent to half the monthly salary of a low-level bureaucrat (then as now, a low-paying job), books came within the purview of a rapidly expanding group of scholars.[10] Books were signs of the promise of examination success. Their desirability inspired stories of individual and family sacrifice. The images of a scholar pawning clothes for books or a family scrimping on food and other expenses to buy books for the education of male family members were celebrated in commemorative biographies. The number of private collections in the Southern Song period grew across the southern territories despite the losses many private collectors had incurred as a result of the Jin invasions and the following unrest and dislocation.[11] Some literati hailed private collectors for their liberal access policies.

---

9. Weng, "Yinshua duiyu shuji chengben de yingxiang," 36–37.

10. Liang Gengyao, "Nan Song jiaoxue hangye xingsheng de beijing," 323–26.

11. From available lists of collectors and major collectors (over 10,000 *juan*), the numbers for Northern Song and Southern Song collectors are about the same. However, the reduction in territory implies that there was a marked increase in the number of collectors in the south. For numbers and the shifting regional distribution of collectors, see Ren Jiyu, *Zhongguo cangshulou*, 1: 750 ff; Fan Fengshu, *Zhongguo sijia cangshu shi*,

The examination genres and examination manuals had taken root in literati culture well before the twelfth century. Anthologies of successful and model essays had been around since at least the Tang dynasty.[12] Encyclopedias, classifying information drawn from a wide variety of sources, were also a feature of examination preparation in the days of the famous poet-statesman Bai Juyi 白居易 (772–846), who compiled an encyclopedia of excerpts of classical and historical texts for examination candidates.[13] The exponential growth in the number of examination candidates and the concomitant proliferation of printed books changed the import of such manuals in twelfth- and thirteenth-century education.[14] Editors and printers thus assisted teachers in promoting their intellectual and political agendas through the distribution of various kinds of printed examination manuals.

## Intellectual History
## and the Examination Field

The shift from state to elite activism in the examination field was apparent not only in the changing authority of the agents but also in the modes of exegesis and discussions of government that teachers promoted, that students adopted, that editors and printers sold, and that examiners and other representatives of the dynastic state reacted to in the twelfth and thirteenth centuries. Both the "Yongjia" and the Learning of the Way

---

62–82; and Fang Jianxin, "Songdai sijia cangshu bulu." Cf. De Weerdt, "The Discourse of Loss."

12. The recently discovered policy questions and answers from the Six Dynasties period may be an even earlier example of the examination anthology genre; see Dien, "Civil Service Examinations."

13. For the most complete overview of relevant titles, see Zhou Yanwen, "Lun lidai shumu zhong de zhiju lei shuji"; see also Liu Hsiang-kwang, "Yinshua yu kaoshi."

14. The impact of an emerging print culture is evident in official prohibitions against certain types of examination cribs. In the eleventh century, government regulations were directed mostly against the practice of carrying hand-copied notes and transcribed books into the examination halls. Edicts from the late Northern Song and later express the government's anxiety over the widespread printing and circulation of commentaries, encyclopedias, and examination guides. See the regulations listed in Chia, *Printing for Profit*, 121–23; cf. Zhu Chuanyu, *Songdai xinwen shi*, 162–69; and Poon, "Books and Printing in Sung China," 106–11.

teachers promoted an intellectual and political agenda that expressed great sensitivity to the interests of local elites, albeit with different emphases.

In the second half of the twelfth century, Yongjia arguments for increasing local wealth and for enhancing the power of the local elite in managing that wealth fell on sympathetic ears among examination candidates from elite families. Likewise, the proposition that non-office-holding literati should be heard in policy debates and imperial decision-making contributed to the authority, first, of a group of Yongjia teachers among the expanding number of Wenzhou examination students, and, second, of the "Yongjia" tradition in the larger examination field comprising the capital and beyond. The Yongjia political program energized Wenzhou elites, but it was not addressed solely to them. The Yongjia teachers' advocacy of customized local solutions and of elite participation in the discussion of current affairs fit into a project aimed at the reform of imperial politics. The "Yongjia" curriculum was imperial in intent. However, it attempted to reshape imperial politics in local elite terms. The Yongjia teachers articulated local interests theoretically. Rather than identifying with the particular interests of their place of origin, they discussed local interests in general and in relationship to centralizing policies. This helps explain the appeal their scholarship enjoyed among students, empire-wide according to their critic Zhu Xi.

Learning of the Way teachers designed new curricula and redesigned old curricula, making moral philosophy the cornerstone of their educational program. As the explication and application of the cosmic imperative of the cultivation of the self, Learning of the Way exegesis and political discourse promoted an elite self-conception that tied moral responsibility to local leadership. As demonstrated in Chapter 7, examination candidates articulated both the metaphysical grounds for the new ethic of self-cultivation and its application in elite-sponsored local welfare programs such as the community granary.

The political meaning of the transition from state to elite activism is, in the case of the civil service examinations, wider than the infusion of the interests of local elites. Examination preparation and examination writing demonstrate that court and imperial politics were of great concern and interest to elites who were developing localist strategies. Arguably, the Song emphasis on policy questions and response essays focused literati on imperial politics in greater numbers than ever before in Chinese history. Official action against authoritative teachers in the examination field, such as the "Yongjia" and Learning of the Way teachers, resulted from reigning court factions' concern about the critical stance teachers and students adopted toward the status quo at court. The burning of the woodblocks of policy essay anthologies and collections of recorded conversations, the frequent bans on the publication of sensitive materials such as archival documents and essays about border affairs, and the calls for court-sponsored anthologies of policy essays testify to the court's attempts to monitor the participation of local literati in court and imperial politics.[15]

During the Southern Song period the examinations were not the precinct for state activism, let alone imperial or bureaucratic absolutism; nor did they become a direct reflection of the interests of participating local literati. The examinations can best be conceived of as a bounded space, with particular conventions and subject to historical change, in which competing groupings of teachers and representatives of the court negotiated standards for the examinations and, by extension, standards for statesmanship and local leadership.

The relative autonomy of the examination field has repercussions for Chinese intellectual history. This study has demonstrated that the dissemination of the Learning of the Way, which became the dominant tradition of Confucian learning in the late imperial period, took place in part in the examination field. It has also demonstrated that Learning of the Way teachers, like other teachers developing examination curricula, adapted their beliefs and methods to the conventions of the examination field. They designed examination questions, produced examination manuals, and trained students in the writing skills appropriate for the examination genres. Both Learning of the Way and "Yongjia" teachers were compelled to reconcile their intellectual agendas with the exigencies of examination preparation. Lü Zuqian and Chen Fuliang, for example, accepted that they had to instruct students in writing, classical philology, and basic knowledge about institutional and political history. At times they expressed the same unease as Learning of the Way

---

15. I discuss the history of prohibitions on sensitive materials in more detail in "What Did Su Che see in the North?" and "Byways."

teachers about the need to prepare students to take the examinations.[16] Further studies are needed to analyze how the continuing adaptation of the Learning of the Way tradition to examination conventions shaped its reception among lesser-known literati. The frequent quoting of the work of the Four Masters of the Learning of the Way and Zhu Xi by the mid-thirteenth century suggests that the canonization of these masters in the examination field significantly changed the reception of their intellectual legacy.

Teachers in the examination field further designed questions, courses, and manuals with an eye to contrasting their approach to that of contemporary competitors in the examination field. Learning of the Way teachers and students in the twelfth century defined exegesis and political analysis based on moral philosophy against the utilitarian approach promoted by "Yongjia" teachers. The Learning of the Way tradition was in part shaped by the rejection or modification of alternative modes of examination preparation. The "Yongjia" teachers defended a historical approach to exegetical and policy questions against exclusively moralistic interpretations. The analysis of the interactions of these two traditions of teachers in the Southern Song examination field thus suggests that such interactions significantly shaped their intellectual production. Given the centrality of the examinations to literati status, intellectual historians cannot afford to ignore shifting trends in the examination field. It was the locus for the formation and transformation of intellectual traditions, and monitoring it provides insight into the changing status and interpretation of intellectual traditions among the literati.

The history of the "Yongjia" tradition and, more clearly, the history of the Learning of the Way suggest that the spread of intellectual traditions in imperial China correlates with their authority in the examination field. The examination field operated according to its own conven-

---

16. After passing the departmental examinations, Chen Fuliang burnt some of his earlier examination compositions and insisted more explicitly on "the learning of moral principles" (Sun Qutian, *Chen Wenjie gong Fuliang nianpu*, 8a–b; cf. Winston Lo, *The Life and Thought of Yeh Shih*, 48n39; and Wang Yu, "Nan Song kechang yu Yongjia xuepai de jueqi," 154–55). In his exchanges with Zhu Xi, Lü Zuqian frequently admitted his discomfort with his examination preparation classes; cf. Lin Sufen, "Lü Zuqian de cizhang zhi xue," *passim*.

tions, but the relationships among agents and their interests related to the broader literati culture and to bureaucratic politics. Specifically, based on the trajectory of the Learning of the Way mode of exegesis and discussion of government traced in this book, I propose the hypothesis that intellectual formations could not gain widespread support among scholar-officials in imperial China unless teachers, students, and court representatives adopted their modes of exegesis and discussing government in the examination field. A brief recapitulation of the trajectory of the Learning of the Way in the last century of Song rule illustrates how acceptance among broader sections of the literati and imperial endorsement resulted from the high profile Learning of the Way teachers and students had assumed in the examination field.

In contrast to Cheng Yi, Zhu Xi upheld the virtues of the civil service examinations. In his proposal for examination reform, "Private Opinion on Schools and Selection," written toward the end of his life, Zhu Xi called for gradual reform of the examinations in various ways; however, he accepted that schools would never replace the examinations, and even that examinations had a place in schools. He comforted frustrated students that the examinations were a necessary evil and maintained that examination essays could be written in accordance with the principles of the Learning of the Way. Zhu Xi's own policy questions, unlike those of Lü Zuqian, Chen Fuliang, or Lu Jiuyuan, were marked by a radical break with contemporary conventions. His short questions about the students' progress in their studies in the 1150s brought out the spirit of his program for moral reformation, but, after this early experiment, he chose to promote the Learning of the Way by means less directly related to examination conventions.

Zhu Xi's disciples integrated the Learning of the Way and examination preparation in their teaching practice. This resulted in the further adaptation of the Learning of the Way to examination conventions and its greater visibility in the examination field. Chen Chun insisted that "the words of our master, Zhu Xi" provided the means "to oppose the enemy in examination writing." In his lectures, published as a lexicon in the mid-1220s, *The Correct Meaning of Terms*, he presented the analytical language and the moral message in the work of the Northern and Southern Song masters of the Learning of the Way as the only proper examination learning fit to replace existing modes of preparation.

The champion of the Learning of the Way in the early decades of the thirteenth century, Zhen Dexiu, taught the informal Learning of the Way canon alongside more conventional courses in composition skills and current examination writing. Zhen's incorporation of authoritative texts of the Learning of the Way into a schedule of tests marked a further move in the establishment of the Learning of the Way as a major force in examination preparation. Zhen Dexiu became well known among examination candidates as a compiler and publisher of examination materials. In his Ancient Prose anthology, *The True Forefathers in Composition*, he used Learning of the Way criteria to redefine the Ancient Prose canon and to adjust the tools used to teach it in contemporary examination preparation manuals. Zhen Dexiu's work absorbed other types of learning, previously cast aside as mere examination learning, into an expanded Learning of the Way ideology. He also promoted the publication of policy essays advocating Learning of the Way values, regardless of the examination success of their authors. In the first decades of the thirteenth century, as the activities of Lin Jiong demonstrate, examination teachers not affiliated with the Learning of the Way also began to cover its main concepts and texts in their courses and textbooks.

The changing positions of the three Learning of the Way teachers, Zhu Xi, Chen Chun, and Zhen Dexiu, were mirrored in the changing perspectives of examination candidates and examiners. The discussion of examination expositions in Chapter 7 shows that examination expositions from the late twelfth century by examinees affiliated with the Learning of the Way adopted an antagonistic stance, which also characterized the intellectual career of Zhu Xi. The authors of the surviving early Learning of the Way expositions articulated their understanding of their master's teachings in an unconventional language of philosophical debate. Examiners at this time perceived the adoption of the style of the recorded conversations in examination essays as a political challenge and a threat to the classical tradition as they saw it.

After the initial rehabilitation of those persecuted, the ennoblement of Zhu Xi and the other masters of the Learning of the Way, and the endorsement of Zhu Xi's commentaries in the 1210s and 1220s, Learning of the Way political theory was acknowledged as a legitimate, albeit not necessarily satisfactory, competitor with other approaches to policy essays and questions. The earlier antagonistic stance gave way to the confident, lengthy elaboration of Learning of the Way tenets in expositions from the 1220s and 1230s. Once Emperor Lizong endorsed its position of authority in the examination field in 1241, the Learning of the Way canon, the legacy of Zhu Xi in particular, became the standard body of reference in exposition and policy essay questions, papers, and evaluations.

The trajectory of the Learning of the Way suggests that literati support for an intellectual formation depended on its transferability to the examination field. The history of the Learning of the Way in the Southern Song examination field traced here is also an essential link between the intellectual histories of its main representatives written by Hoyt Tillman and William Theodore de Bary, on one hand, and the history of its endorsement as the official examination curriculum under the Yuan and Ming dynasties traced by Benjamin Elman on the other hand. Its endorsement by Emperor Lizong powerfully underscores that the examinations connected literati interests and the imperial order. The Southern Song examination field is thus another example of the blurry boundaries between state and (elite) society in imperial China.[17]

---

17. For a succinct discussion of scholarship on Song and later imperial relations between state and society, see R. Bin Wong, "Social Order and State Activism in Sung China."

*Appendixes*

# Appendix A

—

## *Notes on Primary Sources*

### *Secret Tricks for Responding to Policy Questions* (*Dace mijue* 答策秘訣)[1]

An extant copy of this work is appended to a Yuan anthology of policy essays, *The Mirror of Peace: A Collection of Policy Essays* (*Taiping jinjing ce* 太平金鏡策*), preserved at the National Palace Museum Library in Taibei. *Secret Tricks for Responding to Policy Questions* is listed as a separate work in the Siku catalog.[2] The copy appended to *The Mirror of Peace* seems identical to the one described in the Siku catalog. In both cases, Liu Jinwen 劉錦文 is mentioned as the editor. A comment from an editor at the Daily Renewal Printshop (Rixin tang 日新堂) explains that the author is unknown, but that the work has commonly been attributed to Zeng Jian 曾堅. The Siku editor concluded that the attribution to Liu Jinwen was spurious. He cited the manual's interest in the discussion of strategies to invade the north from the south from the Six Dynasties as evidence that it must date from the Southern Song period. According to the Siku editor, this note, which is undated in the version in the copy at the National Palace Museum, was written in 1349.

In a sixteenth-century local gazetteer for Jianyang district, Liu Jinwen is listed as Liu Wenjin 劉文錦. He is named there as the compiler of *Secret Tricks* and praised as an indefatigable teacher, erudite and good at

---

1. I am greatly indebted to Lin Sufen for copying the whole manuscript for me after a series of mishaps prevented me from completing this task and long before the Harvard-Yenching Library obtained a copy.

2. *SKTY*, 40.4398–99.

composition.[3] Other evidence, in addition to the evidence cited by the Siku editor, suggests that Liu Jinwen was not the original compiler. *Secret Tricks* cites examination policy essays by forty-one authors. All twenty-one of the traceable authors cited in *Secret Tricks* obtained the *jinshi* degree in the late twelfth or early thirteenth century. I believe this work dates to some time between 1229 (the last year in which a traceable author obtained his degree) and the end of the Southern Song dynasty. The Daily Renewal Printshop was managed by the Liu family, of which Liu Jinwen was a member.[4] Liu Jinwen was not the compiler of *Secret Tricks*; he may, however, have been responsible for the Rixin tang edition of the work attributed to Zeng Jian.

### The Yongjia Master's "To the Point in All Cases" (Yongjia xiansheng bamian feng 永嘉先生八面鋒)

No Song or Yuan editions have survived. In an early Ming dynasty preface, Zhang Yi 張益 wrote that he had obtained a copy from the court academies' collection from Gao Qi 高啓. Gao Qi had access to the collection while working on the *Yuan History* in 1369–70.[5] Zhang provided little further information about the copy. He claimed that there had been a preface by Chen Fuliang 陳傅良 (1137–1203), which had been lost. Writing in 1503, Du Mu 都穆 claimed that there had been a Song print. He related that the work had conventionally been attributed to Chen Fuliang, although some had suggested that "Yongjia xiansheng" referred to Ye Shi 葉適 (1150–1223). He supported the first case because of similarities, which he did not identify, between Chen's writings and this compilation. In 1778, the Siku editor wrote that there was no clear evidence that Chen Fuliang had authored these essays. He cited internal evidence suggesting that this work dated from the South-

---

3. Feng Jike, ed., *Jiajing Jianyang xian zhi*, 12.18b; cited in *SRZJZL-BB*, 1846.

4. For the Liu family's printing activities during the Song period, see Chia, *Printing for Profit*, 78–87, 111, 128.

5. *MB*, 696–99. This preface, as well as Du Mu's cited below, can be found in an 1844 Japanese edition preserved at Shanghai Library. It also includes a postface by Chen Chun 陳春, who supported a reprint in 1819.

ern Song period.[6] To this evidence I add: (1) In a discussion of the problem of local unrest in Liang Huai and the dislocation of the population, the author referred to the "army on the frontier" as the army stationed on the border between Jin and Song territories.[7] In Chapter 3 of the edition used by the Siku editor, the author referred to "the Xining period in our reigning dynasty."[8] (2) This work cannot date back to the Northern Song period. In his discussions of Wang Anshi 王安石 (1021–86), Su Shi 蘇軾 (1036–1101), Sima Guang 司馬光 (1019–86), and other Northern Song political figures, the author referred to "the past" (*xi* 昔) as opposed to "in recent days" (*jinri* 近日).

### *Detailed Explanations of Institutions Throughout the Ages* (*Lidai zhidu xiangshuo* 歷代制度詳説)

There are two editions, one in twelve chapters and one in fifteen. The SKQS edition has thirteen categories in twelve chapters. The editor noted that the copy he was using was damaged. The sections on currency and famine relief were incomplete; where possible, corrections were made on the basis of quotations from *WXTK*.[9] In the Xu Jinhua congshu, there is a reprint of a fifteen-chapter edition that had been in private possession. This edition may go back to a Song manuscript.[10] The organization is basically the same as in the SKQS edition. Some omissions remain. The chapter on schools, for example, is cut off in the middle of the first section. The last two chapters do not appear in the SKQS edition.[11]

The manual has traditionally been attributed to Lü Zuqian 呂祖謙 (1137–81). Many have questioned the attribution because the title does not appear in established lists of Lü's publications. The fact that

---

6. *SKTY*, 26.2798–99. For a contemporary argument in favor of the attribution of this work to Chen Fuliang, see Chen Zhenbo, "*Yongjia xiansheng Bamian feng tanxi.*"

7. *BMF*, 10.76.

8. *SKTY*, 26.2799.

9. *SKTY*, 26.2798.

10. Lu Xinyuan, *Bisong lou cangshu zhi*, 59.18a.

11. Hu Zongmao's 胡宗楙 preface from 1924 to the Xu Jinhua congshu edition. Hu obtained the source text of this edition from Ding Bing's Eight Thousand Volume Pavilion (Baqian juan lou 八千卷樓).

publishing houses utilized Lü Zuqian's name for examination compilations because of his popularity as a teacher is another factor that makes the attribution suspicious. Others have argued that this title was omitted in Lü's list of publications for other reasons. One nineteenth-century bibliophile argued that this work does not appear in Lü's year-by-year biography because of Zhu Xi's 朱熹 (1130–1200) criticism of Lü's examination preparation manuals. His mistaken argument that this work is the incarnation of *The Skilled Cavalryman* (*Jing qi* 精騎), a work Zhu Xi censured in a letter to Lü Zuqian, is borrowed in Liu Zhaoren's study of Lü's life and work.[12]

Two thirteenth-century encyclopedias attributed the authorship of different passages in *Detailed Explanations of Institutions Throughout the Ages* to Lü Zuqian.[13] Given, moreover, Zhu Xi's complaints about Lü Zuqian's focus on reading history and investigating institutions in his teaching and the compatibility of the contents of this work to other work of his (for instance, his policy questions), I believe that there is a connection between this manual and Lü Zuqian. The latest time reference in *Detailed Explanations of Institutions Throughout the Ages* mentions 1180 or so as the time of writing.[14] Since Lü Zuqian died in 1181, it is not improbable that someone collected and published the lectures after his death.

## *The Key to Ancient Prose*
### (*Guwen guanjian* 古文關鍵)

No Song editions of this work are extant. Song catalogs and an extant Song edition of "The Key to Ancient Prose," instructions for reading and writing preceding the anthology in current editions, suggest that

---

12. Ding Bing, *Shanben shushi cangshu zhi*, 20.16a; excerpted in Hu, *Siku quanshu zongmu tiyao buzheng*, 1062–63. Liu Zhaoren, *Lü Zuqian de wenxue yu shixue*, 54. For more on *The Skilled Cavalryman*, see Bol, "Reading Su Shi in Southern Song Wuzhou."

13. For example, *Yuanliu zhilun*, xuji, 2.9b, citing a passage found in *Xiangshuo*, 2.9b; ibid., 4.1a–5b, citing 6.2a–3a; ibid., 4.16, citing 5.5a–6b; *Bishui*, 79.4, citing 4.8a–b.

14. "Since the Shaoxing period peace treaty with the barbarians [1141] up till now, there has been no major clash for about forty years. Since we concluded a peace treaty with the barbarians in 1164, there has been no skirmish for about ten years" (*Xiangshuo*, 11.6b).

current editions derive from Song editions.[15] The description given in Chen Zhensun's 陳振孫 (ca. 1186–ca. 1262) catalog, *Explanations of Titles in the Catalog of the Upright Studio*, compiled around 1249, matches current editions.[16] Chen Zhensun noted that *The Key to Ancient Prose* was an anthology of the work of Han Yu 韓愈 (768–824), Liu Zongyuan 柳宗元 (773–819), Ouyang Xiu 歐陽修 (1007–1072), the Su family (Su Xun 蘇洵 [1009–66], Su Shi, Su Che 蘇轍 [1039–1112]), and Zeng Gong 曾鞏 (1019–1083), edited and annotated by Lü Zuqian. This anthology was listed under the same title in the bibliographic treatise in *The History of the Song Dynasty*,[17] which also attributed the work to Lü Zuqian. Chen Zhensun recorded that the work had two chapters; current editions have preserved this format. The number of chapters recorded in the bibliographic treatise, twenty, is probably an editorial mistake or a typographical error.[18]

Judging from Qing catalogs[19] and Qing editions of *The Key*, Song editions were still extant in the eighteenth century. Xu Shuping's 徐樹屏 (*js.* 1712) collated edition from the Kangxi period (1662–1722) was based on two Song printed editions. Xu Shuping, whose father, Xu Qianxue 徐乾學 (1631–94), was noted among his contemporaries as an editor with high philological standards and had introduced him to the discipline of collation studies,[20] observed that the two Song editions were annotated in different ways. According to his diagnosis, one edition was a bit earlier than the other and had fewer critical marks in the margins than the slightly later edition. The second edition had

---

15. "The Key," instructions for reading and writing preceding the anthology in current editions, was attached as an appendix to an anonymous edition of anthologies associated with Lü Zuqian. This collection, *Xu zeng "Lidai zouyi" "Lize jiwen"—fulu "Guanjian," zengguang "Lize jiwen"* 續增 "歷代奏議" "麗澤集文"—附錄 "關鍵" 增廣 "麗澤集文," preserved at Beijing Library, has been dated to the Song period. Cf. Liu Zhaoren, *Lü Zuqian de wenxue yu shixue*, 64.

16. Chen Zhensun, *Zhizhai shulu jieti*, 15.451.

17. *SS*, 209.5411.

18. Wang Ruisheng, "Jin cun Songdai zongji kao," 73.

19. Shao, *Zengding Siku jianming mulu biaozhu*, 893; cited in Wang Ruisheng, "Jin cun Songdai zongji kao," 73. Cf. Zhu Shangshu, *Song ren zongji xu lu*, 133–39.

20. *ECCP*, 310–12. Xu Shuping acknowledged his debt to his father in his postface, reproduced at the end of his edition of *The Key*.

philological notes by Cai Wenzi 蔡文子, a Song scholar whose career could not be traced.[21]

Chen Zhensun attributed both the critical comments and the philological comments to Lü Zuqian; it appears from Xu Shuping's description that Lü made only the critical comments, and Cai Wenzi, who may have been commissioned by a commercial publisher to prepare an improved copy from earlier manuscript or printed editions, added philological notes. According to an anonymous postface, Lü's *Key* was based on a prior compilation,[22] and Lü had only added critical comments. Although this may have been the case—Lin Zhiqi 林之奇 (1112–76), Lü's teacher, compiled an Ancient Prose anthology, and Lü was named as its commentator in existing Song prints of this work[23]—there is no further evidence of a similar earlier compilation.

I consider *The Key* as a textbook based on Lü Zuqian's courses in composition. The selection may have been influenced by his teacher. As argued in Chapter 4, an Ancient Prose canon was taking shape in the twelfth century. Lü contributed to this process through his critical analyses of Ancient Prose texts. His use of critical marks and critical comments was recorded in the year-by-year biography written by his brother.[24] Zhu Xi attested and objected to it.[25]

*Ultimate Essays on Origins and Developments (from the Past to the Present)—(Newly Annotated for Examination Success)*
*([Xinjian jueke] [Gujin] yuanliu zhilun*
[新笺决科][古今]源流至論)

The first three installments of this encyclopedia contain essays by Lin Jiong 林駉 (fl. 1210s). Lin Jiong made a career out of private teaching in Longxi (Zhangzhou, Fujian circuit). Another collection of essays on

---

21. Lü Zuqian, *Guwen guanjian*, editorial notes preceding the text in Xu Shuping's edition.

22. The anonymous preface is located between the postfaces by Yu Yue 俞樾 (1821–1906) and Xu Shuping at the end.

23. For a further discussion of this anthology, *Observing Foamy Waves: A Collection of Texts—Organized by Category and with Notes Assembled by Lü Zuqian*, see Chapter 4.

24. Lü Zujian and Lü Qiaonian, "Donglai Lü Taishi nianpu," 233, 235.

25. *ZZYL*, 139.3321, translated in Chapter 4.

government resulting from his lectures, *The Imperial Mirror* (*Huangjian* 皇鑑), was completed around 1216.[26] Essays attributed to Lin Jiong in *Ultimate Essays* contain references to 1219 and 1226.[27] This implies that *Ultimate Essays* was completed after *The Imperial Mirror*. Huang Lüweng's 黃履翁 (*js.* 1232) preface to the first three installments was written in 1237. His preface to his own supplement, the fourth installment in current editions, dates from 1233. In this preface, Huang wrote that Lin Jiong's collected essays—he referred to the collection as *"Ultimate Essays"*—were already circulating widely. Prior to 1237, Lin's papers do not seem to have circulated under the full title quoted above. Huang explained in the 1237 preface that this title, *Ultimate Essays on Origins and Developments from the Past to the Present*, was adopted for the collection of Lin's essays on government "to widen its circulation [even more]." It is possible that some rearrangement, and even some false attributions, went into the 1230s edition of Lin's essays, but there are no indications to this effect in Huang's preface. A further investigation of *The Imperial Mirror* may shed light on this question.

No datable Song editions are extant. There are some references to Song editions in Qing catalogs. Most of these references are vague. Yang Shaohe 楊紹和 (1831–76) listed a Yuan edition he believed went back to an original Song edition. The main reasons for his claim were the blank spaces left after sequences referring to the Song imperial house.[28] Mo Youzhi 莫友芝 (1811–71) in *Record of Old Song and Yuan Titles I Have Seen* (*Song Yuan jiuben jingyan lu* 宋元舊本經眼錄) and Ding Richang 丁日昌 (1823–82) in *Catalog of the Preserving Quietude Studio* (*Chijing zhai xu zeng shumu* 持靜齋續增書目) listed a Song copy of an

---

26. A manuscript copy from the Qing dynasty has been preserved at Beijing University Library, but I have not yet been able to examine it. The table of contents and the use of footnotes to document arguments, as described in an entry on *The Imperial Mirror* in a Qing catalogue, suggest that this work bears close resemblance to *Ultimate Essays*; see Zhang Jinwu, *Airi jinglu cangshu zhi*, 26.9a–10a.

27. In the essay on the Veritable Records, there is a reference to a decree from 1219 ordering the compilation of the Veritable Records of the reign of Emperor Xiaozong (*Yuanliu zhilun*, qianji, 4.12a–b). In the essay "The Teachings of Master Zhu," a footnote refers to a "recent" pronouncement by the emperor in which he expressed his high regard for Zhu Xi's commentaries on the Four Books. The emperor's comment occurred in 1226 (ibid., qianji, 5.2a).

28. Yang Shaohe, *Ying shu yu lu*, 3.50b.

edition of the four collections but gave no further evidence for their claims.[29] In *List of Transmitted Editions Known to Me* (*Lüting zhijian chuanben shumu* 邵亭知見傳本書目), Mo Youzhi provided more detail on another Song edition. According to his notes, this edition, titled *Yuanliu zhilun*, was a Song print dating from the Jiayou 嘉祐 reign (1056–63).[30] The dating is problematic. The year *dingyou* 丁酉 in the Jiayou reign quoted in Mo's catalogue is equivalent to 1057. If we read Jiaxi *dingyou* 嘉熙丁酉, this printed edition would date from 1237, the year in which Huang Lüweng wrote his preface. If, less likely, we read Baoyou dingsi 寶祐丁巳, it could have been an edition from 1257, the earliest edition mentioned by Nakajima Satoshi.[31]

The format of this edition is unlike the format of most editions from the Yuan period and afterward. The edition Mo was reviewing was a mid-size-character edition (*zhongzi ben* 中字本), twenty-two characters per line and twelve lines per page. Most other editions are small-character versions (*xiaozi ben* 小字本), twenty-five characters per line, fifteen lines per page. The National Palace Museum owns a copy of a mid-size-character edition, which its catalogers have dated to the Yuan period.

The copy has been rebound and restored, and, in a few places, fragments from other editions have been copied in. This undated edition reverses the order of the first installment (*qianji*) and the third installment (*xuji*) compared to the Siku edition. The table of contents of each installment corresponds to the arrangement in other extant editions. Like other extant editions, this edition has top-margin comments and bears the full title *Ultimate Essays on Origins and Developments* (*from the Past*

---

29. Ding Richang, *Chijing zhai xu zeng shumu*, 5, zi, 7a; Mo Youzhi, *Song Yuan jiuben jingyan lu*, 1.28a. For more information on Mo Youzhi and his catalogs, see *ECCP*, 582. Since the edition mentioned in the latter catalog was a small-character edition, Mo Youzhi was here not referring to the copy mentioned in *Lüting zhijian chuanben shumu*. See below.

30. Mo Youzhi, *Lüting zhijian chuanben shumu*, 10.20b–21a. Mo Youzhi cited the full title in the case of the 1317 edition he listed below the Song edition. This may be taken as further proof that this was indeed a Song edition. The 1237 edition to which Huang Lüweng wrote a preface was also not upgraded with top margin headings. Note, however, that in his preface Huang cited *Gujin yuanliu zhilun* as the title.

31. Nakajima does not mention his source for this edition. See his review of *Ultimate Essays* in Hervouet, *A Sung Bibliography*, 328.

*to the Present)*—(*Newly Annotated for Examination Success*). Huang Lü-weng's supplement was not part of this edition.

In a recent bibliographical note inserted in front of the first page of the last volume of the third collection, the commentators argue that this edition was an incomplete commercial edition which confused the original order of Lin Jiong's installments. For the two commentators, the sequence in small-character Yuan editions makes more sense. In my view, the fact that this copy does not have the last collection does not mean that it was omitted by error. Huang Lüweng's collection was circulating separately.[32] Moreover, as I have shown in Chapter 6, the sequence in this edition makes more historical sense. As Lin Jiong learned about the growing influence of the Learning of the Way at court, his lectures and essays addressed its major thinkers and their work. No reference is made to Learning of the Way ideology in the third installment in the Siku edition, which came first in this edition. In contrast, the metaphysical foundations of Learning of the Way ethics came first in the first installment in the Siku edition—this was also the collection with the latest time references in Lin Jiong's three installments.

The first installment in the Siku edition has been the object of philological analysis. Qu Yong 瞿鏞 (19th c.) wrote that Ming editors reversed the order of the first and third collections and that they added Zhou Dunyi's 周敦頤 (1017–73) *Diagram of the Supreme Ultimate* and Zhu Xi's explanations of it at the beginning of their first installment. Qu saw this as a complete distortion of the original.[33] However, my review of Yuan editions shows that two types of editions were already circulating in Yuan times. *Diagram of the Supreme Ultimate* and Zhu Xi's commentary on it also appear in all extant Yuan editions.

There are three different Yuan editions. (1) The Dade 大德 edition: There is some confusion concerning this edition. Catalogs of the book collection of Ding Bing 丁丙 (1832–99) list an edition dated Dade 11 (1307). In his descriptive catalog *A Description of Books Preserved in the Rare Book Library* (*Shanben shushi cangshu zhi* 善本書室藏書志), Ding

---

32. There is an early Ming mid-size-character edition of Huang's collection in the Rare Book Room at the Harvard-Yenching Library.

33. Qu, *Tieqin tongjian lou cangshu mulu*, 17.11a. Cf. De Weerdt, "The Encyclopedia as Textbook."

cited an inscription printed at the end of the table of contents of the first volume of his copy. The inscription was written by Liu Kechang 劉克常, a publisher from Shulin in Jianyang. Ding Bing dated it to Dade *dingwei* 丁未, 1307. In the note, as transcribed by Ding, Liu Kechang related that the blocks of this edition had been damaged by fire. He had obtained a copy of the text and had famous scholars check it for errors. Liu Kechang further noted that this edition was divided into four collections as in preceding editions. Yang Shaohe's catalog listed a similar commercial edition from Jianyang. He pointed out that the two characters referring to the reign period had been blotted out. He cited circumstantial evidence from a blurb for a later commercial edition concerning damage on printing blocks caused by fire to argue that this edition dated from Zhizheng 至正 *dingwei*, i.e., 1367. I checked Ding Bing's transcription of the blurb against the inscription in an extant edition dated to 1367 preserved at Beijing Library (see below).[34] Even though some sentences have been left out, the wording is virtually identical. Since it is unlikely that Liu Kechang would have lived long enough to oversee two commercial editions published sixty years apart and since he made no reference to an earlier edition issued by his house, I conclude that Ding Bing's copy dated from 1367.

Ding Bing may have dated his copy back to the Dade reign period because an earlier catalog, *A Comprehensive Record of Remaining Titles Collected in Zhejiang* (*Zhejiang caiji yishu zonglu* 浙江採集遺書總錄), listed a commercial edition from Jianyang printed in the Dade reign period.[35] This edition was, however, published by a different printer, Mr. Zhan's Jianyang Academy Printshop 詹氏建陽書院.[36]

(2) The Yanyou 延祐 edition (Yanyou 3, 1317): Copies of this edition are still extant. Beijing Library, for example, holds one copy. This edition was printed at Yuansha Academy 圓沙書院, a business that specialized in encyclopedic reference materials.[37] It is a small-character edition of all four installments. The top-margin comments contain the

---

34. This inscription is cited in full in Yang Shaohe, *Ying shu yu lu*, 3.50b–51a.

35. Shen Chu, *Zhejiang caiji yishu zonglu*, xin, 10b–11a. Yang Shaohe refers to this edition in *Ying shu yu lu*, 3.50b.

36. For a list of Jianyang publishers, see Chia, *Printing for Profit*, 284.

37. Wu Kwang-tsing, "Chinese Printing Under Four Alien Dynasties," 491.

same headings as in the undated edition at the National Palace Museum in Taibei discussed above, but supplementary headings have been included. This edition also has the first and the third chapter in reversed order compared to the SKQS edition.[38]

(3) The Zhizheng 至正 edition (Zhizheng 27, 1367): A copy of this edition is preserved in the Beijing Library.[39] This is a low-quality edition. There are holes in the paper, and the characters are fading.[40] The publisher, Liu Kechang, wrote that this edition was based on a copy in the private possession of a certain Dong Yong 董鏞, an academician at the capital, and "an original copy," about which he provided no further details. He explained that he had hired several scholars to collate and punctuate the text. This edition features both punctuation (dots) and accentuation marks (jots). The top-margin comments are the same as in the other editions. The fact that the top-margin comments indicating the topics of the paragraphs of the essays are roughly the same in all editions argues that they must go back to an early authoritative edition. This edition has the same layout as the Siku edition; the first installment starts with the entry on the Supreme Ultimate. This means that the rearrangement of the first and the third series dates back to Liu Kechang's edition at the latest and cannot be attributed to Ming editors.

---

38. This edition appears throughout Qing catalogs. For examples, see Miao Quansun, *Yifeng cangshu ji*, 5.14b; Wang Wenjin, *Wenlu tang fangshu ji*, 3.33b; Mo Youzhi, *Lüting zhijian chuanben shumu*, 10.21a; Zhang Jinwu, *Airi jinglu cangshu zhi*, 26.7b–8b; Lu Xinyuan, *Bisong lou cangshu zhi*, 60.14b; and Qu, *Tieqin tongjian lou cangshu mulu*, 17.11a–b. The copy mentioned in Ye Dehui's *Xiyuan dushu zhi* (6.19a–b) is also listed as a Yanyou Yuansha edition. Whereas all other Yanyou copies are in small-character format, the copy described in this catalog was a mid-size-character edition. Ye wrote that the "Song copy" in Ding Richang's *Chijing zhai xu zeng shumu* was in fact a Yuan edition; he referred to the Yuan edition in Qu Yong's catalog to support his point. However, Qu's copy had a Yuansha imprint; this copy does not. Thus, this edition should not be listed as a Yuansha Yanyou edition. As a mid-size-character copy with the first and third series in reversed order (if compared to the Siku edition), it looks identical to the copy preserved at the National Palace Museum and the copy described by Mo Youzhi.

39. Another copy is described in Yang Shaohe, *Ying shu yu lu*, 3.50a–51a. Cf. my discussion of the Dade edition above.

40. On this edition, see also Chia, *Printing for Profit*, 135–36.

*The Epitome of Eminent Men Responding at the Imperial College*
(*Bishui qunying daiwen huiyuan* 璧水群英
待問會元 [選要]) (A) and
*A Net to Unite and Order the Massive Amounts of Information
in All Books* (*Qunshu huiyuan jie jiang wang*
群書匯元截江網) (B)

A. Recent catalogs do not list extant Song editions. One Song edition is listed in a catalog published in 1929 by the Jiangsu Guoxue Library.[41] This catalog also gives reproductions of the first folio of the first chapter and of the last three pages of the last chapter. The note preceding the reproductions argues that this is a typeset edition from the Chunyou period (1241–52).

The inscription on the last page mentions that the characters were written by Hu Sheng 胡昇 from the lower Yangzi region (Gusu 姑蘇), carved by Zhang Feng 章鳳, and typeset by Zhao Ang 趙昂 for the Communicating Lakes Printshop (Lize tang 麗澤堂).[42] External evidence on Hu Sheng reveals that he hailed from Qingjiang 清江 (Linjiang military prefecture 臨江軍) in Jiangnan and obtained the *jinshi* degree in 1268.[43] Both this edition and extant Ming editions contain, moreover, a preface by Chen Zihe 陳子和 dated 1245. Chen was a native of Jian'an and had obtained the *jinshi* degree in 1244.

The original edition was in ninety volumes. The Ming edition held at the National Library in Taibei has ninety volumes if one looks at the chapter numbers in the margins. The chapter numbers at the beginning of each chapter total eighty-two chapters. The discrepancy starts at chapter 34, where the chapter numbering in the margins suddenly jumps to 42. There is no direct evidence of missing chapters.

---

41. Jiangsu guoxue tushuguan, *Boshan shuying*, 1.58–61. Ding Bing also refers to this work in his 1908 catalog, *Shanben shushi cangshu zhi*, 20.11a–b. His comments are cited in the note preceding the reproductions in *Boshan shuying*.

42. "Communicating lakes" is a reference to the *dui* 兌 hexagram in *The Changes*. The image was used to refer to the beneficence of friends learning with and from one another. Cf. Wilhelm, *The I Ching*, 685–86.

43. *SRZJZL-BB*, 689.

In his preface to the Ming edition of 1509, *Selections from "The Epitome of Eminent Men Responding at the Imperial College,"* Wang Chi 王敕 recounted his odyssey in recovering the set and stated briefly that corrections, cuts, and additions were made where necessary. The Ming editors did not make major alterations to the organization and the content of the original. Chen Zihe mentioned the threefold organization scheme used in the Ming edition (subject area [*men* 門], category [*lei* 類], and menu [*jiemu* 節目]) in his 1245 preface. To judge from the reproductions of the alleged Song copy discussed above, it appears that the layout of the reproduced pages is identical to that in the Ming edition. A comparison with *A Net to Unite and Order the Massive Amounts of Information in All Books,* for which a Yuan edition is on record, further illustrates the continuity between pre-Ming and Ming editions.

B. *A Net to Unite and Order the Massive Amounts of Information in All Books* reads like an updated version of *The Epitome of Eminent Men Responding at the Imperial College.* According to the SKQS abstract, the full title of the encyclopedia read *A Net to Unite and Order the Massive Amounts of Information in All Books—An Enlarged Edition from the Imperial College (Taixue zengxiu qunshu huiyuan jie jiang wang* 太學增修群書匯元截江網).[44]

There are many overlaps between *A Net* and *The Epitome.* Besides the layout, the information in many sections is also identical. In a few cases where the information in *A Net* differs from *The Epitome,* the text cites examination essays dated after Chen Zihe's 1245 preface, which can be taken as the date of completion of *The Epitome.* All the dated essays are from either 1249, in the case of the local / lower-level examinations, or 1250, in the case of the departmental examinations.[45] I suspect *A Net* was published around 1250. The earliest known editions date

---

44. *SKTY,* 26.2802. Cf. Chapter 6.

45. Local-level examinations were also held in the Imperial College. At least one of the dated essays was written for the local-level examinations at the Imperial College held in 1249 (*Jie jiang wang,* 12.11b). Another essay was written for the local level examinations of 1249—the place where the examinations were held was not specified in this case (ibid., 17.11b–12a). For examples of passages from essays from the 1250 departmental examinations, see ibid., 3.19a–20a, 12.13a, 12.13b, 23.20a–b, 24.18a–b, 34.21b–22b.

from the late Yuan period. The SKQS edition was based on a Yuan edition that bore a preface by Hu Zhu 胡助.[46] According to the Siku editor, this preface dated from 1347. However, according to the bibliographer of *A Comprehensive Record of Remaining Titles Collected in Zhejiang*, the copy that served as the source text for the Siku edition was printed in 1344.[47]

### *Awesome Expositions from Ten Masters, with Notes* (*Shi xiansheng aolun zhu* 十先生奧論註)

The edition used by the Siku editors was anonymous and contained no references to the date of publication. Judging from the layout and the quality of the print, the editors concluded that this was a commercial edition from Masha, the district in Jianyang notorious for its cheap commercial prints.[48]

The original *Awesome Expositions* dates to the late twelfth century. Lü Zuqian referred to his pieces in *Awesome Expositions* as early compositions he wrote in preparation for the examinations.[49] There are frequent references to Lü Zuqian's "Expositions on the Seven Sages" ("Qi sheng lun" 七聖論) in *Standards for the Study of the Exposition*.[50] In the notes on the examination essays preserved in *Standards*, there are also references to the anonymous collection *Awesome Expositions*.[51] In one instance, the text mentions the author and the title of a piece from the collection. Liu Heqing's 劉和卿 essay on the hexagram *fu* seems to refer to the essay on the same subject by Liu Muyuan 劉穆元 in *Shi xiansheng aolun zhu*.[52] One of the authors, Fang Tian 方恬 (*js.* 1169), had his "awesome expositions" published.[53] In 1196, during the heat of the False Learning campaign, the

---

46. *SKTY*, 26.2802.

47. Shen Chu, *Zhejiang caiji yishu zonglu*, geng, 45a.

48. *SKTY*, 38.4167.

49. Lü Zuqian, *Donglai ji*, bieji, 10.44b; cited in Lin Sufen, "Lü Zuqian de cizhang zhi xue," 158.

50. *LXSC*, 2.1b–2a, 3a, 4a, 5.41a, 42a, 47a, 48b, 6.10a–b, 9.86b.

51. Ibid., 5.17a, 8.33b, 9.45a, 10.28a, 34b.

52. Ibid., 7.52b, and *Shi xiansheng aolun zhu*, qianji, 9.12b–14a. No further biographical information could be found for Liu Muyuan.

53. Wang Ruisheng, "Jin cun Songdai zongji kao," 76.

Directorate of Education reported that it had confiscated materials that propagated Learning of the Way discourse. Among the handful of titles listed in the report was "Awesome Expositions of Seven Masters,"[54] a compilation that was probably a source text for the current edition of *Awesome Expositions from Ten Masters.*

The current edition contains essays by sixteen authors. However, the Siku editors wrote that later additions had been made to their edition. The essays by Zhu Xi, Cheng Yi 程頤 (1033–1107), Zhang Lei 張耒 (1052–1112), and Zhang Shi 張栻 (1133–80) in the second installment were most definitely additions to the twelfth-century work mentioned by Lü Zuqian.

A similar anthology attributed to Chen Yizeng 陳繹曾, the Yuan dynasty author of two works on Ancient Prose stylistics, was probably based on the same source text. This anthology, titled *Awesome Expositions by Various Masters* (*Zhuru aolun* 諸儒奧論), featured the same essays by the same authors, but incorporated work by some Song Ancient Prose masters who had not been featured in *Awesome Expositions from Ten Masters, with Notes,* such as the three Sus and Chen Liang 陳亮 (1143–94). Essays by Learning of the Way authors were added in the fourth and last installment in this edition.[55] The first installment of *Awesome Expositions by Various Masters* is identical to the Yuan dynasty work *Awesome Expositions by Various Masters: The Foundations of the Mastery of the Policy Essay* (*Zhuru aolun cexue tongzong* 諸儒奧論策學統宗). According to a copy of a Yuan edition of the latter work, the editor of this collection was Tan Jinsun 譚金孫. The work may have been attributed to Chen Yizeng later on, because his style manual, *The Writer's Fishing-Net* (*Wen quan* 文筌), was attached to an edition of this work. Such an edition is listed in the catalog of the Library of the Humanities Research Institute at Kyoto University.[56]

---

54. *SHY* (AS), *XF*, 2.127. Cf. Zhu Shangshu, *Song ren zongji xu lu*, 346–48.

55. The National Library in Taibei holds a Ming dynasty print from 1617.

56. Kyōto daigaku, Jinbun kagaku kenkyūjo, *Kyōto daigaku Jinbun kagaku kenkyūjo kanseki mokuroku*, 1164b. Cf. *SKTY*, 38.4246–47.

*The Legitimate Seal of Compositions from the Past and Present—*
*A New Collection with Several Scholars' Criticism*
*(Xinbian zhuru pidian gujin wenzhang zheng yin*
新編諸儒批點古今文章正印*)* and
*A Complete Compilation of Ancient Prose—A Newly Printed*
*Edition with Famous Scholars' Criticisms*
*(Xinkan zhuru pingdian guwen jicheng*
(新刊諸儒評點古文集成)

A Song print of *The Legitimate Seal* with a preface from 1273 is preserved in excellent condition at the Palace Museum Library in Taibei. Liu Zhensun 劉振孫, notary of the administrative assistant to the military commissioner of Wu'an 武安 (Tanzhou 潭州, Jinghu South circuit), collected and organized the texts; Liao Qishan 廖起山, professor at the prefectural school of Raozhou 饒州 (Jiangnan East circuit), did the collation.[57] The anthology was published in four installments, a common technique in the commercial printing of encyclopedias and anthologies in the late Southern Song and Yuan periods.

I have seen no Song editions of *A Complete Compilation of Ancient Prose.* To judge from the use of taboo characters and from the fact that a blank space follows each reference to the Song dynastic house in memorials, the Siku editors argued that their edition was a Southern Song commercial edition.[58] Qing catalogs also list Southern Song copies.[59] Apart from his native place, Luling 盧陵 (Jizhou 吉州, Jiangnan West), no information is available on the compiler, Wang Tingzhen 王霆震. Only the first installment appeared or survived. The virtual identity of the two anthologies raises questions concerning their relationship. In my reading, Wang's edition came after Liu's.

First, in their preface and postface, Liu Zhensun and Liao Qishan claimed credit for the compilation and mention no previous antholo-

---

57. Liu Zhensun, *Gujin wenzhang zheng yin*, prefaces. No further information could be obtained about them. Cf. Zhu Shangshu, *Song ren zongji xu lu*, 429–30.

58. *SKTY*, 38.4163. Cf. Takatsu, "Sō Gen hyōten kō," 136–37.

59. Shen Chu, *Zhejiang caiji yishu zonglu*, xin, 6b; for a full discussion, see Fu Zengxiang, *Cangyuan qunshu tiji*, 927–31. Cf. Zhu Shangshu, *Song ren zongji xu lu*, 262–67.

gies *of this kind*. The term "new compilation" seems to refer to Liu's broader selection of texts and his inclusion of the comments of various previous critical anthologies. As explained in Chapter 6, their anthology drew from the Ancient Prose anthologies previously compiled and annotated by Lou Fang 樓昉 (*js.* 1193), Lü Zuqian, and Zhen Dexiu 眞德秀 (1178–1235).

Second, the selection of the texts for the genres in the existing first series of Wang's work is virtually identical to that in *The Legitimate Seal* (the latter is generally broader); the comments are the same. The organization of Wang's compilation is, however, an improvement over Liu's. For example, Liu's section on expositions is further organized by author, without a strict chronological line. The expositions in *The Complete Collection*, by contrast, are reorganized thematically, starting with all the essays concerning the ruler and continuing with those about ministers and the bureaucracy, the people, regulations, government, finance, border defense, morale, and ethical foundations. Wang's "new print" seems to refer to a reprint of a previously existing collection, probably *The Legitimate Seal*.

# Appendix B

———

# Tables

Southern Song Examination Manuals, by Category

Note: This table illustrates the examination-preparation manuals discussed in the main body of this text. It is not intended as a comprehensive overview of writing manuals for examination candidates. Many extant anthologies are not included.

A. Encyclopedias (*leishu* 類書)

| Title | Author | Date of completion/ first appearance | Date of publication | Place of publication | Earliest pre- and postfaces |
|---|---|---|---|---|---|
| *Lidai zhidu xiangshuo* 歷代制度詳説 | Lü Zuqian 呂祖謙 | early 1180s | Song[a] | | Peng Fei 彭飛 (ca. 1326) |
| (*Yongjia xiansheng*) *Bamian feng* (永嘉[先生])八面鋒 | Anon. (attrib. Chen Fuliang 陳傅良 or Ye Shi 葉適) | 12th c. | Song print?[b] | | early Ming postfaces |
| (*Xinjian jueke*) *Gujin yuanliu zhilun* (新箋決科)古今源流至論—前集、後集、續集、別集 (*qianji, houji, xuji, bieji*) | a. Lin Jiong 林駉 first 3 installments b. Huang Lüweng 黃履翁, 4th installment | a. 1220s–1230s b. ca. 1237 | a. in circulation in the 1230s b. ca. 1237 | Yuan editions: Yuansha Academy(?); Jianyang | a. Huang Lüweng, 1233 b. idem, 1237 |
| *Qunshu huiyuan jie jiang wang* 群書匯元截江網 | | early 1250s[c] | | | Hu Zhu 胡助, 1344 or 1347 |
| *Bishui qunying daiwen huiyuan* (*xuanyao*) 璧水群英待問會元(選要) | Liu Dake 劉達可 | ca. 1245 | Song movable type edition, ca. 1245 | Jianyang?[d] | Chen Zihe 陳子和, 1245 |

a. Lu Xinyuan claimed the existence of a Song edition, and early twelfth-century sources quote it. Cf. the Appendix.
b. Mentioned in Du Mu's preface, 1503.
c. Dated examination essays quoted in this work were written for the examinations of 1249 and 1250. The Siku editors dated it to the reign of Lizong (1225–64), based on references to the Duanping (1234–36) and Chunyou (1241–52) periods. Cf. the Appendix.
d. Both Liu Dake and Chen Zihe hailed from Jianyang.

Table B1 cont., B. Anthologies,
1. General Ancient Prose

| Title | Compiler | Date of completion/first appearance | Date of publication | Place of publication | Earliest pre- and postfaces | Organized by |
|---|---|---|---|---|---|---|
| Guwen guanjian 古文關鍵 | Lü Zuqian | ca. 1160–80 | Song prints[a] | | | author |
| Donglai jizhu leibian guan lan wenji (jiaji, yiji) 東萊集註 類編觀闌文集（甲集、乙集） | Lin Zhiqi 林之奇, Lü Zuqian | 2nd half of 12th c. | Song print[b] | | | genre, then period and author |
| Jingqi 精騎 | Lü Zuqian? | 1160s– early 1170s[c] | by 1173 a printed edition was available in Jianyang | Jianyang, Yongkang | | author and title |
| Yuzhai xiansheng biaozhu chong gu wen jue 迂齋先生標註 崇古文訣 | Lou Fang 樓昉 | 1220s | ca. 1227 | | (Chen Zhensun 陳振孫, 1226) Chen Sen 陳森, 1227[d] | period and author |
| Yuzhai biaozhu zhujia wen ji 迂齋標註 諸家文集[e] | Lou Fang | 1220s | ca. 1226 | | Chen Zhensun, 1226 | period and author |
| Wenzhang zhengzong 文章正宗 | Zhen Dexiu 真德秀 | 1220s–1230s | ca. 1232 | Song editions[f] | (outline and editorial explanations) | mode, then chronological |
| Xu Wenzhang zhengzong 續文章正宗 | Zhen Dexiu et al. | 1230s | ca. 1266 | Song editions[g] | Ni Cheng 倪澄, 1266 | mode, then theme, then author |

| | | | Yuan edition | | |
|---|---|---|---|---|---|
| Wenzhang guifan 文章規範 | Xie Fangde 謝枋得 | mid 13th c. | | Wang Yuanji 王淵濟, late 13th c.[h] | pedagogical stages: elementary to advanced |
| (Xinbian zhuru pidian) gujin wenzhang zhengyin (qianji, houji, xuji, bieji) 新編 諸儒批點 古今文章正印 (前集, 後集, 續集, 列集) | Liu Zhensun 劉振孫 | ca. 1273 | ca. 1273 | Liu Zhensun, 1273; Liao Qishan 廖起山, 1273 | genre, then author or subject and author |
| (Xinkan zhuru pingdian) guwen jicheng (qianji) (新刊諸儒評點) 古文集成—前集[i] | Wang Ting-zhen 王霆震 | 2nd half of the 13th c. | 2nd half of the 13th c.[j] | | genre, then author or subject and author |
| Shi xiansheng aolun zhu (qianji, houji, xuji) 十先生奥論註 (前, 後, 續集) | | 13th c.? | (Song Masha print)[k] | | subject and author |
| Wensui 文髓 | Zhou Yinglong 周應龍 | manuscripts circulated among family members and students from the mid-13th c. on | ca. 1428 | postface from 1428 (biographies of Zhou Yinglong appended) | author |

a. Xu Shuping's edition was based on two Song printed editions. Cf. Appendix.

b. Song print, Beijing Library.

c. Lin Sufen, "Lü Zuqian de cizhang zhi xue," 148; Bol, "Reading Su Shi in Southern Song Wuzhou." The copy at the National Library, Taibei, was printed in Yongkang. Zhu Xi writes that the copy he saw was printed in Jianyang.

Table B1, B. Anthologies

1. General Ancient Prose notes, cont.

d. Takatsu, "Sō Gen hyōten kō," 133.

e. This edition in five *juan* is the one Chen Zhensun and Ma Duanlin referred to. Only three *juan* are preserved in the copy at the Beijing Library. From a comparison of the extant chapters, it appears that only a few additions were made to the edition in 20 *juan*. The layout and the authors and pieces included are basically the same.

f. Several Song editions are mentioned in Qing catalogs. See Wang Wenjin, *Wenlu tang fangshu ji*, 5.29b; Yu Minzhong, *Tianlu linlang shumu*, 3.41a–b; Qu, *Tieqin tongjian lou cangshu mulu*, 24.29a–b.

g. National Library, Taibei. See also Yu Minzhong, *Tianlu linlang shumu*, 3.41b–43a; Peng Yuanrui, *Tianlu linlang shumu houbian*, 7.22a–b.

h. Takatsu, "Sō Gen hyōten kō," 139.

i. This is a re-edition of *Guwen zheng yin*. The internal organization has gone through some alterations.

j. For references to Song copies, see Shen Chu, *Zhejiang caiji yishu zonglu*, xin 辛, 6b; for a full discussion, see Fu Zengxiang, *Cangyuan qunshu tiji*, 927–31. Cf. Appendix.

k. Cf. Appendix.

Table B1, B. Anthologies, cont.
2. Specific authors

| Title | Compiler | Date of completion/ first appearance | Date of publication | Place of publication | Earliest pre- and postfaces | Organized by |
|---|---|---|---|---|---|---|
| Zengguang zhushi yinbian Tang Liu xiansheng ji 增廣註釋音辯唐柳先生集ᵃ | Liu Yuxi 劉禹錫 et al. | ca. 1167 | ca. 1167 | Masha | Lu Zhiyuan 陸之淵, 1167/1168 | genre |
| Donglai biaozhu Laoquan xiansheng wenji 東萊標註老泉先生文集 | Lü Zuqian, compilation and comments Wu Yan 吳炎, collation | ca. 1193 | ca. 1193 | | Wu Yan, 1193 | genre |
| Quandian Longchuan Shuixin xiansheng wencui 圈點龍川水心先生文粹 | | ca. 1212 | ca. 1212 | Jian'an | Rao Hui 饒輝, 1212 | genre and author |
| Pidian fenlei Chengzhai xiansheng wenkuai 批點分類誠齋先生文膾 | Li Chengfu 李誠夫 | ca. 1259 | ca. 1259 | Jian'an | Fang Fengchen 方逢辰, 1259 | subject |
| Xinkan Jiaofeng pidian Zhizhai lunzu 新刊皎峰批點止齋論祖 | Anon., critical comments by Fang Fengchen | ca. 1268 | ca. 1268 | | Fu Canzhi 傅參之, 1268 | essay title |

a. Includes a list of compilers and annotators from Liu Yuxi in the 9th c. to Pan Wei 潘緯 in the 12th c.

Table B1, B. Anthologies, cont.
3. Examination Essays

| Title | Compiler | Date of completion/first appearance | Date of publication | Place of publication | Pre- and postfaces | Organized by |
|---|---|---|---|---|---|---|
| (Pidian fen'ge leiyi jujie) Lunxue Shengchi (批點分格類意句解) 論學繩尺 | Wei Tianying 魏天應 Lin Zizhang 林子長 | 1270s | | | You Cui 游萃, 1330s | style register |
| (Jingxuan) Huang Song cexue shengchi (精選) 皇宋策學繩尺 | | mid 13th c. | | | | essay question |

C. Style Manuals

| Title | Author | Date of completion/first appearance | Date of publication | Place of publication | Earliest pre- and postfaces |
|---|---|---|---|---|---|
| "Guwen guanjian" "Lunjue" 論訣 | Lü Zuqian | ca. 1160–1180 1st half of the 13th c. | Song print | | |
| Fuzao wenzhang baiduan jin 黼藻文章百段錦 | Fang Yisun 方頤孫 | ca. 1249 | ca. 1249 | Jian'an | Chen Yue 陳樾, 1249 |
| Cixue zhinan 辭學指南 | Wang Yinglin 王應麟 Zeng Jian 曾堅? | 1250s | Yuan period[a] | | |
| Dace mijue 答策秘訣 | Liu Jinwen 劉錦文, ed. | 1230s | Yuan period[b] | Jianyang | |

a. Langley, "Wang Ying-Lin," 478. b. Cf. Appendix.

Table B1, cont.,
D. Learning of the Way Manuals

| Title | Compiler | Date of completion/first appearance | Date of publication | Place of publication | Organized by |
|---|---|---|---|---|---|
| Jinsi lu 近思錄 | Zhu Xi 朱熹, Lü Zuqian | 1175 | several Song editions[a] | | Learning of the Way concepts (13) |
| Beixi ziyi 北溪字義 | Chen Chun 陳淳 | 1220s | Song prints in the 1220s and 1240s[b] | | Learning of the Way concepts (26) |
| Wenchang ziyong fenmen Jinsi lu 文場資用分門近思錄 | | | 13th c. | Jian'an | Learning of the Way concepts (121) |
| Hui'an xiansheng yulu da gangling—fulu 晦庵先生語錄大綱領—附錄 | | mid 13th c. | mid 13th c. | | Learning of the Way concepts (25) |
| Zhuzi jingji wenheng leibian (qianji, houji, xuji) 朱子經濟文衡類編 (前集, 後集, 續集) | Teng Gong 滕珙 | 13th–early 14th c. | 1324 | | Learning of the Way concepts (64), historical subjects (75), issues in administration (52), subthemes |
| (Xinbian yindian) Xingli qunshu jujie (qianji, houji) 新編音點性理群書句解—(前集, 後集) | Xiong Jie 熊節, compilation; Xiong Gangda 熊剛大, annotation | mid 13th c. | Song editions[c] | Jian'an | genre, then Learning of the Way canon |

a. For examples, see Jiang Biao, Song Yuan ben bangge biao, 1.12a, 2.40b.
b. Inoue, "Hokki jigi hanpon kō," 14.
c. Several Qing catalogs list Song editions. For examples, see Miao Quansun, Yifeng cangshu ji, xuji, 2.7a–9a; Lu Xinyuan, Bisong lou cangshu zhi, 41.5a–b.

Table B2

Examination Essays in *Standards for the Study of the Policy Response* and
*Standards for the Study of the Exposition* by Examination Level

A. Examination Essays by Examination Level

| Type/institution | Subtype | Standards for the Study of the Policy Response | Standards for the Study of the Exposition[a] |
|---|---|---|---|
| Departmental | | 1 | 44 |
| Local | | | 4 |
| Avoidance | Departmental | | 5 |
| | Local | 1 | 3 |
| | SUBTOTAL | 1 | 8 |
| Imperial College | Monthly | 8 | 45[b] |
| | Yearly | 2 | 18 |
| | Upper hall | 2 | 3 |
| | Local | 1 | 3 |
| | SUBTOTAL | 13 | 69 |
| Local school | | 3 | 2 |
| Imperial clan school | | | 1 |
| Uncertain | | 1 | 27[c] |
| TOTAL | | 19 | 155 |

a. The Yuan edition at the Seikadō in Tokyo mentions the type of examination for which the essay had been written as well as the ranking the authors of the expositions attained. The majority gained first place in the relevant examination. The total number of essays in the collection is sometimes listed as 156; one essay was not counted here because it is incomplete and its status in the original collection uncertain.

b. This number includes four Directorate School (Guoxue) examination essays. For more information on the different kinds of examinations held at the Imperial College, see Wang Jianqiu, *Songdai Taixue yu Taixuesheng*, 60–78; Zhu Zhongsheng, "Songdai Taixue zhi qushi ji qi zuzhi"; *Zhongguo lishi da cidian—Song shi*, 65, "*gongshi*," 210, "*sishi*"; and Chaffee, *The Thorny Gates*, 103, 108.

c. This number includes the model essays of famous authors such as Chen Fuliang and Ouyang Qiming.

Table B2, cont.

B. Learning of the Way Essays in *Standards for the Study of the Exposition* by Examination Level[a]

| Type/institution | Subtype | pre-1200 | 1200–1209 | 1210–19 | 1220–29 | 1230–39 | 1240–49 | 1250–59 | 1260–69 | Undated |
|---|---|---|---|---|---|---|---|---|---|---|
| Departmental | | | | | 1 | 4 | 2 | 8[b] | 4 | 4 |
| Local | | | | | | 1 | | | | |
| Avoidance | Departmental | | | | | 1 | 1 | | 1 | 1 |
| | Local | | | | | | 1 | | 1 | 2 |
| | SUBTOTAL | | | | | 1 | | | | 3 |
| Imperial College | Monthly | | | | | 1 | | | 5 | 8 |
| | Yearly | | | | | | | 1 | 1 | 2 |
| | Upper hall | | | | | | | | | |
| | Local | | | | | | | | | |
| | SUBTOTAL | | | | | 1 | | 1 | 6 | 10 |
| Local school | | | | | | | | | | |
| Imperial clan school | | | | | | | | | | |
| Uncertain | | 2 | | 1 | 2 | | | | | 1 |
| TOTAL | | 2 | 0 | 1 | 3 | 7 | 3 | 9 | 11 | 18 |

a. An essay is considered a Learning of the Way essay if it refers to the ideas or the writings of Zhou Dunyi, the Cheng brothers, Zhang Zai, Zhu Xi, or Zhang Shi. In contrast to policy essays, references in expositions were not explicit; candidates did not cite sources. In *Standards for the Study of the Exposition*, direct citations and borrowings from Learning of the Way textual sources were indicated in the commentary and footnotes.

b. The higher number of departmental examinations in this decade is impacted by the fact that four departmental examinations were held in the 1250s, whereas only three departmental examinations were held during the 1230s, 1240s, and 1260s.

Table B2, cont.
C. Learning of the Way Essays in *Standards for the Study of the Policy Response* by Examination Level

| Type/institution | Subtype | Learning of the Way | Zhu Xi |
|---|---|---|---|
| Departmental Local | | 0 | 0 |
| Avoidance | Departmental | | |
| | Local | 1 | 1 |
| | SUBTOTAL | 1 | 1 |
| Imperial College | Monthly | 3 | 3 |
| | Yearly | 2 | 0 |
| | Upper hall | 1 | 0 |
| | Local | 1 | 1 |
| | SUBTOTAL | 7 | 4 |
| Local school Imperial clan school | | 1 | 1 |
| Uncertain | | 1 | 1 |
| TOTAL | | 10 | 7 |

Table B3

Dates of the Essays in *Standards for the Study of the Exposition* and the
Impact of the Learning of the Way in the Thirteenth Century

| Period | Total number of essays[a] | Number of Learning of the Way essays[b] | Number of essays referring to Zhu Xi |
|---|---|---|---|
| pre-1200 | 27 | 2 | 1 |
| 1200–1209 | 3 | 0 | 0 |
| 1210–19 | 3 | 2 | 1 |
| 1220–29 | 8 | 3 | 1 |
| 1230–39 | 9 | 7 | 2 |
| 1240–49 | 7 | 3 | 1 |
| 1250–59 | 15 | 9 | 5 |
| 1260–69 | 15 | 11 | 8 |
| Uncertain | 68 | 32 | 17 |

a. The numbers listed here differ from those compiled by Zhang Haiou and Sun Yaobin ("Lunxue shengchi yu Nan Song luntiwen," 91). The larger number of essays of unknown date in this table derives in part from the fact that I listed all essays for which no exact date can be estimated in the "uncertain" column, whereas Zhang and Sun roughly divide essays up into those written before and those written after 1213. Their list of essays of unknown date (44) does not include essays that can be roughly dated before 1213. Their article appeared as this book went to press, and I was therefore unable to double check all areas of disagreement.

b. See Table B2B, note a, for the criteria for determining Learning of the Way essays.

Table B4
Sources of Exposition Topics

A. In *Standards for the Study of the Exposition* (*Lunxue shengchi*);
*Chen Fuliang, the Founding Father of the Exposition* (*Zhizhai lunzu*);
and *Ouyang's Model Expositions* (*Ouyang lunfan* 歐陽論範)[a]

| Source | *Lunxue shengchi*[b] | *Zhizhai lunzu* | *Ouyang lunfang* |
|---|---|---|---|
| *Records of the Historian* | 3 | | |
| *History of the Han Dynasty* | 54 | 7 | 15 |
| *History of the Later Han Dynasty* | 5 | 2 | 2 |
| *History of the Jin Dynasty* | 2 | | |
| *New History of the Tang Dynasty* | 13 | 5 | 5 |
| *Zuo's Commentary*[c] | 2 | | 1 |
| *Guliang's Commentary*[c] | | 1 | |
| *Gongyang's Commentary*[c] | | | 1 |
| *Classic of Filial Piety* | 1 | | |
| *Analects* | 11 | 8 | 7 |
| *Mencius* | 23 | 8 | 9 |
| *Laozi* | 1 | | |
| *Zhuangzi* | | 2 | 1 |
| *Xunzi* | 10 | 2 | 6 |
| Works of Yang Xiong | 14 | | 6 |
| Xun Yue's *Extended Reflections* (*Shenjian lun*) | | | 1 |
| *Wenzhongzi* | 7 | 1 | 4 |
| *Lu Zhi's Memorials* | 4 | | 1 |
| Works of Han Yu | 3 | 3 | 1 |
| Works of Liu Zongyuan | | | 1 |
| *Selections of Texts* (*Wenxuan*) | 2 | | |
| TOTAL | 155 | 39 | 61 |

a. *Ouyang's Model Expositions* is a collection of sample examination essays by Ouyang Qiming 歐陽 起鳴 (fl. late Southern Song-Yuan times). Like Chen Fuliang, Ouyang was an authority on exposition writing. Three of Ouyang's model essays were selected in *Standards for the Study of the Exposition*; his comments on exposition writing were included in *The Art of the Exposition* (*Lun jue*). In addition to the Siku quanshu cunmu congshu edition (reprint of a 1471 edition), I used the similarly dated copy preserved at the Naikaku bunko, Tokyo.

b. Zhang Haiou and Sun Yaobin list slightly higher numbers for the historical sources ("*Lunxue shengchi* yu Nan Song luntiwen," 95). I found this article as the book went to press and was therefore unable to determine the reasons behind the discrepancies.

c. On the *Spring and Autumn Annals*.

Table B4, cont.

B. In *Standards for the Study of the Exposition*, 1150–1200

| Source | Number |
| --- | --- |
| *Records of the Historian* | 1 |
| *History of the Han Dynasty* | 10 |
| *Zuo's Commentary* | 1 |
| *New History of the Tang Dynasty* | 2 |
| *Analects* | 3 |
| *Mencius* | 4 |
| *Xunzi* | 1 |
| Works of Yang Xiong | 3 |
| Works of Han Yu | 1 |
| TOTAL | 26 |

Table B5

Subjects Covered in Twelfth- and Thirteenth-Century Policy Questions

A. In Zeng Jian's 曾堅 *Secret Tricks for*
*Responding to Policy Questions* (*Dace mijue* 答策秘訣)

---

1. The way of government (*zhidao* 治道)
   The situation of the state (*guoshi* 國勢)
   Decrees (*haoling* 號令)
   Using penal law (*yong xing* 用刑)
   Rewards and punishments (*shangfa* 賞罰)
   Success and failure in government throughout history (*gujin zhidao deshi* 古今治道得失)
   The greatness and decadence of the ruler's virtue (*junde longti* 君德隆替)
2. The learning of the sages (*shengxue* 聖學)
   The Classics and the histories (*jingshi* 經史)
   The Precious Instructions[a] (*baoxun* 寶訓)
   Lecturing at court (*jinjiang* 進講)
   Reading at court (*jindu* 進讀)
3. Regulations (*zhidu* 制度)
   Ritual utensils (*qiyong* 器用)
   The honorary guard (*yiwei* 儀衛)
   Military organization (*bingzhi* 兵制)
   Carriages and robes (*jufu* 車服)[b]
   Palace buildings (*gongshi* 宮室)
   The bureaucracy (*guanzhi* 官制)
   Sacrifices (*jisi* 祭祀)
   Ritual observances (*liyi* 禮儀)
   Ritual and music (*liyue* 禮樂)
   Land management (*tianzhi* 田制)
      Corvee regulations (*yifa* 役法)
      The well-fields (*jingtian* 井田)
      Estates (*yuantian* 園田)
      Water conservation (*shuili* 水利)
   Schools (*xuexiao* 學校)
      Establishing schools and regulations (*li xue zhidu* 立學制度)
      Rise and decline through history (*gujin xingfei* 古今興廢)
      The ceremonies of the biannual sacrificial offerings and the regular food offerings (*shi dian zhi yi* 釋奠之儀)[c]
      Methods of instruction (*jiaoyang zhi fa* 教養之法)
4. The learning of the nature (*xingxue* 性學)
   The unity of the Way (*daotong* 道統)
   The transmission through famous scholars (*zhuru chuanshou* 諸儒傳授)
5. Recruiting talent (*qu cai* 取材)
   Selection by recommendation (*jianju* 薦舉)
   Selection for promotion and placement (*quanxuan* 銓選)

Table B5, A cont.

---

Examination tracks (*kemu* 科目)

Evaluations (*kaoke* 考課)

Long tenure (*jiu ren* 久任)

Hits and misses in the ruler's employment of men (*renjun yong ren deshi* 人君用人 得失)

Generals on the frontiers (*bianfang jiangshuai* 邊防將帥)

The mores of the scholar-officials (*shidafu fengsu* 士大夫風俗)

6. Human talent (*rencai* 人才)

Capacities (*caipin* 才品)

Learning (*xueshu* 學術)

Integrity (*jieyi* 節義)

Factionalism (*pengdang* 朋黨)

Talented men living in retirement (*yinyi* 隱逸)

Scholarship (*rushu* 儒術)

The morally superior and the morally inferior (*junzi xiaoren* 君子小人)

Talented men from the period of the Three Kingdoms and the Six Dynasties (*sanguo liuchao rencai* 三國六朝人才)

Inconsistencies between Confucius' words and his actions (*Kongzi yanxing bu tong* 孔子言行不同)

7. Writing (*wenzhang* 文章)

Genres (*wenti* 文體)

The ranking of literature from the past and the present (*gujin wenzhang gaoxia* 古今文章高下)

Compositions of famous authors from the past and present (*gujin mingru zhushu* 古今名儒著述)

8. Geopolitics (*xingshi* 形勢)

Victory and defeat (*shengbai* 勝敗)

Reunification (*huifu* 恢復)

9. Disasters (zaiyi 災異)

Astronomical anomalies (*xing bian* 星變)

Meteorological anomalies (*lei bian* 雷變)

Droughts (*shuihan* 水旱)

Locusts (*minghuang* 蝗蝗)

10. Remonstration (*jianyi* 諫議)

Soliciting opinion (*qiu yan* 求言)

Receptivity (*tingna* 聽納)

Presenting criticism (*jinjian* 進諫)

Organizing debates (*jiyi* 集議)

Public debate (*gongyi* 公議)

11. Questions about the Classics (*jingyi* 經疑)

Questions about the Six Classics (*liujing yi* 六經疑)

Discrepancies between the commentaries of famous scholars (*zhuru zhushu yitong* 諸儒注疏異同)

(*cont. on next page*)

Table B5, A, cont.

---

(11. *Questions about the Classics, cont.*)

  *Records of the Historian* (*Shiji* 史記)

  Historiography in the past and the present (*gujin zuo shi* 古今作史)

12. Astronomical observations (*lixiang* 曆象)

  Astronomy (*tianwen* 天文)

  Instruments used in the calculation of the calendar (*tuibu zhi qi* 推步之器)

  Differences in the nomenclature and depiction of the celestial bodies (*xiangwei mingming quxiang butong* 象緯命名取象不同)

---

a. The Precious Instructions were compilations of an emperor's sayings and decrees.

b. The carriages and robes refer to the types of carriages an official was allowed to use or the kind of robes one was allowed to wear according to one's rank. The locus classicus is in *The Book of Documents*, "The Canon of Shun," see Legge, *The Chinese Classics*, vol. 3, *The Shoo King*, 37.

c. Zeng mentions the *shidian* 釋奠 and *shicai* 釋菜 ceremonies in his discussion of questions on regulations. For a short description of these ceremonies, see Neskar, "The Cult of Worthies," 175.

Table B5, cont.
B. In *The Epitome of Eminent Men Responding at the Imperial College* (A)
and *A Net to Unite and Order the Massive Amounts of Information in All Books* (B)[a]

| A | B |
|---|---|
| Reform (*gexin* 革新)[b] | |
| The most urgent matters in current affairs | |
| Establishing the root of the state | |
| The recruitment of high officials | |
| The imposition of administrative punishments | |
| Eliminating disasters | |
| The teaching of the sages (*shengxue* 聖學) | |
| The teaching of the sages | |
| The Classics mat | |
| The emperor's writings | The emperor's writings (*shengzhi* 聖製) (chap. 1) |
| The emperor's calligraphy | The emperor's calligraphy (*shenghan* 聖翰) (chap. 2) |
| The way of the ruler (*jundao* 君道) | |
| The ruler's mind | |
| The ruler's judgment | |
| The ruler's virtue | |
| Revering heaven | Revering heaven (*jingtian* 敬天) (chap. 3) |
| Taking the imperial ancestors as the model | Taking the imperial ancestors as the model (*fazu* 法祖) (chap. 4) |
| Cherishing the people | |
| Employing people | |
| Listening to advice | |
| The way of government (zhidao 治道) | |
| The essence of government | |
| Government by law | The guidelines of the dynasty (*jigang* 紀綱) (chap. 17) |
| | Decrees and orders (zhaoling 詔令) (chap. 18) |
| Calling to account | [Regulations (fadu 法度)] (chap. 19) |
| The situation of the state (*guoshi* 國勢) | |
| The court line | The court line (*guolun* 國論);[c] [war, self-protection and peace (*zhan, shou, he* 戰守和)] (chap. 24) |
| Rewarding achievements | |
| Penal law | |

Table B5, B cont.

| A | B |
| --- | --- |
| The way of the minister (*chendao* 臣道) | |
|     The ethos of the scholar-officials | |
|     The evil and the good | |
|     Reputation and integrity | |
|     Corruption | |
|     Serving and withdrawing | |
| Officials and clerks (*guanli* 官吏) | |
|     The administrative system | The bureaucracy (*guanzhi* 官制) (chap. 8) |
|     The councilors | |
|     The censors | |
|     The drafters | |
| Selection (*xuanju* 選舉) | |
|     Human talent | |
| The occupations of the scholar (*rushi* 儒事) | |
|     The learning of the scholar | Heterodoxy (*yiduan* 異端) (chap. 34) |
|     Literary genres | |
|     Points of doubt in the Classics | |
|     The Four Books | |
|     The philosophers | The philosophers (*zhuzi* 諸子) (chap. 35) |
|     The histories | The histories (*zhushi* 諸史) (chap. 29) |
|     The history of the reigning dynasty | The history of the reigning dynasty (*guoshi* 國史) (chap. 30) |
| The learning of the Way (Daoxue 道學) | The transmission of the teachings of the Cheng brothers (*Yiluo chuanshou* 伊洛傳授) (chap. 31) |
|     The lineage of the Way | Zhu Xi (Huian 晦庵) (chap. 33); Lu Jiuyuan (Xiangshan 象山) (chap. 32) |
|     The learning of the Way | |
| Nature and coherence (*xingli* 性理) | |
|     The learning of coherence | |
|     The learning of the mind | |
|     The Supreme Ultimate | |
|     The Five Constants | |
| Matters relating to the people (*minshi* 民事) | [The population (*hukou* 戶口] (chap. 26) |
|     The people's customs | Customs (*fengsu* 風俗) (chap. 27) |
|     Land distribution | |
|     Corvée regulations | Corvée regulations (*yifa* 役法) (chap. 28) |
|     Famine relief | |
|     Vagrants | |

Table B5, B cont.

| A | B |
|---|---|
| Military affairs (*wushi* 武事) | |
|     Generals | Generals (*jiangshuai* 蔣帥) (chaps. 21–22) |
|     The military organization | [State armies (*junguo bing* 郡國兵)] (chap. 13); [the infantry and the cavalry (*buji* 步騎)] (chap. 15) |
|     Local militia | Local militia (*minbing* 民兵) (chap. 14) |
|     The administration of the armies | |
|     The navy | |
|     The horse administration | The horse administration (*mazheng* 馬政) (chap. 25) |
|     Provisions | |
|     Camps | |
|     Training | |
|     Fortifications | |
|     Border defense | |
|     Strategic locations | [Military strategy (*bingfa* 兵法)] (chap. 16) |
|     The treatment of barbarians | |
|     Surrender | |
|     Quelling banditry | [Pacification campaigns (*pingshu* 平戍)] (chap. 23) |
| Accounting (caiji 財計) | accounting (*kuaiji* 會計) (chap. 9) |
|     Expenditure | [The treasury (*fuku* 府庫)] (chap. 10) |
|     Storehouses | Storing [grain] (*chuji* 儲積) (chap. 5) |
|     Grain transport | Grain transport (*caoyun* 漕運) (chap. 6) |
|     Taxation | |
|     Equitable marketing regulations | Equitable marketing (Hedi 和糴) (chap. 7) |
|     Paper money | Paper money (*chubi* 楮幣) (chap. 12) |
|     Metal currency | Money (*qianbo* 錢帛) (chap. 11) |
|     Commercial regulations | |
|     Monopolies | |
| Ritual codes (*lidian* 禮典) | |
|     Sacrifices | |
| The numerological sciences (*shuxue* 數學) | |
| Numerology | |
| The Chart from the River and the Book from Luo | |
| (Yang Xiong's) *The Ultimate Origin* and (Sima Guang's) *The Hidden Beginning* | |
| Medicine (the alternation of the five phases and the six types of energy) | |

*Appendix B*

Table B5, B cont.

| A | B |
| --- | --- |
| Music and the calendar | |
| The calendar | |

a. B is based on A; see Appendix. The categories in B are listed next to the corresponding categories in A. Categories in B with no exact equivalents in A are listed next to the categories to which they are most closely related and enclosed in square brackets. The original chapter numbers are shown in parentheses.

b. The text reads *cuixin* 萃新, which one could translate as "gathering the new." However, this is not a compound. *Cui* probably stands for *ge* 革, since both characters look very similar. *Gexin* is a compound meaning "changing and renovating" or "change and renovation." The subcategories indicate that the term refers to a reformation package and not just to "the recruitment of new people," which would be the most plausible translation for *cuixin*. Throughout the reigns of Ningzong and Lizong, several programs of political reinvigoration (*genghua*) were announced. See Chapter 6.

c. I translate *guolun* as "court line" in accordance with Yü Yingshi's discussion of the history of the term in the twelfth century. Rather than "the discussion of state affairs" broadly defined, the term was used to refer to whatever policy set the basic guidelines for court policy. Court lines were at the center of factional struggles and functioned in ways similar to "the party line" in twentieth-century factional politics; see Yü Yingshi, *Zhu Xi de lishi shijie*, vol. 1, chaps. 5, 7.

*Reference Matter*

# Works Consulted

*Primary Sources*

This list includes all Chinese-language titles published before the twentieth century. In the publication data, the first year is the date of publication of the edition used. Dates in parentheses refer to the first publication or indicate the era in which the work was completed. For works in the Siku quanshu series (SKQS), the 1983 edition by Taiwan Shangwu yinshuguan, Taibei, has been used. For most sources, the year of completion and the date of the first publication are unknown; in many cases, exact dates cannot be determined. For the collected works of many Song writers, centuries and decades have been used; when Song prints of their collected writings are documented, this has been mentioned. Dates of authors are given only in the main body of the text.

Ban Gu 班固. *Han shu* 漢書. Beijing: Zhonghua shuju, 1962 (1st c. CE).

Bao Hui 包恢. *Bizhou gao lue* 敝帚藁略. SKQS (13th c.).

Bi Yuan 畢沅. *Xu Zizhi tongjian* 續資治通鑒. 2 vols. Beijing: Zhongguo guoji guangbo chubanshe, 1993 (1801).

Buyeshanren 不夜山人 and Ni Shiyi 倪士毅. *Song ren jingyi yue chao* 宋人經義約鈔. 2 vols. Kaifeng: n.p., 1898.

Chen Chun 陳淳. *Beixi daquan ji* 北溪大全集. SKQS (ca. 1220s).

———. *Beixi ziyi* 北溪字義. Lixue congshu. Beijing: Zhonghua shuju, 1983 (1220s).

———. *Beixi ziyi* 北溪字義. SKQS (1220s).

Chen Fuliang 陳傅良. *Chen Fuliang xiansheng wenji* 陈傅良先生文集. Ed. Zhou Mengjiang 周梦江. Hangzhou: Zhejiang daxue chubanshe, 1999 (1213).

———. *Lidai bingzhi* 歷代兵制. SKQS (late 12th c.).

———. *Zhizhai xiansheng wenji* 止齋先生文集. Sibu congkan chubian. Shanghai: Shangwu yinshuguan, 1929? (reprint of a Ming Hongzhi [1488–1505] edition; 1213).

———. *Zhizhai ji* 止齋集. SKQS (1213).

Chen Liang 陳亮. *Chen Liang ji (zengding ben)* 陳亮集 (增訂本). 2 vols. Beijing: Zhonghua shuju, 1987 (1204; ca. 1214 print).

Chen Yizeng 陳繹曾. *Wenzhang ouye* 文章歐冶 (*Wen quan* 文筌). Wakokuhon kanseki zuihitsu shū. Tokyo: Kyūko shoin, 1977 (reprint of a 1688 Japanese reprint of a 1552 Korean edition; 1332).

Chen Zao 陳藻. *Lexuan ji* 樂軒集. Ed. Lin Xiyi 林希逸. SKQS (first half 13th c.).

Chen Zhensun 陳振孫. *Zhizhai shulu jieti* 直齋書錄解題. Shanghai: Shanghai guji chubanshe, 1987 (ca. 1249).

Cheng Bi 程珌. *Mingshui ji* 洺水集. SKQS (13th c.).

Cheng Duanli 程端禮. *Dushu fennian richeng* 讀書分年日程. Shandong: Shangzhi tang, 1871 (1310s).

Cheng Hao 程顥 and Cheng Yi 程頤. *Er Cheng ji* 二程集. Beijing: Zhonghua shuju, 1981 (mid-11th c.).

———. *Er Cheng quan shu* 二程全書. Sibu beiyao. Shanghai: Zhonghua shuju, 1927–36 (mid-11th c.).

*Chong guang hui shi* 重廣會史. Beijing: Zhonghua shuju, 1986 (1060s).

Ding Du 丁度 et al., eds. *Fu shiwen huzhu Libu yunlue—fu gongju tiaoshi* 附釋文互註禮部韻略—附貢舉條式. Sibu congkan xubian. Shanghai: Shangwu yinshuguan, 1934? (ca. 1230).

Ding Richang 丁日昌. *Chijing zhai shumu* 持靜齋書目. N.p., 1870.

———. *Chijing zhai xu zeng shumu* 持靜齋續增書目. Ca. 1870.

Fan Ye 范曄. *Hou Han shu* 後漢書. Beijing: Zhonghua shuju, 1965 (1st half 5th c.).

Fang Dacong 方大琮. *Tiean ji* 鐵庵集. SKQS (13th c.).

Fang Yisun 方頤孫. *Fuzao wenzhang baiduan jin* 黼藻文章百段錦. Early Ming ed. (ca. 1249). National Library, Taibei.

———. *Taixue xinbian Fuzao wenzhang baiduan jin* 太學新編黼藻文章百段錦. Siku quanshu cunmu congshu. Ji'nan: Qi Lu shushe chubanshe, 1997 (ca. 1249).

Feng Jike 馮繼科, ed. *Jiajing Jianyang xian zhi* 嘉靖建陽縣志. Tianyige cang Mingdai fangzhi xuankan. Shanghai: Shanghai guji shudian, 1964 (16th c.).

Feng Mengzhen 馮夢禎. *Lidai gongju zhi* 歷代貢舉志. Changsha: Shangwu yinshuguan, 1937 (16th c.–early 17th c.).

Feng Qi 馮琦 and Chen Bangzhan 陳邦瞻. *Song shi jishi benmo* 宋史紀事本末. 3 vols. Beijing: Zhonghua shuju, 1977 (1605).

Gu Yanwu 顧炎武. *Rizhilu jishi* 日知錄集釋. 3 vols. Shanghai: Shanghai guji chubanshe, 1984 (1670, 1695).

*Guanzi* 管子. Guoxue jiben congshu. Shanghai: Shangwu yinshuguan, 1934 (5th c.–1st c. BCE).

He Yan 何晏 and Xing Bing 邢昺, annots. *Lunyu zhushu* 論語注疏. In *Shisan jing zhushu*. Beijing: Beijing daxue chubanshe, 2000 (3rd c.; 999).

Hong Zikui 洪咨夔. *Pingzhai wenji* 平齋文集. SKQS (13th c. print).

Huang Gan 黃榦. *Mianzhai ji* 勉齋集. SKQS (13th c. print).

Huang Lüweng 黃履翁. *(Xinjian jueke) Gujin yuanliu zhilun—bieji* (新箋決科)古今源流至論—別集. SKQS (1230s).

Huang Yuji 黃虞稷. *Qianqing tang shumu* 千頃堂書目. Shiyuan cong shumu. Shanghai: Shanghai guji chubanshe, 1990 (17th c.; first printed in 1916).

Huang Zongxi 黃宗羲, Quan Zuwang 全祖望, et al. *Song Yuan xuean* 宋元學案. Taibei: Huashi chubanshe, 1987 (17th–18th c., first printed in 1846).

*Huian xiansheng yulu da gangling—Fulu* 晦菴先生語錄大綱領—附錄. Song ed., mid-13th c. Beijing Library.

Ji Yun 紀昀 et al., eds. *Siku quanshu zongmu tiyao* 四庫全書總目提要. In *Heyin Siku quanshu zongmu tiyao ji Siku weishou shumu jinhui shumu* 合印四庫全書總目提要及四庫未收書目禁毀書目, ed. Wang Yunwu 王雲五. Taibei: Taiwan Shangwu yinshuguan, 1985 (1782).

Jiang Biao 江標. *Song Yuan ben hangge biao* 宋元本行格表. Siku weishou shu jikan. Beijing: Beijing chubanshe, 1997 (1897).

*Jing qi* 精騎. Wuzhou: Chenzhai, 1160s–early 1170s. National Library, Taibei.

*(Jingxuan) Huang Song Cexue shengchi* (精選)皇宋策學繩尺. Qing copy of a Song edition (mid-13th c.). Beijing Library.

*Jingyi mofan* 經義模範. SKQS (16th c).

Li Chengfu 李誠父. *Pidian fenlei Chengzhai xiansheng wenkuai (qianji, houji)* 批點分類誠齋先生文膾 (前、後集). Siku quanshu cunmu congshu. Ji'nan: Qi Lu shushe chubanshe, 1997 (ca. 1259).

Li Maoying 李昴英. *Wenxi ji* 文溪集. SKQS (late 13th c.).

Li Xinchuan 李心傳. *Daoming lu* 道命錄. Congshu jicheng chubian. Beijing: Zhonghua shuju, 1985 (1935; 1239).

———. *Jianyan yilai chaoye zaji* 建炎以來朝野雜記. Lidai shiliao biji congkan. Beijing: Zhonghua shuju, 2000 (1202–16).

———. *Jianyan yilai xinian yao lu* 建炎以來系年要錄. Beijing: Zhonghua shuju, 1988 (ca. 1208).

Liang Kejia 梁克家. *Chunxi Sanshan zhi* 淳熙三山志. SKQS (1182).

Liang Zhangju 梁章鉅. *Zhiyi cong hua* 制藝叢話. Shanghai: Shanghai shudian chubanshe, 2001 (1851).

Lin Jiong 林駉 and Huang Lüweng 黃履翁. *(Xinjian jueke) Gujin yuanliu zhilun—qianji, houji, xuji, bieji* (新箋決科)古今源流至論—前、後、續、別集. SKQS (1220s–1230s).

———. (*Xinjian jueke*) *Gujin yuanliu zhilun—qianji, houji, xuji, bieji* (新笺决科)古今源流至論—前、後、集、別集. Jianyang?: Yuansha Academy, 1317. Beijing Library.

———. (*Xinjian jueke*) *Gujin yuanliu zhilun—qianji, houji, xuji, bieji* (新笺决科)古今源流至論—前、後、續、別集. 1367. Beijing Library.

Lin Jizhong 林季仲. *Zhuxuan zazhu* 竹軒雜著. SKQS (13th c.).

Lin Xiyi 林希逸. *Zhuxi yan zhai shiyi gao xuji* 竹溪鬳齋十一藁續集. SKQS (ca. 1269).

Lin Zhiqi 林之奇. (*Donglai jizhu leibian*) *Guan lan wenji* (*jiaji, yiji*) (東萊集註類編)觀瀾文集(甲、乙集). Ed. Lü Zuqian 呂祖謙. 2nd half 12th c. Beijing Library.

Liu, Anjie 劉安節. *Liu Zuoshi ji* 劉左史集. SKQS (early 13th c. print).

Liu Chenweng 劉辰翁. *Xuxi sijing shiji* 須溪四景詩集. SKQS (13th c.).

Liu Dake 劉達可. *Bishui qunying daiwen huiyuan (xuanyao)* 璧水群英待問會元 (選要). 1509 / early Ming ed. (ca. 1245). Naikaku bunko, Tokyo; National Library, Taibei.

———. *Bishui qunying daiwen huiyuan* 璧水羣英待問會元. Xuxiu Siku quanshu. Shanghai: Shanghai guji chubanshe, 1995 (ca. 1245).

Liu Kezhuang 劉克莊. *Houcun ji* 后村集. SKQS (1270s print).

Liu Xiang 劉向. *Xin xu* 新序. Sibu congkan chubian. Shanghai: Shangwu yinshuguan, 1929? (1st c. BCE).

Liu Xie 劉勰. *Wen xin diao long* 文心雕龍 [electronic database]. Ed. Zhongyang yanjiuyuan, Zixun kexue yanjiusuo 中央研究院資訊科學研究所. Taibei: Zhongyang yanjiuyuan, Jisuan zhongxin, 2000.

———. *Wen xin diao long yi zheng* 文心雕龍義証. Ed. Zhan Ying 詹鍈. Zhongguo gudian wenxue congshu. Shanghai: Shanghai guji chubanshe, 1989 (ca. 500).

Liu Zhensun 劉振孫. (*Xinbian zhuru pidian*) *Gujin wenzhang zheng yin (qianji, houji, xuji, bieji*) (新編諸儒批點)古今文章正印 (前、後、續、別集). 1273. Palace Museum Library, Taibei.

Liu, Zongyuan 柳宗元. *Liu Hedong ji* 柳河东集. 2 vols. Shanghai: Renmin chubanshe, 1974.

———. *Zengguang zhushi yinbian Tang Liu xiansheng ji* 增廣註釋音辯唐柳先生集. Comp. Liu Yuxi 劉禹錫 et al. 1167. Palace Museum Library, Taibei.

Lou Fang 樓昉. *Yuzhai biaozhu zhujia wen ji* 迂齋標註諸家文集. 1220s. Beijing Library.

———. (*Yuzhai xiansheng biaozhu*) *Chonggu wen jue* (迂齋先生標註)崇古文訣. 1220s. Beijing Library.

Lou Yue 樓鑰. *Gongkui ji* 攻媿集. SKQS (1st half 13th c. print).

Lu Jiuyuan 陸九淵. *Lu Jiuyuan ji* 陸九淵集. Ed. Zhong Zhe 鍾哲. Beijing: Zhonghua shuju, 1980 (1205).

———. *Xiangshan ji* 象山集 (1205). Sibu congkan. Shanghai: Shangwu yinshuguan, 1922.

———. *Xiangshan yulu* 象山语录. Ed. Wang Dianli 王佃利 et al. Ji'nan: Shandong youyi chubanshe, 2001.

Lu Xinyuan 陸心源. *Bisong lou cangshu zhi* 皕宋樓藏書志. *Qingren shumu tiba congkan*. Beijing: Zhonghua shuju, 1990 (1882).

Lü Zujian 呂祖儉 and Lü Qiaonian 呂喬年. "Donglai Lü Taishi nianpu" 東萊呂太史年譜. In *Songren nianpu jimu—Song ren nianpu xuankan* 宋人年譜集目—宋人年譜選刊, ed. Wu Hongze 吳洪澤, 229–35. Chengdu: Bashu, 1995 (1204).

Lü Zuqian 呂祖謙. *Donglai bo yi* 東萊博議. Beijing: Zhongguo shudian, 1986 (1936; 1168).

———. *Donglai ji—bieji, waiji, fulu* 東萊集—別集外集附錄. SKQS (1204 print).

———. *Donglai Lü Taishi wenji—bieji, waiji, fulu, buyi, kaoyi* 東萊呂太史文集—別集外集附錄補遺考異. Xu Jinhua congshu. [China]: Yongkang Hu shi Meng xuan lou, 1924? (1204 print).

———. *Donglai xiansheng Guwen guanjian* 東萊先生古文關鍵. Annot. Cai Wenzi 蔡文子; collated by Xu Shuping 徐樹屏. 2 vols. [Fujian]: Zhang shi Li zhi shuwu, 1870.

———. *Guwen guanjian* 古文關鍵. Taibei: Hongxue, 1989 (ca. 1898; mid 12th–mid 13th c.).

———. *Lidai zhidu xiangshuo* 歷代制度詳説. Xu Jinhua congshu. [China]: Yongkang Hu shi Meng xuan lou, 1924? (late 12th–early 13th c.).

———. *Lü Zuqian quanji* 呂祖謙全集. Ed. Huang Linggeng 黃靈庚. Zhejiang guji chubanshe, 2005.

———. *Song wen jian* 宋文鑑. 3 vols. Beijing: Zhonghua shuju, 1992 (1179).

———. *Zuoshi bo yi* 左氏博議. SKQS (1168).

Lü Zuqian 呂祖謙 and Lü Qiaonian 呂喬年. *Lize lunshuo jilu* 麗澤論説集錄. Xu Jinhua congshu. [China]: Yongkang Hu shi Meng xuan lou, 1924? (late 12th c.–13th c.).

*Lun jue* 論訣. In *Lunxue shengchi* 論學繩尺. Mid-13th c.

Ma Duanlin 馬端臨. *Wenxian tongkao* 文獻通考. 2 vols. Beijing: Zhonghua shuju, 1986 (late 13th–early 14th c.).

Mo Youzhi 莫友芝. *Chijing zhai cangshu jiyao* 持靜齋藏書紀要. Suzhou: Wenxue shanfang, 1870.

———. *Song Yuan jiuben jingyan lu* 宋元舊本經眼錄. Yingshan caotang liuzhong. Shanghai, 1873.

*Nan Song guan'ge xu lu* 南宋館閣續錄. In Chen Kui 陳騤 et al., *Nan Song guan-ge lu, xu lu* 南宋館閣錄, 續錄. Beijing: Zhonghua shuju, 1998 (1178–13th c.).

Ouyang Qiming 歐陽起鳴. *Ouyang lun fan* 歐陽論範. Siku quanshu cunmu congshu. Ji'nan: Qi Lu shushe chubanshe, 1997 (1471; 13th c.).

Ouyang Shoudao 歐陽守道. *Xunzhai wenji* 巽齋文集. SKQS (13th c.).

Ouyang Xiu 歐陽修 and Song Qi 宋祁. *(Xin) Tang shu* 新唐書. Beijing: Zhonghua shuju, 1975 (1060).

Peng Guinian 彭龜年. *Zhitang ji* 止堂集. Congshu jicheng chu bian. Shanghai: Shanghai yinshuguan, 1935 (early 13th c.).

Peng Yuanrui 彭元瑞 et al. *Tianlu linlang shumu houbian* 天祿琳琅書目後編. Qingren shumu tiba congkan. Beijing: Zhonghua shuju, 1995 (1798).

Qiao chuan qiao sou 樵川樵叟. *Qingyuan dangjin* 慶元黨禁. Congshu jicheng chubian. Shanghai: Shangwu yinshuguan, 1939 (ca. 1245).

*Qingyuan tiaofa shilei* 慶元條法事類. Ed. Yang Yifan 楊一凡 and Tian Tao 田濤. 10 vols. Zhongguo zhenxi falü dianji xubian. Haerbin: Heilongjiang renmin chubanshe, 2002 (ca. 1202).

Qu Yong 瞿鏞. *Tieqin tongjian lou cangshu mulu* 鐵琴銅鑑樓藏書目錄. Qingren shumu tiba congkan. Beijing: Zhonghua shuju, 1990 (1860; rev. ed., 1897–98).

*Quandian Longchuan Shuixin xiansheng wen cui (qianji, houji)* 圈點龍川水心先生文粹 (前、後集). Jian'an, 1240s or 1250s? (1212). National Library, Taibei.

*Qunshu huiyuan jie jiang wang* 群書匯元截江網. SKQS (early 1250s).

Ruan Yuan 阮元. *Siku weishou shumu tiyao* 四庫未收書目提要. In *Heyin Siku quanshu zongmu tiyao ji Siku weishou shumu Jinhui shumu* 合印四庫全書總目提要及四庫未收書目禁毀書目, ed. Wang Yunwu 王雲五. Taibei: Taiwan Shangwu yinshuguan, 1985 (1820s).

Shen Chu 沈初, ed. *Zhejiang caiji yishu zonglu* 浙江採集遺書總錄. Hangzhou: Zhejiang buzheng shi si, 1774.

*Sheng Song wenxuan* 聖宋文選. 1165–73. National Library, Taibei.

*Shi xiansheng aolun zhu (qianji, houji, xuji)* 十先生奧論註 (前、後 、續集). SKQS (13th c.).

Sima Qian 司馬遷. *Shi ji* 史記. Beijing: Zhonghua shuju, 1959 (90 BCE).

*Song shi quanwen* 宋史全文. SKQS (14th c.).

Su Shi 蘇軾. *Su Shi wenji* 蘇軾文集. 6 vols. Beijing: Zhonghua shuju, 1986 (12th c. prints).

Su Xun 蘇洵. *Jiayou ji* 嘉祐集. SKQS (2nd half 11th c. print).

———. *Donglai biaozhu Laoquan xiansheng wenji* 東萊標註老泉先生文集. Annot. Lü Zuqian 呂祖謙; collated by Wu Yan 吳炎. 1193. Beijing Library.

Tan Jinsun 譚金孫. (*Xinkan jingxuan*) *Zhuru aolun cexue tongzong—qianji* (新刊精選)諸儒奧論策學統宗—前集. Wanwei biecang. Shanghai: Shangwu yinshuguan, 1935 (late 13th–early 14th c.).

Teng Gong 滕珙, comp. *Zhuzi jingji wenheng leibian (qianji, houji, xuji)* 朱子經濟文衡類編 (前、後、續集). SKQS (13th–14th c.).

Tuo-tuo 脫脫, ed. *Song shi* 宋史. Beijing: Zhonghua shuju, 1977 (1345).

Wang Dingbao 王定保. (*Tang*) *Zhiyan* 唐摭言. SKQS (mid-10th c.).

Wang Fuzhi 王夫之, annot. *Zhangzi Zheng meng zhu* 張子正蒙注. Beijing: Zhonghua shuju, 1975 (17th c.).

Wang Gou 王構. *Xiuci jianheng* 修辭鑑衡. Guoxue jiben congshu. Shanghai: Shangwu yinshuguan, 1937 (1330s).

Wang Mai 王邁. *Quxuan ji* 臞軒集. SKQS (13th C).

Wang Maohong 王懋竑. *Zhu Xi nianpu* 朱熹年譜. Annot. He Zhongli 何忠禮. Nianpu congkan. Beijing: Zhonghua shuju, 1998 (18th c.).

Wang Tingzhen 王霆震. (*Xinkan zhuru pingdian*) *Guwen jicheng—qianji* (新刊諸儒評點)古文集成—前集. SKQS (1270s?).

Wang Yinglin 王應麟. *Cixue zhinan* 辭學指南. SKQS (1250s?).

———. *Yuhai* 玉海. SKQS (1250s).

Wei Liaoweng 魏了翁. *Haoshan ji* 鶴山集. SKQS (1259 print).

Wei Tianying 魏天應, ed.; Lin Zizhang 林子長, annot. (*Pidian fenge leiyi jujie*) *Lunxue shengchi* (批點分格類意句解)論學繩尺. Yuan ed. (1270s). Seikadō bunko, Tokyo.

*Wenchang ziyong fenmen Jinsi Lu* 文場資用分門近思錄. 13th c. National Library, Taibei.

Wu Yong 吳泳. *Haolin ji* 鶴林集. SKQS (13th c).

Wu Zeng 吳曾. *Nenggai zhai man lu* 能改齋漫錄. Beijing: Zhonghua shuju, 1960 (1157).

Wu Ziliang 吳子良. *Lin xia ou tan* 林下偶談. Congshu jicheng chubian. Shanghai: Shangwu yinshuguan, 1936 (mid-13th c.).

Wu Zimu 吳自牧. *Meng liang lu* 夢梁錄. Hangzhou: Zhejiang renmin chubanshe, 1981 (1274).

Xie Fangde 謝枋德. *Wenzhang guifan* 文章軌範. SKQS (mid-13th c.).

———. *Xie Dieshan quanji jiaozhu* 謝疊山全集校注. Annot. Xiong Fei 熊飛 et al. Shanghai: Huadong shifan daxue chubanshe, 1995 (13th c.).

(*Xin kan Jiaofeng pidian*) *Zhizhai lunzu* (新刊蛟峰批點)止齋論祖. With commentary by Chen Fuliang 陳傅良 and Fang Fengchen 方逢辰. Ming ed. (ca. 1268). Naikaku bunko, Tokyo.

Xing Bing 邢昺, *see* He Yan.

Xiong Jie 熊節, comp.; Xiong Gangda 熊剛大, annot. (*Xinbian yindian*) *Xingli qunshu jujie* (*qianji, houji*) (新編音點)性理群書句解(前、後集). Yuan ed. (first half 13th c.). National Library, Taibei.

Xu Shizeng 徐師曾. *Wenti ming bian xushuo* 文體明辯序說. Beijing: Renmin wenxue chubanshe, 1962 (16th c.).

Xu Song 徐松, ed. *Song huiyao jigao* 宋會要輯稿. 8 vols. Beijing: Zhonghua shuju, 1957 (ca. 1809).

———. *Song huiyao* 宋會要 [electronic database]. Ed. Zhongyang yanjiuyuan 中央研究院 and Harvard University. Zhongyang yanjiuyuan, Jisuan zhongxin, 2003.

Xu Yinglong 許應龍. *Dongjian ji* 東澗集. SKQS (13th c.).

Xu Yuanjie 徐元杰. *Meiye ji* 木芙埜集. SKQS (13th c.).

Xue Jixuan 薛季宣. *Langyu ji* 浪語集. SKQS (late 12th c.).

Yang Fang 陽枋. *Zixi ji* 字溪集. SKQS (13th c.).

Yang Shaohe 樣紹和. *Ying shu yu lu* 楹書隅錄. Qingren shumu tiba congkan. Beijing: Zhonghua shuju, 1990 (ca. 1871).

Yang Shi 楊時. *Yang Guishan ji* 楊龜山集. 2 vols. Shanghai: Shangwu yinshuguan, 1936 (1590 edition; 12th c.).

Yang Wanli 楊萬里. *Chengzhai ji* 誠齋集. SKQS (1170s–1180s prints).

———. (*Pidian fenlei*) *Chengzhai xiansheng wenkuai* (批點分類)誠齋先生文膾. Ed. Li Chengfu 李誠父. Yuan edition (ca. 1259). Palace Museum Library, Taibei.

Yao Mian 姚勉. *Xuepo ji* 雪坡集. SKQS (13th c.).

Ye Shaoweng 葉紹翁. *Sichao wenjian lu* 四朝聞見錄. Tang Song shiliao biji congkan. Beijing: Zhonghua shuju, 1989 (ca. 1225).

Ye Shi 葉適. *Ye Shi ji* 葉適集. Beijing: Zhonghua shuju, 1961 (ca. 1230).

(*Yongjia* [*xiansheng*]) *Bamian feng* (永嘉[先生])八面鋒. Congshu jicheng chubian. Shanghai: Shanghai yinshuguan, 1936 (12th c.).

Yu, Changcheng 俞長城. *Keyi tang yibaiershi mingjia zhiyi* 可儀堂一百二十名家制義. Ca. 1699.

Yu Minzhong 于敏中 et al. *Tianlu linlang shumu* 天祿琳琅書目. Qingren shumu tiba congkan. Beijing: Zhonghua shuju, 1995 (1775).

Zeng Jian 曾堅 and Liu Jinwen 劉錦文, eds. *Dace mijue* 答策秘訣. Yuan ed. (mid-13th c.). Palace Museum Library, Taibei.

Zhang Jinwu 張金吾. *Airi jinglu cangshu zhi* 愛日精盧藏書志. Qingren shumu tiba congkan. Beijing: Zhonghua shuju, 1990 (1826).

Zhang Ruyu 章如愚. *Qunshu kaosuo* 群書考索. Beijing: Zhonghua shuju, 1992 (reprint of a Yuan ed.; 13th c. print).

Zhang Shi 張栻. *Zhang Nanxuan ji* 張南軒集. Fuzhou: Zhengyi tang, 1709 (1184).

Zhen Dexiu 眞德秀. *Wenzhang zhengzong* 文章正宗. Yuan ed. (1232). Palace Museum Library and National Library, Taibei.

———. *Xishan wenji* 西山文集. SKQS (mid-13th c.).

———. *Zhen Wenzhong gong Xu wenzhang zhengzong* 眞文忠公續文章正宗. 1266 (1230s?). National Library, Taibei.

———. *Zhen Xishan xiansheng ji* 眞西山先生集. Congshu jicheng chubian. Shanghai: Shangwu yinshuguan, 1937.

Zhongyang yanjiuyuan, Lishi yuyan yanjiusuo 中央研究院歷史語言研究所, ed. Ershiwu shi 二十五史 [electronic database]. Taibei: Zhongyang yanjiuyuan, Jisuan zhongxin, 2000.

———. Hanji quanwen ziliaoku 漢籍全文資料庫 [electronic database]. Taibei: Zhongyang yanjiuyuan, Jisuan zhongxin, 2000.

———. Shisan jing 十三經 [electronic database]. Taibei: Zhongyang yanjiuyuan, Jisuan zhongxin, 2000.

Zhou Bida 周必大. *Wenzhong ji* 文忠集. SKQS (1206 print).

Zhou Dunyi 周敦頤. *Zhou Lianxi xiansheng quanji* 周濂溪先生全集. Fuzhou: Zhengyi tang, 1707–13 (ca. 1241).

Zhou Mi 周密. *Guixin zashi* 癸辛雜識. Tang Song shiliao biji congkan. Beijing: Zhonghua shuju, 1988 (1290s).

———. *Qidong yeyu* 齊東野語. Tang Song shiliao biji congkan. Beijing: Zhonghua shuju, 1983 (1291).

Zhou Nan 周南. *Shanfang ji* 山房集. SKQS (13th c.).

Zhou Xingji 周行己. *Fuzhi ji* 浮沚集. SKQS (12th c.).

Zhou Yinghe 周應合. *Jingding Jiankang zhi* 景定建康志. Song Yuan fangzhi congkan. Beijing: Zhonghua shuju, 1990 (1261).

Zhou Yinglong 周應龍. *Wen sui* 文髓. 1428 (mid-13th c.). National Library, Taibei.

Zhu Mu 祝穆 and Fu Dayong 富大用. *(Xin bian) Gujin shiwen leiju* (新編)古今事文類聚. 3 vols. Beijing: Shumu wenxian chubanshe, 1991 (1246).

Zhu Xi 朱熹. *Sishu zhangju jizhu* 四書章句集註. Xinbian Zhuzi jicheng. Beijing: Zhonghua shuju, 1983 (1182–92).

———. *Zhu Xi ji* 朱熹集. 10 vols. Chengdu: Sichuan jiaoyu chubanshe, 1996 (early 13th c.).

———. *Zhuzi da quan* 朱子大全. Sibu beiyao. Shanghai: Zhonghua shu ju, 1927–36 (early 13th c.).

———. *Zhuzi wenji* 朱子文集. Guoxue jiben congshu. 3 vols. Shanghai: Shangwu yinshuguan, 1937 (1708; early 13th c.).

———. *Zhuzi wenji* 朱子文集. Ed. Chen Junmin 陳俊民 and Yu Yingshi 余英時. 10 vols. Taibei: De fu wenjiao jijinhui, 2000 (early 13th c.).

———. *Zhuzi yulei* 朱子語類. Ed. Li Jingde 黎靖德. Beijing: Zhonghua shuju, 1986 (1270).

Zhu Xi 朱熹 and Li Youwu 李幼武. *Song mingchen yanxing lu* 宋名臣言行錄. Song shi ziliao cui bian. Taibei: Dahai, 1967 (late 12th–mid-13th c.).

Zhu Xi 朱熹 and Lü Zuqian 呂祖謙, comps. *Jinsi lu* 近思錄. Zhuzi baijia congshu. Shanghai: Shanghai guji chubanshe, 1994 (1178).

Zhu Xi 朱熹 et al. *Zizhi tongjian gangmu* 資治通鑑綱目. SKQS (ca. 1172).

Zhu Xizhao 朱希召. *Song li ke zhuangyuan lu* 宋歷科狀元錄. Song shi ziliao cuibian. Taibei: Wenhai chubanshe, 1981 (Ming).

*Secondary Sources*

Adler, Joseph A. "Chu Hsi and Divination." In *Sung Dynasty Uses of the I Ching*, ed. Kidder Smith, Jr., et al., 169–205. Princeton: Princeton University Press, 1990.

Althusser, Louis. "Ideology and Ideological State Apparatuses (Notes Towards an Investigation)." In *Mapping Ideology*, ed. Slavoj Žižek, 100–140. London and New York: Verso, 1994.

Angle, Stephen C. "The Possibility of Sagehood: Reverence and Ethical Perfection in Zhu Xi's Thought." *Journal of Chinese Philosophy* 25, no. 3 (1998): 281–303.

Araki Kengo 荒木見悟. *Chūgoku shisōshi no shosō* 中国思想史の諸想. Fukuoka: Chūgoku shoten, 1989.

———. "Rin Ki'eki no tachiba" 林希逸の立場. *Chūgoku tetsugaku ronshū* 中国哲学論集 7 (1981): 48–61.

Araki Toshikazu 荒木敏一. *Sōdai kakyo seido kenkyū* 宋代科舉制度研究. Kyoto: Dōhōsha, 1969.

Association for Asian Studies. Ming Biographical History Project Committee; L. Carrington Goodrich; and Chao-ying Fang, eds. *Dictionary of Ming Biography, 1368–1644*. New York: Columbia University Press, 1976.

Azuma Hidetoshi 東英壽. "Kōkan yori mita Hoku Sō shoki kobun undō ni tsuite—Ō Ushō o tegakari to shite" 行卷よりみた北宋初期古文運動について—王禹稱を手がかりとして. *Chūgoku bungaku ronshū* 中国文学論集, no. 12 (1993): 29–48.

———. "'Taigakutai' kō—sono Hoku Sō kobun undō ni okeru ichi kōsatsu" "太学体"考—その北宋古文運動における一考察. *Nihon Chūgaku gakkai hō* 日本中国学会報 40 (1988): 94–108.

Bao, Weimin 包偉民. Review: *Songdai jiaoyu: Zhongguo gudai jiaoyu de lishixing zhuanzhe* by Yuan Zheng. *JSYS* 24 (1994): 321–25.

———. "Zhongguo jiu dao shisan shiji shehui shizilü tigao de jige wenti" 中国九到十三世纪社会识字率提高的几个问题. *Hangzhou daxue xuebao* 杭州大学学报 1992, no. 4: 79–87.

Beattie, Hilary. *Land and Lineage in China: A Study of T'ung-Ch'eng, Anhwei, in the Ming and Ch'ing Dynasties*. Cambridge: Cambridge University Press, 1979.

Beijing tushuguan 北京圖書館. *Beijing tushuguan guji shanben shumu* 北京圖書館古籍善本書目. 5 vols. Beijing: Shumu wenxian chubanshe, 1987.

Berkey, Jonathan Porter. *The Transmission of Knowledge in Medieval Cairo: A Social History of Islamic Education*. Princeton Studies on the Near East. Princeton: Princeton University Press, 1992.

Berthrong, John. "Glosses on Reality: Chu Hsi as Interpreted by Ch'en Ch'un." Ph.D. diss., University of Chicago, 1979.

———. "To Catch a Thief: Chu Hsi (1130–1200) and the Hermeneutic Art." *Journal of Chinese Philosophy* 18 (1991): 195–212.

Bol, Peter K. "Ch'eng Yi as a Literatus." In *The Power of Culture: Studies in Chinese Cultural History*, ed. Willard J. Peterson, Andrew H. Plaks, and Yü Ying-shih, 172–94. Hong Kong: Chinese University Press, 1994.

———. "Chu Hsi's Redefinition of Literati Learning." In *Neo-Confucian Education: The Formative Years*, ed. Wm. Theodore de Bary and John Chaffee, 151–87. Berkeley: University of California Press, 1989.

———. "Culture and the Way in Eleventh-Century China." Ph.D. diss., Princeton University, 1982.

———. "Examinations and Orthodoxies: 1070 and 1313 Compared." In *Culture and State in Chinese History: Conventions, Accomodations, and Critiques*, ed. Theodore Huters, R. Bin Wong, and Pauline Yu, 29–57. Stanford: Stanford University Press, 1997.

———. "The Examination System and Sung Literati Culture." In *La Société civile face à l'etat: dans les traditions chinoise, japonaise, coréenne et vietnamienne*, ed. Léon Vandermeersch, 55–75. Paris: Ecole française d'Extrême-Orient, 1994.

———. "Intellectual Culture in Wuzhou Ca. 1200—Finding a Place for Pan Zimu and the *Complete Source for Composition*." In *Proceedings of the Second Symposium on Sung History—Di erjie Song shi xueshu yantaohui lunwen ji* 第二屆宋史學術研討會論文集, ed. Di er jie Song shi xueshu yantaohui mishuchu 第二屆宋史學術研討會秘書處, 788–38. Taibei: Chinese Culture University, 1996.

———. "Neo-Confucianism and Chinese History: Position, Identity, and Movement." Paper presented at the Conference on Sung-Yuan-Ming Transitions, University of California at Los Angeles, June 5–11, 1997.

———. "Neo-Confucianism and History—Some Issues for Historians." *Chūgoku shigaku* 中国史学 6 (1996): 1–22.

———. "Neo-Confucianism and Local Society, Twelfth to Sixteenth Century: A Case Study." In *The Song-Yuan-Ming Transition in Chinese History*, ed. Paul J. Smith and Richard von Glahn, 241–83. Cambridge: Harvard University Asia Center, 2003.

———. "Reading Su Shi in Southern Song Wuzhou." *East Asian Library Journal* 8, no. 2 (1998): 69–102.

———. "Reconceptualizing the Nation in Southern Song: Some Implications of Ye Shi's Statecraft Learning." In *Thought, Political Power, and Social Forces*, ed. Ko-wu Huang, 33–64. Taibei: Institute of Modern History, Academia Sinica, 2002.

———. "Reflections on Sung Literati Thought." *Bulletin of Sung-Yuan Studies* 18 (1986): 88–97.

———. "The Sung Examination System and the *Shih*." *Asia Major*, 3rd series, 3, no. 2 (1990): 149–71.

———. *"This Culture of Ours": Intellectual Transitions in T'ang and Sung China*. Stanford: Stanford University Press, 1992.

———. "Zhang Ruyu, the *Qunshu kaosuo*, and Diversity in Intellectual Culture—Evidence from Dongyang County in Wuzhou." In *Qingzhu Deng Guangming jiaoshou jiushi huadan lunwen ji* 慶祝鄧廣銘教授九十華誕論文集, 644–73. Shijiazhuang: Hebei jiaoyu chubanshe, 1997.

Bossler, Beverly Jo. *Powerful Relations: Kinship, Status, & the State in Sung China (960–1279)*. Cambridge: Council on East Asian Studies, Harvard University, 1998.

Bourdieu, Pierre. *Distinction: A Social Critique of the Judgement of Taste*. London: Routledge & Kegan Paul, 1986.

Bourdieu, Pierre, and Randal Johnson. *The Field of Cultural Production: Essays on Art and Literature, European Perspectives*. New York: Columbia University Press, 1993.

Bourdieu, Pierre, and John B. Thompson. *Language and Symbolic Power*. Cambridge: Harvard University Press, 1991.

Brooks, E. Bruce, and A. Taeko Brooks, *see* Confucius.

Cai Zongyang 蔡宗陽. *Chen Kui Wenze xin lun* 陳騤文則新論. Taibei: Wenshizhe, 1993.

Carruthers, Mary. *The Book of Memory: A Study of Memory in Medieval Culture*. Cambridge: Cambridge University Press, 1990 (1992).

Chaffee, John W. *Branches of Heaven: A History of the Imperial Clan of Sung China*. Cambridge, MA: Harvard University Asia Center, 1999.

———. "Chao Ju-Yü, Spurious Learning, and Southern Sung Political Culture." *JSYS* 22 (1990–92): 23–61.

————. "Chūgoku shakai to kakyo: Ōbei ni okeru kenkyū dōkō" 中国社会と 科挙: 欧米における研究動向. *Chūgoku shakai to bunka* 中国社会と文化 17 (2002): 174–85.

————. "Chu Hsi and the Revival of the White Deer Grotto Academy, 1179– 1181 A.D." *T'oung Pao* 71 (1985): 40–62.

————. "Chu Hsi in Nan-K'ang: Tao-Hsüeh and the Politics of Education." In *Neo-Confucian Education: The Formative Stage*, ed. Wm. Theodore de Bary and John W. Chaffee, 414–31. Berkeley: University of California, 1989.

————. "Education and Examinations in Sung Society (960–1279)." Ph.D. diss., University of Chicago, 1979.

————. "Examinations During Dynastic Crisis: The Case of the Early Southern Song." Paper presented at the International Conference on the Imperial Examination System and the Study of Imperial Examinations, Xiamen, September 2–4, 2005.

————. "The Historian as Critic: Li Hsin-Ch'uan and the Dilemmas of Statecraft in Southern Sung China." In *Ordering the World: Approaches to State and Society in Sung Dynasty China*, ed. Robert P. Hymes and Conrad Schirokauer, 310–35. Berkeley, CA: University of California Press, 1993.

————. *The Thorny Gates of Learning in Sung China: A Social History of the Examinations*. Albany, NY: State University of New York Press, 1995 (1985).

Chan, Hok-lam. *Control of Publishing in China, Past and Present*. Canberra: Australian National University Press, 1983.

Chan, Wing-tsit 陳榮捷. *Chu Hsi: New Studies*. Honolulu: University of Hawaii Press, 1989.

————. "Chu Hsi's Completion of Neo-Confucianism." In idem, *Chu Hsi, Life and Thought*, 103–138. Hong Kong, 1987 (1973).

————. "The Principle of Heaven vs. Human Desires." In *Chu Hsi: New Studies*, ed. Wing-tsit Chan, 197–211. Honolulu: University of Hawaii Press, 1989.

————. *Xin ruxue lun ji* 新儒學論集. Taibei: Zhongyang yanjiuyuan, Zhongguo wenzhe yanjiusuo, 1995.

————. *Zhuzi menren* 朱子門人. Taibei: Xuesheng shuju, 1982.

Chan, Wing-tsit 陳榮捷, comp. and trans. *A Sourcebook in Chinese Philosophy*. Princeton: Princeton University Press, 1963.

Chan, Wing-tsit 陳榮捷, trans. *Reflections on Things at Hand, the Neo-Confucian Anthology*. New York: Columbia University Press, 1967.

Chan, Wing-tsit 陳榮捷, trans. and ed. *Neo-Confucian Terms Explained (the Pei-Hsi Tzu-I)*. New York: Columbia University Press, 1986.

Chang Bide 昌彼得, et al., eds. *Song ren zhuanji ziliao suoyin* 宋人傳記資料索引 = *Index to Biographical Materials of Sung Figures*. Taibei: Dingwen shuju, 1974.

Che Jixin 車吉心 and Liu Dezeng 劉德增. *Zhongguo zhuangyuan quan zhuan* 中國狀元全傳. Ji'nan: Shandong meishu chubanshe, 1993.

Chen Deyun 陳德芸. "Bagu wenxue" 八股文學. *Lingnan xuebao* 嶺南學報 4 (1941): 17–21.

Chen Jo-shui. *Liu Tsung-Yüan and Intellectual Change in T'ang China, 773–819*. Cambridge: Cambridge University Press, 1992.

Chen Lai 陈来. "Lue lun *Zhuru mingdao ji*" 略论《诸儒鸣道集》. *Beijing daxue xuebao* 北京大学学报 (1986): 30–38.

———. *Zhu Xi zhexue yanjiu* 朱熹哲学研究. Beijing: Zhonghua shuju, 1988.

———. *Zhuzi shuxin biannian kaozheng* 朱子书信编年考证. Shanghai: Shanghai renmin chubanshe, 1989.

Chen Renhua 陳仁華. *Zhuzi dushufa* 朱子讀書法. Taibei: Yuanliu chuban, 1991.

Chen Wenyi 陳雯怡. *You guanxue dao shuyuan: cong zhidu yu linian de hudong kan Songdai jiaoyu de yanbian* 由官學到書院—從制度與理念的互動看宋代教育的演變. Taibei: Lianjing, 2004.

Chen, Yu-shih. *Images and Ideas in Chinese Classical Prose: Studies of Four Masters*. Stanford: Stanford University Press, 1988.

Chen Zhenbo 陈镇波. "*Yongjia xiansheng Bamian feng* tanxi" 永嘉先生八面锋探析. In *Chen Fuliang danchen babai liushi zhounian jinian ji* 陈傅良诞辰八百六十周年纪念集, ed. Jinian Chen Fuliang danchen babai liushi zhounian chouweihui 纪念陈傅良诞辰八百六十周年筹委会, 39–55. N.p., 1997?

Cherniack, Susan. "Book Culture and Textual Transmission in Sung China." *HJAS* 54, no. 1 (1994): 5–125.

Chia, Lucille. "Book Emporium: The Development of the Jianyang Book Trade, Song-Yuan." *Late Imperial China* 17, no. 1 (1996): 10–48.

———. "Printing for Profit: The Commercial Printers of Jianyang, Fujian (Song-Ming)." Ph.D. diss., Columbia University, 1996.

———. *Printing for Profit: The Commercial Publishers of Jianyang, Fujian (11th–17th Centuries)*. Cambridge: Harvard University Asia Center for Harvard-Yenching Institute, 2002.

Chow Kai-wing. "Discourse, Examination and Local Elite: The Invention of the T'ung-Ch'eng School in Ch'ing China." In *Education and Society in Late Imperial China, 1600–1900*, ed. Benjamin A. Elman and Alexander Woodside, 183–219. Berkeley: University of California Press, 1994.

———. *Publishing, Culture, and Power in Early Modern China*. Stanford: Stanford University Press, 2004.

———. Review: Benjamin A. Elman. *A Cultural History of Civil Examinations in Late Imperial China. American Historical Review* 107, no. 1 (2002): 168.

————. "Writing for Success: Printing, Examinations, and Intellectual Change in Late Ming China." *Late Imperial China* 17, no. 1 (1996): 120–57.

Chu, Ping-tzu. "Tradition Building and Cultural Competition in Southern Song China (1160–1220): The Way, the Learning, and the Texts." Ph.D. diss., Harvard University, 1998.

Chu, Ron-Guey. "Chen Te-Hsiu and the 'Classic on Governance': The Coming of Age of Neo-Confucian Statecraft." Ph.D. diss., Columbia University, 1988.

Confucius, E. Bruce Brooks, and A. Taeko Brooks. *The Original Analects: Sayings of Confucius and His Successors, 0479–0249*. Translations from the Asian Classics. New York: Columbia University Press, 1998.

Cui Fuzhang 崔富章. *Siku tiyao buzheng* 四库提要补正. Hangzhou: Hangzhou daxue chubanshe, 1990.

Cui Ji 崔驥. "Xie Fangde nianpu" 謝枋得年譜. *Jiangxi jiaoyu yuekan* 江西教育月刊 4 (1935): 34–45; 7 (1935): 27–44.

Davis, Richard L. *Court and Family in Sung China, 960–1279: Bureaucratic Success and Kinship Fortunes for the Shih of Ming-Chou*. Durham: Duke University Press, 1986.

————. "Custodians of Education and Endowment at the State Schools of Southern Sung." *JSYS* 25 (1995): 95–119.

————. *Wind Against the Mountain: The Crisis of Politics and Culture in Thirteenth-Century China*. Cambridge: Council on East Asian Studies, Harvard University, 1996.

de Bary, Wm. Theodore. *Learning for One's Self: Essays on the Individual in Neo-Confucian Thought*. New York: Columbia University Press, 1991.

————. *The Liberal Tradition in China*. New York: Columbia University Press, 1983.

————. *The Message of the Mind in Neo-Confucianism*. New York: Columbia University Press, 1989.

————. *Neo-Confucian Orthodoxy and the Learning of the Mind-and-Heart*. New York: Columbia University Press, 1981.

————. "Some Common Tendencies in Neo-Confucianism." In *Confucianism in Action*, ed. David S. Nivison and Arthur F. Wright, 25–49. Stanford: Stanford University Press, 1959.

————. "The Uses of Confucianism: A Response to Professor Tillman." *Philosophy East and West* 43, no. 3 (1993): 541–55.

de Bary, Wm. Theodore, Irene Bloom, et al., eds. *Sources of Chinese Tradition*. 2nd ed. Introduction to Asian Civilizations. New York: Columbia University Press, 1999.

de Bary, Wm. Theodore, and John W. Chaffee, eds. *Neo-Confucian Education: The Formative Stage*. Berkeley: University of California Press, 1989.

Deng Guangming 邓广铭 and Cheng Yingliu 程应镠. *Zhongguo lishi da cidian—Song shi* 中国历史大辞典—宋史. Shanghai: Shanghai cishu chubanshe, 1984.

Deng Guangming 邓广铭 and Li Jiaju 郦家驹, eds. *Song shi yanjiu lunwen ji* 宋史研究论文集. Zhengzhou: Henan renmin chubanshe, 1984.

Deng Hongbo 邓洪波, Peng Mingzhe 彭明哲, and Gong Kangyun 龚抗云. *Zhongguo lidai zhuangyuan dianshi juan* 中国历代状元殿试卷. Haikou: Hainan chubanshe, 1993.

Deng Xiaonan 邓小南. "Guanyu 'daoli zui da'—jian tan Song ren duiyu 'zuzong' xingxiang de suzao" 关于 "道理最大"—兼谈宋人对于 "祖宗" 形象的塑造. *Ji'nan xuebao* 暨南学报 25, no. 2 (2003): 116–26.

———. *Songdai wenguan xuanren zhidu zhu cengmian* 宋代文官选任制度层面. Song shi yanjiu congshu. Baoding: Hebei jiaoyu chubanshe, 1993.

Des Rotours, Robert. *Le traité des examens, traduit de la "Nouvelle histoire des T'ang."* Paris: Librairie Ernest Leroux, 1932.

De Weerdt, Hilde. "Amerika no Sōdaishi kenkyū ni okeru kinnen no dōkō: chihō shūkyō to seiji bunka アメリカの宋代史研究における近年の動向：地方宗教と政治文化." Trans. Kenji Ueuchi 上内健司. *Ōsaka shiritsu daigaku Tōyōshi ronsō* 大阪市立大学東洋史論叢 15 (2006): 121–38.

———. "Aspects of Song Intellectual Life: A Preliminary Inquiry into Some Southern Song Encyclopedias." *Papers on Chinese History* 3 (1994): 1–27.

———. "Byways in the Imperial Chinese Information Order: The Dissemination and Commercial Publication of State Documents." *HJAS* 66, no. 1 (2006): 145–88.

———. "Canon Formation and Examination Culture: The Construction of *Guwen* and *Daoxue* Canons," *JSYS* 29 (1999): 91–134.

———. "Changing Minds Through Examinations: Examination Critics from Medieval Through Late Imperial Times." *Journal of the American Oriental Society*, forthcoming 2007.

———. "The Composition of Examination Standards: *Daoxue* and Southern Song Dynasty Examination Culture." Ph.D. diss., Harvard University, 1998.

———. "Content and Composition: An Investigation of 'Lü Zuqian' Examination Teaching." In *Jiangnan wenhua yanjiu (di yi ji)* 江南文化研究 (第一辑), ed. Zhejiang shifan daxue renwen xueyuan 浙江师范大学人文学院, 92–109. Beijing: Xueyuan chubanshe, 2006.

———. "'Court Gazettes' and 'Short Reports': The Blurry Boundaries Between Official News and Rumor." Paper presented at the Annual Meeting of the Association for Asian Studies, San Francisco, April 9, 2006.

————. "The Discourse of Loss in Private and Court Book Collecting in Imperial China." *Library Trends* 55, no. 3 (2007): 404–20.

————. "The Empire-Wide Significance of Local Intellectual Traditions: Yongjia Scholarship in the Twelfth Century." In *Education and Local Development in Late Imperial China*, ed. Liu Hsiang-kwang, forthcoming.

————. "The Encyclopedia as Textbook: Selling Private Chinese Encyclopedias in the Twelfth and Thirteenth Centuries." *Extrême-orient, Extrême-occident-*, special volume, 2007, 77–102.

————. Review: *The Class of 1761: Examinations, State, and Elites in 18th Century China* by Iona Man-Cheong. *JAS* 64, no. 2 (2005): 453–54.

————. Review: *Publishing, Culture, and Power in Early Modern China* by Kai-Wing Chow. *Technology and Culture* 47 (2006): 192–93.

————. "The Ways of the Teacher: Commemorating the Administrator and the Transmitter in Twelfth-Century Teacher Biographies." *Chūgoku shigaku / Studies in Chinese History* 16 (2006): 1–24.

————. "What Did Su Che See in the North? Publishing Laws, State Security, and Political Culture in Song China." *T'oung Pao: International Journal of Chinese Studies* 92, no. 4–5 (2006): 466–94.

DeWoskin, Kenneth. "Lei-shu." In *The Indiana Companion to Traditional Chinese Literature*, ed. William Nienhauser, Jr., 526–29. Bloomington: Indiana University Press, 1986.

Dien, Albert. "Civil Service Examinations: Evidence from the Northwest." In *Culture and Power in the Reconstruction of the Chinese Realm, 200–600*, edited by Scott Pearce, Audrey Spiro, and Patricia Ebrey, 99–121. Cambridge: Harvard University Asia Center, 2001.

Diény, Jean-Pierre. "Les encyclopédies chinoises." In *L'encyclopédisme: actes du colloque de Caen, 12–16 Janvier 1987*, ed. Annie Becq, 195–200. Paris: Editions aux amateurs de livres, 1991.

Ding Bing 丁丙. *Shanben shushi cangshu zhi* 善本書室藏書志. Qingren shumu tiba congkan. Beijing: Zhonghua shuju, 1990 (1908).

Ditmanson, Peter Brian. "Contesting Authority: Intellectual Lineages and the Chinese Imperial Court from the Twelfth to the Fifteenth Centuries." Ph.D. diss., Harvard University, 1999.

Drège, Jean-Pierre. "Des effets de l'imprimerie en Chine sous la dynastie des Song." *Journal Asiatique* 282, no. 2 (1994): 409–42.

Du Haijun 杜海军. *Lü Zuqian wenxue yanjiu* 吕祖谦文学研究. Beijing: Xueyuan chubanshe, 2003.

————. "Lü Zuqian yu *Jinsi lu* de bianzuan" 吕祖谦与近思录的编纂. *Zhongguo zhexue shi* 中国哲学史 2003, no. 4: 43–49.

Eagleton, Terry. *Ideology: An Introduction*. London; New York: Verso, 1991.

Ebrey, Patricia Buckley. *Confucianism and Family Rituals in Imperial China: A Social History of Writing About Rites*. Princeton: Princeton University Press, 1991.

―――. "The Dynamics of Elite Domination in Sung China." *HJAS* 48 (1988): 493–519.

Ebrey, Patricia Buckley, trans. and annot. *Chu Hsi's Family Rituals*. Princeton: Princeton University Press, 1991.

Edgren, Sören. "Southern Song Printing at Hangzhou." *Bulletin of the Museum of Far Eastern Antiquities*, no. 62 (1989): 1–212.

Egan, Ronald C. *The Literary Works of Ou-Yang Hsiu (1007–72)*. Cambridge Studies in Chinese History, Literature and Institutions. Cambridge: Cambridge University Press, 1984.

Elman, Benjamin A. "Changes in Confucian Civil Service Examinations from the Ming to the Ch'ing Dynasty." In *Education and Society in Late Imperial China, 1600–1900*, ed. idem and Alexander Woodside, 111–49. Berkeley: University of California Press, 1994.

―――. "The Changing Role of Historical Knowledge in Southern Provincial Civil Examinations During the Ming and Ch'ing." *Zhongyang yanjiuyuan Zhongshan renwen shehui kexue yanjiusuo renwen ji shehui kexue jikan* 中央研究院中山人文社會科學研究所人文及社會科學集刊 (*The Journal of Social Sciences and Philosophy*) 5, no. 1 (1992): 265–319.

―――. "The 'Chinese Sciences' in Policy Questions from Confucian Civil Examinations During the Late Ming." In *Western Learning and Christianity in China: The Contribution and Impact of Johann Adam Schall Von Bell, S.J. (1592–1666)*, ed. Roman Malek, 619–66: Sankt Augustin: China-Zentrum, 1998.

―――. *A Cultural History of Civil Examinations in Late Imperial China*. Berkeley: University of California Press, 2000.

―――. "An Early Ming Perspective on Song-Jin-Yuan Civil Service Examinations." Paper presented at the conference The Song-Yuan-Ming Transition: A Turning Point in Chinese History? UCLA, June 5–11, 1997.

―――. "The Evolution of Civil Service Examinations in Late Imperial China." *Newsletter for Modern Chinese History* 11 (1991): 65–88.

―――. "The Formation of Dao Learning as Imperial Ideology During the Early Ming Dynasty." In *Culture & State in Chinese History: Conventions, Accommodations, and Critiques*, ed. Theodore Huters, R. Bin Wong, and Pauline Yu, 58–82. Stanford: Stanford University Press, 1997.

―――. "Political, Social, and Cultural Reproduction Via Civil Service Examinations in Late Imperial China." *JAS* 50, no. 1 (1991): 7–28.

―――. "Rethinking 'Confucianism' and 'Neo-Confucianism' in Modern Chinese History." In *Rethinking Confucianism: Past and Present in China, Japan, Ko-*

*rea, and Vietnam*, ed. idem, John B. Duncan, and Herman Ooms, 518–54. Los Angeles: UCLA Asian Pacific Monograph Series, 2002.

Elman, Benjamin A.; John B. Duncan; and Herman Ooms. "Introduction." In *Rethinking Confucianism: Past and Present in China, Japan, Korea, and Vietnam*, ed. Elman, Duncan, and Ooms, 1–29. Los Angeles: UCLA Asian Pacific Monograph Series, 2002.

————, eds. *Rethinking Confucianism: Past and Present in China, Japan, Korea, and Vietnam*. Los Angeles: UCLA Asian Pacific Monograph Series, 2002.

Elman, Benjamin A., and Alexander Woodside, eds. *Education and Society in Late Imperial China, 1600–1900*. Berkeley: University of California Press, 1994.

Ephrat, Daphna. *A Learned Society in a Period of Transition: The Sunni "Ulama" of Eleventh Century Baghdad*. SUNY Series in Medieval Middle East History. Albany: State University of New York Press, 2000.

Fan Fengshu 范凤书. *Zhongguo sijia cangshu shi* 中国私家藏书史. Zhengzhou: Daxiang chubanshe, 2001.

Fang Jianxin 方建新. "Nan Song sijia cangshu zai bulu" 南宋私家藏书再补录. *Song shi yanjiu jikan* 宋史研究集刊, 1989, no. 2.

————. "Songdai sijia cangshu bulu" 宋代私家藏书补录. 2 pts. *Wenxian* 文献 1988, no. 1: 220–39; 1988, no. 2: 229–43.

Fang Yanshou 方彦寿. *Zhu Xi shuyuan menren kao* 朱熹书院门人考. Shanghai: Huadong shifan daxue chubanshe, 2000.

Feng Xiaoting 馮曉庭. *Song chu jingxue fazhan shulun* 宋初經學發展述論. Taibei: Wanjuan lou tushu, 2001.

Foster, Robert Wallace. "Differentiating Rightness from Profit: The Life and Thought of Lu Jiuyuan (1139–1193)." Ph.D. diss., Harvard University, 1997.

————. "Seeking a Tradition: Lu Jiuyuan's Attempt to Define *Ru* Approaches to Education, Politics, and Philosophy in Southern Song." *Papers on Chinese History* 1 (1992): 1–21.

Frank, Paulo. "National Security in Northern Sung: A Subject for the Civil Service Examination." *Papers on Chinese History* 1 (1992): 22–38.

Franke, Herbert. *Sung Biographies*. Wiesbaden: Steiner, 1976.

Franke, Herbert, and Denis Crispin Twitchett, eds. *Cambridge History of China*, vol. 6, *Alien Regimes and Border States, 907–1368*. Cambridge: Cambridge University Press, 1994.

Franke, Wolfgang. *The Reform and Abolition of the Traditional Chinese Examination System*. Cambridge: Center for East Asian Studies, Harvard University, 1968.

Fu Xuanzong 傅璇琮. *Tangdai keju yu wenxue* 唐代科舉與文學. Taibei: Wenshizhe, 1994 (1986).

Fu Zengxiang 傅增湘. *Cangyuan qunshu jingyan lu* 藏園群書經眼錄. 5 vols. Beijing: Zhonghua shuju, 1983.

———. *Cangyuan qunshu tiji* 藏園群書題記. Shanghai: Shanghai guji chubanshe, 1989 (1930s–1940s).

Gao Lingyin 高令印. *Zhu Xi shiji kao* 朱熹事迹考. Shanghai: Shanghai renmin chubanshe, 1987.

Gao Mingshi 高明士. *Sui Tang gongju zhidu* 隋唐貢舉制度. Sui Tang wenhua yanjiu congshu. Taibei: Wenjin chubanshe, 1999.

———. *Zhongguo chuantong zhengzhi yu jiaoyu* 中國傳統政治與教育. Taibei: Wenjin chubanshe, 2003.

Gardner, Daniel K. *Chu Hsi and the Ta-Hsueh: Neo-Confucian Reflection on the Confucian Canon.* Cambridge: Council on East Asian Studies, Harvard University, 1986.

———. "Confucian Commentary and Chinese Intellectual History." *JAS* 57, no. 2 (1998): 397–422.

———. *Learning to Be a Sage.* Berkeley: University of California Press, 1990.

———. "Modes of Thinking and Modes of Discourse in the Sung: Some Thoughts on the Yü Lu ('Recorded Conversations') Texts." *JAS* 50, no. 3 (1991): 574–603.

———. "Principle and Pedagogy: Chu Hsi and the Four Books." *HJAS* 44, no. 1 (1984): 57–81.

———. "Transmitting the Way: Chu Hsi and His Program of Learning." *HJAS* 49, no. 1 (1989): 141–72.

Ge Shao'ou 葛邵歐. "Songdai Fuzhou de gongyuan" 宋代福州的贡院. In *Guoji Song shi yantaohui lunwen xuanji* 国际宋史研讨会论文选集, ed. Deng Guangming 邓广铭 and Qi Xia 漆侠, 304–19. Baoding: Hebei daxue chubanshe, 1993.

Gong Wei Ai. "The Consolidation of Southern Sung China: The Reign of Hsiao-Tsung (1162–1189)." In *Cambridge History of China*, vol. 5, pt. I, *The Five Dynasties and Sung China, 960–1279 A.D.*, ed. Paul Smith and John W. Chaffee. Cambridge: Cambridge University Press, forthcoming.

———. "Emperor Hsiao-Tsung and the Consolidation of Southern Sung China. 2 pts. "Court Politics During the Ch'ien-Tao Era, 1165–1173" and "Court Politics During the Ch'un-Hsi Era, 1174–1189." *Chinese Culture* 28, no. 3 (1987): 47–78; 28, no. 4 (1987): 35–65.

———. "Ideal and Reality: Student Protests in Southern Sung China, 1127–1279." In *Proceedings of the Second Symposium on Sung History* 第二屆宋史學術研討會論文集, ed. Di er jie Song shi xueshu yantaohui mishuchu 第二屆宋史學術研討會秘書處, 720–696. Taibei: Chinese Culture University, 1996.

Gong Yanming 龔延明. *Songdai guanzhi cidian* 宋代官制辞典. Beijing: Zhonghua shuju, 1997.

Graham, A. C. *Two Chinese Philosophers.* London: Lund Humphries, 1958.

————. "What Was New in the Ch'eng-Chu Theory of Human Nature?" In *Chu Hsi and Neo-Confucianism*, ed. Wing-tsit Chan, 138–57. Honolulu: University of Hawaii Press, 1986.

Gu Hongyi 顾宏义. *Jiaoyu zhengce yu Songdai Liang Zhe jiaoyu* 教育政策与宋代兩浙教育. Wuhan: Hubei jiaoyu chubanshe, 2003.

Guan Changlong 关长龙. *Liang Song daoxue mingyun de lishi kaocha* 兩宋道学命运的历史考察. Qiu shi cong shu. Shanghai: Xuelin chubanshe, 2001.

Guan Minyi 管敏义. *Zhedong xueshu shi* 浙东学术史. Shanghai: Huadong shifan daxue chubanshe, 1993.

Guan Xihua 管錫華. *Zhongguo gudai biaodian fuhao fazhan shi* 中國古代標點符號發展史. Sichuan shifan daxue wenxueyuan xueshu congshu. Chengdu: Ba Shu shushe, 2002.

Guarino, Marie. "Learning and Imperial Authority in Northern Sung China (960–1126): The Classics Mat Lectures." Ph.D. diss., Columbia University, 1994.

Guillory, John. "Canon." In *Critical Terms for Literary Study*, ed. Frank Lentricchia and Thomas McLaughlin, 233–49. Chicago: University of Chicago Press, 1995 (1990).

————. *Cultural Capital: The Problem of Literary Canon Formation*. Chicago: University of Chicago Press, 1993.

Guo Qijia 郭齊家. *Zhongguo gudai kaoshi zhidu* 中國古代考試制度. Ed. Ren Jiyu 任繼愈. Zhongguo wenhuashi zhishi congshu. Taibei: Taiwan Shangwu yinshuguan, 1994.

Guo Shaoyu 郭紹虞. *Zhongguo wenxue piping shi* 中國文學批評史. Taibei: Wenshizhe chubanshe, 1990 (1934).

Guoli zhongyang tushuguan 國立中央圖書館. *Guoli zhongyang tushuguan shanben shumu* 國立中央圖書館善本書目. 2d ed., rev. and enl. 4 vols. Taibei: Guoli zhongyang tushuguan,1986.

————. *Guoli zhongyang tushuguan Songben tulu* 國立中央圖書館宋本圖錄. Taibei: Zhonghua congshu weiyuanhui, 1958.

Guy, R. Kent. "Fang Pao and the *Ch'in-Ting Ssu-Shu-Wen*." In *Education and Society in Late Imperial China, 1600–1900*, edited by Benjamin A. Elman and Alexander Woodside, 150–82. Berkeley: University of California Press, 1994.

Haeger, John Winthrop. "The Intellectual Context of Neo-Confucian Syncretism." *JAS* 31, no. 3 (1972): 499–513.

Haeger, John Winthrop, ed. *Crisis and Prosperity in Sung China*. Tucson: University of Arizona Press, 1975.

Hansen, Valerie. *Changing Gods in Medieval China, 1127–1276*. Princeton: Princeton University Press, 1990.

Hartman, Charles. "Bibliographical Notes on Sung Historical Works: The Original *Record of the Way and Its Destiny (Tao-Ming Lu)* by Li Hsin-Ch'uan." *JSYS* 30 (2000): 1–61.

———. *Han Yü and the T'ang Search for Unity*. Princeton: Princeton University Press, 1986.

———. "The Making of a Villain: Ch'in Kuei and Tao-Hsueh." *HJAS* 58, no. 1 (1998): 59–146.

———. "Preliminary Bibliographical Notes on the Sung Editions of Han Yü's *Collected Works.*" In *Critical Essays on Chinese Literature*, ed. William H. Nienhauser, Jr., 89–100. Hong Kong: Chinese University of Hong Kong, 1976.

———. "Zhu Xi and His World." *JSYS*, forthcoming.

Hartwell, Robert M. "Financial Expertise, Examinations, and the Formulation of Economic Policy in Northern Sung China." *JAS* 30 (1971): 281–314.

———. "Historical Analogism, Public Policy, and Social Science in Eleventh and Twelfth Century China." *American Historical Review* 76, no. 3 (1971): 690–727.

Hatch, George Cecil, Jr. "Su Hsun's Pragmatic Statecraft." In *Ordering the World. Approaches to State and Society in Sung Dynasty China*, ed. Robert P. Hymes and Conrad Schirokauer, 59–75. Berkeley: University of California Press, 1993.

———. "The Thought of Su Hsun (1009–1066): An Essay in the Social Meaning of Intellectual Pluralism in Northern Sung." Ph.D. diss., University of Washington, 1972.

He Huaihong 何怀宏. *Xuanju shehui ji qi zhongjie: Qin Han zhi wan Qing lishi de yizhong shehuixue chanshi* 选举社会及其终结: 秦汉至晚清历史的一种社会学阐释. Sanlian, Hafo Yanjing xueshu congshu. Beijing: Shenghuo dushu xin zhi sanlian shudian, 1998.

He Jipeng 何寄澎. *Bei Song de guwen yundong* 北宋的古文運動. Taibei: Youshi, 1992.

———. *Tang Song guwen xin tan* 唐宋古文新探. Taibei: Da'an, 1990.

———. "Zhuzi de wen lun" 朱子的文論. In *Guoji Zhuzixue huiyi wenji* 國際朱子學會議文集, ed. Zhong Caijun 鍾彩鈞, 1213–32. Taibei: Zhongyangyuan wenzhesuo, 1993.

He Jun 何俊. *Nan Song ruxue jian'gou* 南宋儒学建构. Shanghai: Shanghai renmin chubanshe, 2004.

He Yousen 何祐森. "Liang Song xuefeng zhi dili fenbu" 兩宋學風之地理分布. *Xinya xuebao* 新亞學報 1 (1955): 331–79.

He Zhongli 何忠礼. *Song shi "Xuanju zhi" buzheng* 宋史选举志补正. Song shi buzheng. Hangzhou: Zhejiang guji chubanshe, 1992.

He Zhongli 何忠礼 and Xu Jijun 徐吉军. *Nan Song shi gao: zhengzhi, junshi, wenhua* 南宋史稿: 政治・军事・文化. Hangzhou: Hangzhou daxue chubanshe, 1999.

Hervouet, Yves, ed. *A Sung Bibliography (Bibliographie Des Sung)*. Hong Kong: Chinese University Press, 1978.

Hirata Shigeki 平田茂樹. "Gen'yū jidai no seiji ni tsuite: senkyo rongi o tegakari ni shite" 元祐時代の政治について: 選舉論議を手掛りにして. In *Sōdai no chishikijin* 宋代の知識人, 109–36. Tokyo: Kyūko shoin, 1993.

———. *Kakyo to kanryōsei* 科舉と官僚制. Sekaishi riburetto. Tokyo: Yamakawa shuppansha, 1997.

Ho, Ping-ti. *The Ladder of Success in Imperial China: Aspects of Social Mobility, 1368–1911*. New York: Columbia University Press, 1962.

Hong Dexuan 洪德旋. *Zhongguo kaoshi zhidu shi* 中國考試制度史. Kaoquan congshu. Taibei: Kaoshi yuan (Zhengzhong shuju), 1983.

Honma Tsugihiko 本間次彦. "Yomigaeru Shu Ki" 甦る朱熹. *Chūgoku tetsugaku kenkyū* 中国哲学研究, no. 5 (1993): 1–42.

Hou Shaowen 侯紹文. *Tang Song kaoshi zhidu shi* 唐宋考試制度史. Taibei: Taiwan Shangwu yinshuguan, 1973.

Hou Wailu 侯外庐, Qiu Hansheng 邱汉生, and Zhang Qizhi 张岂之. *Song Ming lixue shi* 宋明理学史. 2 vols. Beijing: Renmin chubanshe, 1984.

Hsu, Yeong-huei. "Song Gaozong (r. 1127–1162) and His Chief Councilors: A Study of the Formative State of the Southern Song Dynasty (1127–1279)." Ph.D. diss., University of Arizona, 2000.

Hu Yujin 胡玉缙. *Siku quanshu zongmu tiyao buzheng* 四庫全書總目提要補正. Taibei: Muduo, 1981 (first half 20th c., first printed in 1964).

Huang, Chün-chieh. "The Synthesis of Old Pursuits and New Knowledge: Chu Hsi's Interpretation of Mencian Morality." *Xinya xueshu jikan* 新亞學術集刊 3 (1982): 197–222.

Huang, Chün-chieh, and Erik Zürcher. *Norms and the State in China*. Leiden: E. J. Brill, 1993.

Huang Jinjun 黄锦君. *Er Cheng yulu yufa yanjiu* 二程语录语法研究. Chengdu: Sichuan daxue, 2005.

Huang Kuanchong 黄寛重. "Keju, jingji, yu jiazu xingshuai: yi Songdai Dexing Zhang shi jiazu wei li" 科舉經濟與家族興衰: 以宋代德興張氏家族爲例. In *Proceedings of the Second Symposium on Sung History—第二屆宋史學術研討會論文集*, ed. Di er jie Song shi xueshu yantaohui mishuchu 第二屆宋史學術研討會秘書處, 127–46. Taibei: Chinese Culture University, 1996.

————. "Lue lun Nan Song shidai de guizheng ren" 略論南宋時代的歸正人. In idem, *Nan Song shi yanjiu ji* 南宋史研究集, 185–231. Taibei: Xin wenfeng, 1985.

————. *Nan Song shi yanjiu ji* 南宋史研究集. Taibei: Xin wenfeng, 1985.

————. "Wan Song chaochen dui guoshi de zhengyi—Lizong shidai de hezhan, bianfang yu liumin" 晚宋朝臣對國是的爭議—理宗時代的和戰、邊防與流民. Master's thesis, National Taiwan University, 1975.

Huang Qingwen 黃晴文. "Zhongguo gudai shuyuan zhidu ji qi keshu tanyan" 中國古代書院制度及其刻書探研. M.A. thesis, Zhongguo wenhua daxue, 1984.

Huff, Toby E. *The Rise of Early Modern Science: Islam, China and the West.* Cambridge: Cambridge University Press, 1993.

Huters, Theodore; R. Bin Wong; and Pauline Yu, eds. *Culture and State in Chinese History: Conventions, Accommodations, and Critiques.* Stanford: Stanford University Press, 1997.

Hymes, Robert P. "Lu Chiu-Yuan, Academies, and the Problem of the Local Community." In *Neo-Confucian Education: The Formative Stage,* ed. Wm. Theodore de Bary and John W. Chaffee, 432–56. Berkeley: University of California Press, 1989.

————. "Moral Duty and Self-Regulating Process in Southern Sung Views of Famine Relief." In *Ordering the World. Approaches to State and Society in Sung Dynasty China,* ed. idem and Conrad Schirokauer, 280–309. Berkeley: University of California Press, 1993.

————. *Statesmen and Gentlemen: The Elite of Fu-Chou, Chiang-Hsi, in Northern and Southern Sung.* Cambridge: Cambridge University Press, 1986.

————. *Way and Byway: Taoism, Local Religion, and Models of Divinity in Sung and Modern China.* Berkeley: University of California Press, 2002.

Hymes, Robert P., and Conrad Schirokauer, eds. *Ordering the World. Approaches to State and Society in Sung Dynasty China.* Berkeley: University of California Press, 1993.

Ichikawa Mototarō 市川本太郎. *Bunchū shi* 文中子. Chūgoku koten shinsho. Tokyo: Meitoku shuppansha, 1970.

————. "*Shu Ki gorui* zakki" 朱熹語類雜記. *Jinbun kagakuka kiyō* 人文科學科紀要 21 (1959): 137–84.

Ichikawa Yasuji 市川安司. *Shushi tetsugaku ronkō* 朱子哲學論考. Tokyo: Kyūko shoin, 1985.

Ichiki Tsuyuhiko 市来津由彦. "Shin Jun ron josetsu—Shushigaku keisei no shiten kara" 陳淳論序説—朱子学形成の視点から. *Tōyō kotengaku kenkyū* 東洋古典學研究 15 (2003): 143–58.

————. *Shu Ki monjin shūdan keisei no kenkyū* 朱熹門人集團形成の研究. Tōyōgaku sōsho. Tokyo: Sōbunsha, 2002.

————. "Shu Ki no Rikuchō hyō—dō bun itchi no ron kara mita chūsei shō" 朱熹の六朝評—道文一致の論からみた中世像. In *Sōgō kenkyū chūsei no bunka* 綜合研究中世の文化, ed. Katano Tatsurō 片野達郎, 465–83. Tokyo: Kadokawa shoten, 1988.

Ihara Hiroshi 伊原弘. "Chūgoku chishikijin no kisō shakai—Sōdai Onshū Eika gakuha o rei to shite" 中国知識人の基層社會—宋代温州永嘉学派を例として. *Shisō* 思想 802, no. 4 (1991): 82–103.

————. "Chūgoku shomin kyōiku kenkyū no tame no joshō" 中国庶民教育研究のための序章. *Tōyō kyōikushi kenkyū* 東洋教育史研究 11 (1987): 61–77.

————. "Sōdai no shitaifu oboegaki" 宋代の士大夫覚え書き. In *Sōdai no shakai to shūkyō* 宋代の社會と宗教, 257–95. Tokyo: Kyūko shoin, 1985.

Inoue Susumu 井上進. *Chūgoku shuppan bunkashi: shomotsu sekai to chi no fūkei* 中国出版文化史: 書物世界と知の風景. Nagoya: Nagoya daigaku shuppankai, 2002.

————. "*Hokki jigi* hanpon kō" 北溪字義版本考. *Tōhōgaku* 東方学 80 (1990): 111–25.

————. "Zōsho to dokusho" 藏書と讀書. *Tōhō gakuhō* 東方學報 62 (1990): 409–45.

Ishida Hajime 石田肇. "Shū Mi to Dōgaku" 周密と道学. *Tōyōshi kenkyū* 東洋史研究 49, no. 2 (1990): 25–47.

Ivanhoe, Philip J. "Reflections on the *Chin-Ssu Lu*." *Journal of the American Oriental Society* 108, no. 2 (1988): 269–75.

Ji Xiao-bin. "Inward-Oriented Ethical Tension in Lü Tsu-Ch'ien's Thought." Master's thesis, Arizona State University, 1991.

Jia Guirong 賈貴榮 and Wang Guan 王冠. *Song Yuan ban shumu tiba jikan* 宋元版書目題跋輯刊. 4 vols. Beijing: Beijing tushuguan chubanshe, 2003.

Jiang Guanghui 姜广辉. *Lixue yu Zhongguo wenhua* 理学与中国文化. Shanghai: Shanghai renmin chubanshe, 1994.

Jiang Shuge 姜书阁. *Pianwen shi lun* 骈文史论. Beijing: Renmin wenxue chubanshe, 1986.

Jiangsu guoxue tushuguan 江蘇國學圖書館. *Boshan shuying* 盋山書影. 2 vols. Nanjing: Jiangsu guoxue tushuguan, 1929.

Jin Zheng 金诤. *Keju zhidu yu Zhongguo wenhua* 科举制度与中国文化. Shanghai: Shanghai renmin chubanshe, 1990.

Jin Zhongshu 金中樞. "Bei Song keju zheng cidi renyuan renyong zhi zhi xingcheng kao" 北宋科举正賜第人员任用之制形成考. In *Guoji Song shi yantaohui lunwen xuanji* 国际宋史研讨会论文选集, ed. Deng Guangming 鄧广铭 and Qi Xia 漆侠, 281–303. Baoding: Hebei daxue chubanshe, 1993.

———. "Bei Song keju zhidu yanjiu" 北宋科舉制度研究. In *Song shi yanji ji* 宋史研究集, 1–72, 31–112. Taibei: Guoli bianyiguan, 1979–80 (1964).

———. "Bei Song keju zhidu yanjiu xu" 北宋科舉制度研究續. In *Song shi yanjiu ji* 宋史研究集, 61–189, 53–90. Taibei: Guoli bianyiguan, 1981, 1983 (1978–79).

———. "Bei Song keju zhidu yanjiu zai xu" 北宋科舉制度研究再續. In *Song shi yanjiu ji* 宋史研究集, 125–88, 53–90. Taibei: Guoli bianyiguan, 1984, 1986 (1980, 1982).

Jinian Chen Fuliang danchen babai liushi zhounian chouweihui 纪念陈傅良诞辰八百六十周年筹委会, ed. *Chen Fuliang danchen babai liushi zhounian jinian ji* 陈傅良诞辰八百六十周年纪念集. N.p., 1997.

Kakiuchi Keiko 垣内景子. "Shushi gorui yakuchū ken 113–114 'Kun monjin' (1–5)" 朱子語類訳注巻 113–114 訓門人 (1–5). 5 pts. *Meiji daigaku kyōyō ronshū* 明治大学教養論集, no. 318 (1999): 101–17; no. 327 (2000): 43–57; 339 (2001): 73–94; 358 (2002): 65–86; 376 (2004): 29–50.

Kasoff, Ira E. *The Thought of Chang Tsai (1020–1077)*. Cambridge Studies in Chinese History, Literature, and Institutions. Cambridge: Cambridge University Press, 1984.

Kawakami Kyōji 川上恭司. "Sōdai no toshi to kyōiku—Shūkengaku o chūshin ni" 宋代の都市と教育—州県学を中心に. In *Chūgoku kinsei no toshi to bunka* 中国近世の都市と文化, ed. Umehara Kaoru, 359–87. Kyoto: Kyōto daigaku Jinbun kagaku kenkyūjo, 1984.

Ke Dunbo 柯敦伯. *Song wenxue shi* 宋文學史. Guoxue xiao congshu. Shanghai: Shangwu yinshuguan, 1934.

Kieschnick, John. *The Eminent Monk: Buddhist Ideals in Medieval Chinese Hagiography*. Studies in East Asian Buddhism. Honolulu: University of Hawaii Press, 1997.

Kinney, Anne Behnke. *The Art of the Han Essay: Wang Fu's "Ch'ien-Fu Lun."* Monograph Series / Arizona State University, Center for Asian Studies, 26. Tempe: Center for Asian Studies, Arizona State University, 1990.

Kinugawa Tsuyoshi 衣川強. "Kaishi yōhei o megutte" 開禧用兵をめぐって. *Tōyōshi kenkyū* 東洋史研究 36, no. 3 (1977): 128–51.

Kinugawa Tsuyoshi 衣川強 and Chikusa Masaaki 竺沙雅章. *Shu Ki* 朱熹. Chūgoku rekishi jinbutsu sen. Tokyo: Hakuteisha, 1994.

Kiyomizu Yasuyoshi 清水靖義. "Sōdai kakyo seido kenkyū no kaiko to sono mondaiten" 宋代科挙制度研究の回顧とその問題点. *Hiroshima daigaku Tōyōshi kenkyūshitsu hōkoku* 広島大学東洋史研究室報告 14 (1992): 10–14.

Kohn, Livia, and Harold David Roth. *Daoist Identity: History, Lineage, and Ritual.* Honolulu: University of Hawaii Press, 2002.

Kojima Kenkichirō 兒島獻吉郎. *Shina bungaku kō*, vol. 1, *Sanbun kō* 支那文學考: 散文考. Tokyo: Meguro shoten, 1919.

Kojima Tsuyoshi 小島毅. "Riben Song xue yanjiu de xin shidian" 日本宋学研究的新视点. *Song shi yanjiu tong xun* 宋史研究通訊 36 (2000): 37–40.

———. "Shisō dentatsu baitai to shite no shomotsu—Shushigaku no 'bunka no rekishigaku' josetsu" 思想伝達媒体としての書物—朱子学の「文化の歴史学」序説. In *Sōdai shakai no nettowāku* 宋代社会のネットワーク, ed. Sōdaishi kenkyūkai 宋代史研究会, 47–75. Tokyo: Kyūko shoin, 1998.

———. "Shushigaku no dempo, teichaku to shomotsu" 朱子学の伝播・定着と書物. In *Ajia yūgaku—Sōdai chishikijin no shoshō—hikaku no shuhō ni yoru mondai teiki* アジア遊学—宋代知識人の諸相—比較の手法による問題提起, 17–27. Tokyo: Bensei shuppan, 1999.

———. "Shushigaku no hatten to insatsu bunka" 朱子学の展開と印刷文化. In *Chishikijin no shosō: Chūgoku Sōdai o kiten to shite* 知識人の諸相: 中国宋代を基点として, ed. Kojima Tsuyoshi 小島毅 and Ihara Hiroshi 伊原弘, 192–202. Tokyo: Bensei shuppan, 2001.

———. "*Shushi gorui* Gan En mon jin shō yakuchū kō" (1) 朱子語類顔淵問仁章訳注稿. *Tokushima daigaku sōgō kagakubu ningen shakai bunka kenkyū* 德島大学綜合科学部人間社会文化研究 2 (1995): 1–22.

———. *Sōgaku no keisei to tenkai* 宋学の形成と展開. Chūgoku gakugei sōsho. Tokyo: Sōbunsha, 1999.

———. "Songxue zhupai zhong zhi Zhuxue de diwei" 宋学诸派中之朱学的地位. In *Song shi yanjiu lunwen ji: Guoji Song shi yantaohui ji Zhongguo Song shi yanjiuhui di jiu jie nianhui biankan* 宋史研究论文集: 国际宋史研讨会暨中国宋史研究会第九届年会编刊, ed. Qi Xia 漆侠, 516–28. Baoding: Hebei daxue, 2002.

Kondō Haruo 近藤春雄. *Chūgoku gakugei dai jiten* 中國学芸大事典. Tokyo: Taishūkan shoten, 1978.

Kondō Kazunari 近藤一成. "Dōgakuha no keisei to Fukken—Yō Shi no keizai seisaku o megutte" 道学派の形成と福建—楊時の経済政策をめぐって. In *Chūgoku zen kindaishi kenkyū: Kurihara Tomonobu hakushi tsuitō kinen* 中国前近代史研究: 栗原朋信博士追悼記念, ed. Waseda daigaku Bungakubu Tōyōshi kenkyūshitsu 早稲田大学文学部東洋史研究室, 154–69. Tokyo: Yūzankaku shuppan, 1980.

———. "Nan Sō shoki no Ō Anseki hyōka ni tsuite" 南宋初期の王安石評價について. *Tōyōshi kenkyū* 東洋史研究 38, no. 3 (1979): 26–51.

————. "Ō Anseki no kakyo kaikaku o megutte" 王安石の科挙改革を
めぐって. *Tōyōshi kenkyū* 東洋史研究 46, no. 3 (1987): 483–508.

————. "Sōdai Eika gakuha no risai ron" 宋代永嘉学派の理財論. *Shikan*
史観 92 (1975): 40–53.

————. "Sōdai Eika gakuha Yō Teki no kai kan" 宋代永嘉学派葉適の華夷
観. *Shigaku zasshi* 史学雑誌 88, no. 6 (1979): 51–79.

Kōzen Hiroshi 興膳宏, Kizu Yūko 木津祐子, and Saitō Mareshi 齋藤希史.
"*Shushi gorui* 'Dokushohō hen' yakuchū" 朱子語類讀書法篇訳注. 6 pts.
*Chūgoku bungaku hō* 中国文学報, no. 48 (1994): 109–38; no. 49 (1994): 19–52;
no. 50 (1995): 58–80; no. 51 (1995): 48–71; no. 52 (1996): 4–25; no. 53 (1996):
23–43.

————. "*Shushi gorui* 'Ronbun hen' yakuchū" 朱子語類論文篇訳注. 8 pts.
*Chūgoku bungaku hō* 中国文学報, no. 55 (1997): 127–51; no. 56 (1998): 19–46;
no. 57 (1998): 95–20; no. 58 (1999): 15–40; no. 59 (1999): 62–75; no. 60
(2000): 65–82; no. 61 (2000): 92–10; no. 62 (2001): 75–96.

Kracke, Edward, Jr. "Change Within Tradition." *Far Eastern Quarterly* 14 (1954–
55): 479–88.

————. *Civil Service in Early Sung China, 960–1067*. Cambridge: Harvard University Press, 1953.

————. "The Expansion of Educational Opportunity in the Reign of Hui-
Tsung of the Sung and Its Implications." *Sung Studies Newsletter* 13 (1977):
6–30.

————. "Family Versus Merit in Chinese Civil Service Examinations Under
the Empire." *HJAS* 10 (1947): 105–23.

————. "Region, Family and Individuals in the Examination System." In *Chinese Thought and Institutions*, ed. John King Fairbank, 251–68. Chicago: University of Chicago Press, 1957.

Kuhn, Dieter. *Die Song-Dynastie (960 Bis 1279): eine neue Gesellschaft im Spiegel ihrer
Kultur*. Weinheim: Acta Humaniora and VCH, 1987.

Kuhn, Dieter, and Ina Asim. *Beamtentum und Wirtschaftspolitik in der Song-Zeit*.
Würzburger Sinologische Schriften. Heidelberg: Edition Forum, 1995.

Kyōto daigaku. Jinbun kagaku kenkyūjo 京都大学人文科学研究所. *Kyōto
daigaku Jinbun kagaku kenkyūjo kanseki mokuroku* 京都大学人文科学研究所
漢籍目録. 2 vols. Kyoto: Kyōto daigaku Jinbun kagaku kenkyūjo, 1979,
1980.

Lackner, Michael. "Argumentation par diagrammes: Le Ximing depuis Zhang
Zai jusqu'au *Yanjitu*." *Extrême Orient–Extrême Occident* 14 (1992): 131–68.

————. "Zur Verplanung des Denkens am Beispiel der T'u." In *Lebenswelt und
Weltanschauung im frühneuzeitlichen China*, ed. Hellwig Schmidt-Glintzer, 133–
56. Stuttgart: Franz Steiner, 1990.

Lai, T. C. *A Scholar in Imperial China*. Hong Kong: Kelly & Walsh, 1970.

Langley, C. Bradford. "Wang Ying-Lin (1223–1296): A Study in the Political and Intellectual History of the Demise of the Sung." Ph.D. diss., Indiana University, 1980.

Langlois, John D., Jr. Review: Benjamin A. Elman, *A Cultural History of Civil Examinations in Late Imperial China. HJAS* 61, no. 1 (2001): 216–30.

Lau, D. C. *The Analects*. Harmondsworth, Eng.: Penguin Books, 1979.

———. *Mencius*. Harmondsworth, Eng.: Penguin Books, 1970.

Lau, Nap-yin. "The Absolutist Reign of Sung Hsiao-Tsung (r. 1163–1189)." Ph.D. diss., Princeton University, 1986.

Lee, Thomas H. C. 李弘祺. "Academies: Official Sponsorship and Suppression." In *Imperial Rulership and Cultural Change in Traditional China*, ed. Frederic Brandauer and Chün-chieh Huang, 117–43. Seattle: University of Washington Press, 1994.

———. "Books and Bookworms in Song China: Book Collection and the Appreciation of Books." *JSYS* 25 (1995): 193–218.

———. "Chu Hsi, Academies and the Tradition of Private *Chiang-Hsüeh*." *Chinese Studies* 2 (1986): 301–30.

———. *Education in Traditional China, a History*. Handbook of Oriental Studies. Leiden: Brill, 2000.

———. *Government Education and Examinations in Sung China*. Hong Kong: Chinese University of Hong Kong Press, 1985.

———. "Jiang zhang de yifeng: siren jiangxue de chuantong" 降帳的遺風: 私人講學的傳統. In *Haohan de xuehai* 浩瀚的學海, ed. Lin Qingzhang 林慶彰, 343–410. Taibei: Lianjing, 1982.

———. "Jingshe yu shuyuan" 精舍與書院. *Chinese Studies* 10, no. 2 (1992): 307–32.

———. "Neo-Confucian Education in Chien-Yang, Fu-Chien, 1000–1400: Academies, Society and the Development of Local Culture." In *Guoji Zhuzixue huiyi lunwen ji* 國際朱子學會議論文集, 945–96. Taibei: Zhongyang yanjiuyuan Zhongguo wenzhe yanjiusuo choubeichu, 1993.

———. "The Social Significance of the Quota System in Song Civil Service Examinations." *Journal of the Institute of Chinese Studies* 13 (1982): 287–317.

———. "Songdai de juren" 宋代的舉人. In *Guoji Song shi yantaohui lunwen ji* 國際宋史研討會論文集, 297–314. Taibei: Zhongguo wenhua daxue, 1989.

———. "Songdai guanyuan shu de tongji" 宋代官員數的統計. *Shihuo yuekan fukan* 食貨月刊復刊 14, no. 5–6 (1984): 227–39.

———. *Songdai jiaoyu sanlun* 宋代教育散論. Taibei: Dongsheng, 1979.

Legge, James. *The Chinese Classics*. 5 vols. Taibei: SMC Publishing, 1991 (1893–95).

Levine, Ari Daniel. "A House in Darkness: The Politics of History and the Language of Politics in the Late Northern Song, 1068–1104." Ph.D. diss., Columbia University, 2002.

Li Guoling 李國玲, ed. *Song ren zhuanji ziliao suoyin bubian* 宋人傳記資料索引補編. Chengdu: Sichuan daxue chubanshe, 1994.

Li Huarui 李华瑞. *Wang Anshi bianfa yanjiu shi* 王安石变法研究史. Beijing: Renmin chubanshe, 2004.

Li Jixiang 李紀祥. *Liang Song yilai daxue gaiben zhi yanjiu* 兩宋以來大學改本之研究. Taibei: Xuesheng shuju, 1988.

Li Ruiliang 李瑞良. *Zhongguo muluxue shi* 中國目錄學史. Taibei: Wenjin chubanshe, 1993.

Li Tie 李铁. *Kechang fengyun* 科场风云. Beijing: Zhongguo qingnian chubanshe, 1991.

Li Xinda 李新達. *Zhongguo keju zhidu shi* 中國科舉制度史. Taibei: Wenjin chubanshe, 1995.

Li Xuezhi 李學智. "Taida cang Song ban *Xishan xiansheng Zhen Wenzhong gong wenzhang zhengzong*" 臺大藏宋版西山先生眞文重公文章正宗. *Tushuguanxue kan* 圖書館學刊 1 (1967): 77–79.

Li Zhengfu 李正富. "Songdai keju zhidu zhi yanjiu" 宋代科舉制度之研究. Master's thesis, Guoli zhengzhi daxue, 1963.

Li Zhizhong 李致忠. *Lidai keshu kaoshu* 历代刻书考述. Chengdu: Ba Shu shushe, 1989.

Liang Gengyao 梁庚堯. "Nan Song jiaoxue hangye xingsheng de beijing" 南宋教學行業興盛的背景. *Song shi yanjiu ji* 宋史研究集 no. 30, 317–43. Taibei: Guoli bianyiguan, 2000.

Library of Congress, Asian Division, and Arthur William Hummel, eds. *Eminent Chinese of the Ch'ing Period (1644–1912)*. Taibei: Literature House, 1964 (1943).

Lin Ruihan 林瑞翰. "Songdai zhike kao" 宋代制科考. In *Song shi yanjiuji* 宋史研究集 no. 16, 127–53. Taibei: Guoli bianyiguan, 1986.

Lin Sufen 林素芬. "Boshi yi zhiyong—Wang Yinglin xueshu de zai pingjia" 博識以致用—王應麟學術的再評價. Master's thesis, National Taiwan University, 1994.

———. "Lü Zuqian de cizhang zhi xue yu guwen yundong" 呂祖謙的辭章之學與古文運動. *Guoli zhongyang tushuguan guankan* 國立中央圖書館館刊 28, no. 2 (1995): 145–61.

Lin Yisheng 林益勝. "Qingyuan dang'an zhi yanjiu" 慶元黨案之研究. Master's thesis, National Taiwan University, 1970.

Liu Boji 劉伯驥. *Songdai zhengjiao shi* 宋代政教史. 2 vols. Taibei: Zhonghua shuju, 1971.

Liu Hong 刘虹. *Zhongguo xuanshi zhidu shi* 中国选士制度史. N.p.: Hunan jiaoyu chubanshe, 1992.

Liu Hsiang-kwang 劉祥光. "Education and Society: The Development of Public and Private Institutions in Hui-Chou, 960–1800." Ph.D. diss., Columbia University, 1996.

———. "Yinshua yu kaoshi: Songdai kaoshi shiyong cankaoshu chu tan" 印刷與考試: 宋代考試用參考書初探. *Song shi yanjiu ji* 宋史研究集 no. 31, 151–200. Taibei: Guoli bianyiguan, 2001.

Liu Hsieh. *The Literary Mind and the Carving of Dragons: A Study of Thought and Pattern in Chinese Literature.* Trans. and annot. Vincent Yu-chung Shih. Chinese Classics. Hong Kong: Chinese University Press, 1983.

Liu, James T. C. 劉子健. *China Turning Inward: Intellectual-Political Changes in the Early Twelfth Century.* Cambridge: Harvard University Press, 1988.

———. "How Did a Neo-Confucian School Become the State Orthodoxy?" *Philosophy East and West* 23 (1973): 483–505.

———. "Lue lun Songdai difang guanxue he sixue de xiaozhang" 略論宋代地方官學和私學的消長. *Lishi yuyan yanjiusuo jikan* 歷史語言研究所集刊 36 (1965): 237–48.

———. *Ou-yang Hsiu: An Eleventh-Century Neo-Confucianist.* Stanford: Stanford University Press, 1967.

———. "Songdai kaochang biduan—jian lun shifeng wenti" 宋代考場弊端—兼論士風問題. In idem, *Liang Song shi yanjiu huibian* 兩宋史研究彙編, 229–47. Taibei, 1987 (1965).

———. "Songmo suowei daotong de chengli" 宋末所謂道統的成立. In idem, *Liang Song shi yanjiu huibian* 兩宋史研究彙編, 249–82. Taibei: Lianjing, 1987 (1979).

———. "Wei Liaoweng's Thwarted Statecraft." In *Ordering the World. Approaches to State and Society in Sung Dynasty China*, ed. Robert P. Hymes and Conrad Schirokauer, 336–48. Berkeley: University of California Press, 1193.

Liu, James T. C., and Peter J. Golas, eds. *Change in Sung China: Innovation or Renovation?* Problems in Asian Civilizations. Boston: D. C. Heath & Co., 1969.

Liu, James T. C. 劉子健, and Liu Zijian boshi song shou jinian Song shi yanjiu lunji kanxinghui 劉子健博士頌壽紀念宋史研究論集刊行會. *Liu Zijian boshi song shou jinian Song shi yanjiu lunji* 劉健博士頌壽紀念宋史研究論集. Tokyo: Dōhōsha, 1989.

Liu Lin 劉琳 and Shen Zhihong 沈治宏. *Xiancun Song ren zhushu zonglu* 現存宋人著述總錄. Chengdu: Ba Shu shushe, 1995.

Liu, Shi Shun. *Chinese Classical Prose: The Eight Masters of the T'ang-Sung Period.* Hong Kong: Chinese University of Hong Kong Press, 1979.

Liu Zhaoren 劉昭仁. *Lü Zuqian de wenxue yu shixue* 呂祖謙的文學與史學. Taibei: Wenshizhe chubanshe, 1986.

Liu Zhenlun 劉真倫. *Han Yu ji Song Yuan chuanben yanjiu* 韓愈集宋元傳本研究. Tang yanjiu jijinhui congshu. Beijing: Zhongguo shehui kexue chubanshe, 2004.

Liu Zijian 劉子健, *see* Liu, James T. C.

Lo Wing Kwai. "Chen Fuliang (1137–1203) yanjiu" 陳傅良 (1137–1203) 研究. Ph.D. diss., Hong Kong University, 2004.

Lo, Winston Wan. *An Introduction to the Civil Service of Sung China: With Emphasis on Its Personnel Administration.* Honolulu: University of Hawaii Press, 1987.

———. *The Life and Thought of Yeh Shih.* Hong Kong: Chinese University of Hong Kong Press, 1974.

Loewe, Michael. *A Biographical Dictionary of the Qin, Former Han and Xin Periods, 221 BC–AD 24.* Handbuch der Orientalistik. Leiden: Brill, 2000.

———. *The Origins and Development of Chinese Encyclopaedias.* China Society Occasional Papers. London: China Society, 1987.

Loewe, Michael, ed. *Early Chinese Texts: A Bibliographical Guide.* Early China Special Monograph Series, 2. [Berkeley]: Society for the Study of Early China: Institute of East Asian Studies, University of California Berkeley, 1993.

Loewe, Michael, and Edward L. Shaughnessy, eds. *The Cambridge History of Ancient China: From the Origins of Civilization to 221 B.C.* Cambridge: Cambridge University Press, 1999.

Lu Zizhen 卢子震. "Lun Daoxue zhi ming de xingcheng ji qi hanyi de fazhan" 论道学之名的形成及其含义的发展. *Hebei daxue xuebao (zhexue shehui kexue ban)* 河北大学学报 (哲学社会科学版) 24, no. 1 (1999): 30–34.

Luo Genze 羅根澤. *Zhongguo wenxue piping shi* 中國文學批評史. Shanghai: Guji chubanshe, 1984 (1961).

Magone, Rui. Review: Benjamin A. Elman, *A Cultural History of Civil Examinations in Late Imperial China. JAS* 63, no. 4 (2004): 1097–99.

Mai Zhonggui 麥仲貴. *Song Yuan Lixuejia zhushu shengzu nianbiao* 宋元理學家著述生卒年表. Hong Kong: Xinya yanjiusuo, 1968.

Man-Cheong, Iona. *The Class of 1761: Examinations, State, and Elites in Eighteenth-Century China.* Stanford: Stanford University Press, 2004.

Mao Lirui 毛礼锐. *Zhongguo jiaoyu tongshi* 中国教育通史. 6 vols. Ji'nan: Shandong jiaoyu chubanshe, 1985–89.

Marchal, Kai 馬凱之. "Lun Zhu Xi, Lü Zuqian de jian jun sixiang yu qi zhengzhi zhexue neihan" 论朱熹吕祖谦的谏君思想与其政治哲学内涵. In *Lü Zuqian ji Zhedong xueshu wenhua guoji yantaohui lunwen huibian* 呂祖謙暨浙东

学术文化国际研讨会论文汇编 (*Collected Papers from the International Conference on Lü Zuqian and the Intellectual Culture of Eastern Zhejiang*), ed. Lü Zuqian ji Zhedong xueshu wenhua guoji yantaohui dahui mishu zu 吕祖谦暨浙东学术文化国际研讨会大会秘书组, 207–26. Jinhua: Zhejiang shifan daxue, 2005.

Margouliès, Georges. *Le kou-wen chinois: receuil de textes avec introduction et notes*. Paris: Librairie orientaliste, 1926.

McMullen, David. *State and Scholars in T'ang China*. Cambridge: Cambridge University Press, 1988.

McRae, John R. *Seeing Through Zen: Encounter, Transformation, and Genealogy in Chinese Chan Buddhism*. Berkeley: University of California Press, 2003.

Meng Shuhui 孟淑慧. *Zhu Xi ji qi menren de jiaohua linian yu shijian* 朱熹及其門人的教化理念與實踐. Guoli Taiwan daxue wenshi congkan. Taibei: Guoli Taiwan daxue chuban weiyuanhui, 2003.

Menzel, Johanna M., ed. *The Chinese Civil Service: Career Open to Talent?* Problems in Asian Civilizations. Boston: D. C. Heath & Co., 1963.

Miao Chunde 苗春德. *Songdai jiaoyu* 宋代教育. Songdai yanjiu congshu. Kaifeng: Henan daxue chubanshe, 1992.

Miao Quansun 繆荃孫. *Yifeng cangshu ji* 藝風藏書記. N.p., 1901.

———. *Yifeng cangshu zai xu ji* 藝風藏書再續記. Beijing: Yanjing daxue tushuguan, 1940.

Min, Tu-gi (Min Tu-ki). *National Polity and Local Power: The Transformation of Late Imperial China*. Ed. Philip A. Kuhn and Timothy Brook. Cambridge: Council on East Asian Studies, Harvard University, 1989.

———. "The Sheng-Yuan-Chien-Sheng Stratum (Sheng-Chien) in Ch'ing Society." In idem, *National Polity and Local Power: The Transformation of Late Imperial China*, 21–49. Cambridge: Council on East Asian Studies, Harvard University, 1989.

Miyakawa, Hisayuki. "An Outline of the Naitō Hypothesis and Its Effects on Japanese Studies of China." *Far Eastern Quarterly* 14, no. 4 (1955): 533–52.

Miyazaki Ichisada 宮崎市定. *China's Examination Hell*. New Haven: Yale University Press, 1981.

———. *Kakyo* 科舉. Tokyo: Akitaya, 1946.

———. "Sōdai no shifū" 宋代の士風. *Shigaku zasshi* 史学雜誌 62, no. 2 (1953): 139–69.

Mo Yanshi 莫雁诗 and Huang Ming 黄明. *Zhongguo zhuangyuan pu* 中国状元谱. Keju wenhua xilie shu. Guangzhou: Guangzhou chubanshe, 1993.

Mo Youzhi 莫友芝. *Lüting zhijian chuanben shumu* 邵亭知見傳本書目. Shanghai: Guoxue fulun she, ca. 1911 (1909).

Moore, Oliver J., and Wang Dingbao. *Rituals of Recruitment in Tang China: Reading an Annual Programme in the Collected Statements by Wang Dingbao (870–940)*. Sinica Leidensia, 65. Leiden and Boston: Brill, 2004.

Morohashi Tetsuji 諸橋轍次 and Yasuoka Masahiro 安岡正篤, eds. *Shushigaku taikei* 朱子學大系. 14 vols. Tokyo: Meitoku shuppansha, 1974–83.

———. *Shushigaku taikei*, vol. 6, *Shushi gorui* 朱子語類. Tokyo: Meitoku shuppansha, 1981.

Munro, Donald J. *Images of Human Nature: A Sung Portrait*. Princeton: Princeton University Press, 1988.

Murakami Tetsumi 村上哲見. *Kakyo no hanashi* 科挙の話. Tokyo: Kōdansha, 1980.

Nakajima Satoshi 中島敏. *"Sōshi" 'Senkyoshi' yakuchū* 宋史選挙志譯註 (*Treatise on the Civil Service Examinations of the Sung—Translation with Annotation of Chapters 155 and 156 of the "Sung-Shih"*). 3 vols. Tokyo: Tōyō bunko, 1992–2000.

Nakasuna Akinori 中砂明德. "Ryū Kōson to Nan Sō shijin shakai" 劉後村と南宋士人社會. *Tōhō gakuhō* 東方学報 66 (1994): 63–158.

———. "Shitaifu no norumu no keisei—Nan Sō jidai" 士大夫のノルムの形成—南宋時代. *Tōyōshi kenkyū* 東洋史研究 54, no. 3 (1995): 86–117.

Neskar, Ellen. "The Cult of Worthies: A Study of Shrines Honoring Local Confucian Worthies in the Sung Dynasty (960–1279)." Ph.D. diss., Columbia University, 1993.

Nie Chongqi 聶崇岐. "Song cike kao" 宋詞科考. *Yanjing xuebao* 燕京學報 25 (1939): 107–52.

———. "Songdai zhiju kaolue" 宋代制舉考略. *Shixue nianbao* 史學年報 2, no. 5 (1938): 17–37.

———. *Song shi congkao* 宋史叢考. 2 vols. Beijing: Zhonghua shuju, 1980.

Nienhauser, William, Jr., ed. *The Indiana Companion to Traditional Chinese Literature*. Bloomington: Indiana University Press, 1986.

Niida Noboru 仁井田陞. "Keigen jōhō jirui to Sōdai no shuppanhō" 慶元條法事類と宋代の出版法. In *Chūgoku hōseishi kenkyū* 中國法制史研究, 445–65. Tokyo: Tōkyō daigaku shuppankai, 1981 (1935).

———. "Sō kaiyō to Sōdai no shuppanhō" 宋會要と宋代の出版法. In *Chūgoku hōseishi kenkyū* 中國法制史研究, 466–91. Tokyo: Tōkyō daigaku shuppankai, 1981 (1938).

Ning Huiru 甯慧如. "Bei Song jinshi ke kaoshi neirong zhi yanbian" 北宋進士科考試內容之演變. Master's thesis, Guoli Taiwan daxue, 1993.

———. *Bei Song jinshi ke kaoshi neirong zhi yanbian* 北宋進士科考試內容之演變. Taibei xian: Zhi shufang chubanshe, 1996.

———. "Songdai gongju dianshi ce yu zhengju" 宋代貢舉殿試策與政局. *Zhongguo lishi xuehui shixue jikan* 中國歷史學會史學集刊 28 (1996): 143–66.

———. "Zhu Xi lun keju" 朱熹論科舉. *Song shi yanjiu ji* 宋史研究集, no. 33, 125–65. Taibei: Guoli bianyiguan, 2003.

Niu Pu. "Confucian Statecraft in Song China: Ye Shi and the Yongjia School." Ph.D. diss., Arizona State University, 1998.

Nivison, David S. "Protest Against Conventions and Conventions of Protest." In *The Confucian Persuasion*, ed. Arthur Wright, 177–201. Stanford: Stanford University Press, 1960.

Nylan, Michael. *The Five "Confucian" Classics*. New Haven: Yale University Press, 2001.

Oberst, Zhihong Liang. "Chinese Economic Statecraft and Economic Ideas in the Sung Period (960–1279)." Ph.D. diss., Columbia University, 1996.

Oka Motoshi 岡元司. "Nan Sō ki no chiiki shakai ni okeru 'yū'" 南宋期の地域社会における"友." *Tōyōshi kenkyū* 東洋史研究 61, no. 4 (2004): 36–75.

———. "Nan Sō ki Onshū no chihō gyōsei o meguru jinteki ketsugō" 南宋期温州の地方行政をめぐる人的結合. *Shigaku kenkyū* 史学研究, no. 212 (1996): 25–48.

———. "Nan Sō ki Onshū no meizoku to kakyo" 南宋期温州の名族と科舉. *Hiroshima daigaku Tōyōshi kenkyūshitsu hōkoku* 広島大学東洋史研究室報告 17 (1995): 1–23.

———. "Yō Teki no Sōdai zaiseikan to zaisei kaikaku an" 葉適の宋代財政観と財政改革案. *Shigaku kenkyū* 史学研究, no. 197 (1992): 35–55.

Okada Takehiko 岡田武彦. *Chūgoku shisō ni okeru risō to genjitsu* 中国思想における理想と現実. Tokyo: Mokujisha, 1983.

———. "Shushi no fu to shi" 朱子の父と師. In *Chūgoku shisō ni okeru risō to genjitsu* 中国思想における理想と現実, ed. idem, 331–415. Tokyo: Mokujisha, 1983.

Ouyang Guang 欧阳光. "Song Yuan keju yu wenren huishe" 宋元科举与文人会社. In idem, *Song Yuan shishe yanjiu conggao* 宋元诗社研究丛稿, 15–28. Guangzhou: Guangdong gaodeng jiaoyu chubanshe, 1996.

Owen, Stephen. *Readings in Chinese Literary Thought*. Cambridge: Council on East Asian Studies, Harvard University, 1992.

Ozaki Yasushi 尾崎康. *Seishi Sō Gen han no kenkyū* 正史宋元版の研究. Tokyo: Kyūko shoin, 1989.

———. "Songdai diaoban yinshua de fazhan" 宋代雕版印刷的發展. *Gugong xueshu jikan* 故宮學術季刊 20, no. 4 (2003): 167–90.

Pan Fuen 潘富恩 and Xu Yuqing 徐余庆. *Lü Zuqian pingzhuan* 吕祖谦评传. Zhongguo sixiangjia pingzhuan congshu. Ed. Kuang Yaming 匡亚明. Nanjing: Nanjing daxue chubanshe, 1992.

Pan Meiyue 潘美月. *Songdai cangshujia kao* 宋代藏書家考. Taibei: Xuehai chubanshe, 1980.

Parkes, Malcolm Beckwith. "The Influence of the Concepts of *Ordinatio* and *Compilatio* on the Development of the Book." In *Medieval Learning and Literature: Essays Presented to Richard William Hunt*, ed. J. J. G. Alexander and Margaret T. Gibson, 115–41. Oxford, Eng.: Clarendon Press, 1976.

———. *Pause and Effect: An Introduction to the History of Punctuation in the West*. Aldershot, Eng.: Scolar Press, 1992.

Pearce, Scott; Audrey G. Spiro; and Patricia Buckley Ebrey. *Culture and Power in the Reconstitution of the Chinese Realm, 200–600*. Cambridge: Harvard University Asia Center, 2001.

Pelliot, Paul. *Les debuts de l'imprimerie en Chine*. Paris: Impr. nationale Librairie d'Amérique et d'Orient, 1953.

Peterson, Willard J. "Another Look at *Li*." *BSYS* 18 (1986): 13–31.

Peterson, Willard J., Andrew H. Plaks, and Yü Ying-shih, eds. *The Power of Culture: Studies in Chinese Cultural History*. Hong Kong: Chinese University of Hong Kong Press, 1994.

Plaks, Andrew. "The Prose of Our Time." In *The Power of Culture: Studies in Chinese Cultural History*, ed. Willard J. Peterson, Andrew H. Plaks, and Yü Ying-shih, 206–17. Hong Kong: Chinese University Press, 1994.

Poon, Ming-sun. "Books and Printing in Sung China (960–1279)." Ph.D. diss., University of Chicago, 1979.

———. "The Printer's Colophon in Sung China, 960–1279." *Library Quarterly* 43 (1973): 39–52.

Pu Yanguang 蒲彥光. "Songdai keju shiwen yanjiu—'jingyi' wenti chu tan" 宋代科舉時文研究—"經義"文體初探. *Zhongguo haishi shangye zhuanke xuexiao xuebao* 中國海事商業專科學校學報 90 (2002): 153–88.

Qi Gong 启功, Zhang Zhongxing 张中行, and Jin Kemu 金克木. *Shuo bagu* 说八股. Beijing: Zhonghua shuju, 1994.

Qian Mu 錢穆. *Zhuzi xin xue'an* 朱子新學案. 5 vols. Taibei: Sanmin shuju, 1989 (1980).

Qiao Yanguan 喬衍琯. *Songdai shumu kao* 宋代書目考. Wenshizhe xue jicheng. Taibei: Wenshizhe chubanshe, 1987.

Qiu Hansheng 邱漢生 and Xiong Chengdi 熊承滌. *Nan Song jiaoyu lunzhu xuan* 南宋教育論著選. Zhongguo gudai jiaoyu lunzhu congshu. Beijing: Renmin jiaoyu chubanshe, 1992.

Rawski, Evelyn Sakakida. *Education and Popular Literacy in Ch'ing China*. Ann Arbor: University of Michigan Press, 1979.

Ren Jiyu 任继愈. *Zhongguo cangshulou* 中國藏書樓. Shenyang: Liaoning renmin chubanshe, 2001.

Ren Yuan 任远. "Songdai jingdu zhi chuxin yu biduan" 宋代经读之出新与弊端. *Kongzi yanjiu* 孔子研究 2 (1995): 56–61.

Rickett, W. Allyn. *Guanzi: Political, Economic, and Philsophical Essays from Early China.* Vol. 1. Princeton: Princeton University Press, 1985.

Ridley, Charles Price. "Educational Theory and Practice in Late Imperial China: The Teaching of Writing as a Specific Case." Ph.D. diss., Stanford University, 1973.

Ropp, Paul S. *Dissent in Early Modern China: "Ju-Lin Wai-Shih" and Ch'ing Social Criticism.* Ann Arbor: University of Michigan Press, 1981.

Sano Kōji 佐野公治. *Shishogaku shi no kenkyū* 四書學史の研究. Tokyo: Sōbunsha, 1988.

Sariti, Anthony. "Monarchy, Bureaucracy and Absolutism in the Political Thought of Ssu-Ma Kuang." *JAS* 32, no. 1 (1972): 53–76.

Satō Takanori 佐藤隆則. "Shin Jun no gakumon to shisō—Shu Ki jūgaku izen" 陳淳の学問と思想—朱熹從學以前. *Daitō bunka daigaku Kangakkai shi* 大東文化大學漢學會誌 28 (1989): 44–64.

———. "Shin Jun no gakumon to shisō—Shu Ki jūgaku ki" 陳淳の学問と思想—朱熹從學期. *Daitō bunka daigaku kangakkai shi* 大東文化大學漢學會誌 29 (1990): 138–52.

Schaberg, David. *A Patterned Past: Form and Thought in Early Chinese Historiography.* Cambridge: Harvard University Asia Center, 2001.

Schirokauer, Conrad. "Ch'en Fu-Liang." In *Sung Biographies*, ed. Herbert Franke, 103–7. Wiesbaden: Steiner, 1976.

———. "Chu Hsi's Political Career: A Study in Ambivalence." In *Confucian Personalities*, ed. Arthur F. Wright and Denis Crispin Twitchett, 162–88. Stanford: Stanford University Press, 1962.

———. "Chu Hsi's Political Thought." *Journal of Chinese Philosophy* 5, no. 2 (1978): 127–59.

———. "Chu Hsi's Sense of History." In *Ordering the World. Approaches to State and Society in Sung Dynasty China*, ed. Robert P. Hymes and Conrad Schirokauer, 193–220. Berkeley: University of California Press, 1993.

———. "Neo-Confucians Under Attack: The Condemnation of Wei-Hsueh." In *Crisis and Prosperity in Sung China*, ed. John Winthrop Haeger, 163–98. Tucson: University of Arizona Press, 1975.

———. "The Political Thought and Behavior of Chu Hsi." Ph.D. dissertation, Stanford University, 1960.

Schwartz, Benjamin Isadore. *The World of Thought in Ancient China.* Cambridge: Belknap Press of Harvard University Press, 1985.

Shang Yanliu 商衍鎏. *Kechang anjian yu kechang yiwen* 科場案件與科場軼聞. Taibei: Zhongshan tushu, 1972 (1956).

Shanghai Xin si jun lishi yanjiuhui, Yinshua yinchao fenhui 上海新四军历史研究会印刷印钞分会. *Lidai keshu gaikuang* 历代刻书概况. Zhongguo yinshua shiliao xuanji. Beijing: Yinshua gongye chubanshe, 1991.

Shao Yichen 紹懿辰. *Zengding Siku jianming mulu biaozhu* 增訂四庫簡明目錄標注. Beijing: Zhonghua shuju, 1959 (1911).

Shen Jianshi 沈兼士. *Zhongguo kaoshi zhidu shi* 中國考試制度史. Renren wenku. Ed. Wang Yunwu 王雲五. Taibei: Shangwu yinshuguan, 1969.

Shen Songqin 沈松勤. *Nan Song wenren yu dangzheng* 南宋文人与党争. Beijing: Renmin chubanshe, 2005.

Shen Zhong 沈重. *Tangdai mingren keju kaojuan yiping* 唐代名人科举考卷译评. Nanchang: Jiangxi gaoxiao chubanshe, 1995.

Shih, Vincent Yu-chung, *see* Liu Hsieh.

Shimada Kenji 島田虔次. *Chūgoku shisōshi no kenkyū* 中国思想史の研究. Tōyōshi kenkyū sōkan. Kyoto: Kyōto daigaku gakujutsu shuppankai, 2002.

———. *Daigaku, Chūyō* 大学・中庸. Shintei Chūgoku kotensen. Tokyo: Asahi shinbunsha, 1967.

Shiomi Kunihiko 塩見邦彦. *"Shushi gorui" kōgo goyi sakuin* 朱子語類口語語彙索引. Kyoto: Chūbun shuppansha, 1988.

Shu Jingnan 束景南. *Zhu Xi nianpu changbian* 朱熹年譜長編. Shanghai: Huadong shifan daxue chubanshe, 2001.

———. *Zhu Xi yiwen ji kao* 朱熹佚文辑考. Nanjing: Jiangsu guji chubanshe, 1991.

———. *Zhuzi da zhuan* 朱子大传. Fuzhou: Fujian jiaoyu chubanshe, 1992.

Sichuan daxue. Guji zhengli yanjiusuo 四川大學古籍整理研究所. *Xiancun Song ren bieji banben mulu* 現存宋人別集版本目錄. Quan Song wen yanjiu ziliao congkan. Chengdu: Ba Shu shushe, 1990.

Smith, Paul J., and Richard von Glahn. *The Song-Yuan-Ming Transition in Chinese History*. Cambridge: Harvard University Asia Center, 2003.

Soejima Ichirō 副島一郎. "Sō jin no mita Ryū Shūgen" 宋人の見た柳宗元. *Chūgoku bungaku hō* 中国文学報 47 (1993): 103–45.

Song Dingzong 宋鼎宗. *Chunqiu Song xue fawei* 春秋宋學發微. Taibei: Wenshizhe, 1986 (1983).

Song Jaeyoon. "Tensions and Balance: Changes of Constitutional Schemes in Song Commentaries on 'the Office of Heaven,' the *Rituals of Zhou*." Paper presented at The *Rituals of Zhou (Zhouli)* in East Asian History: Premodern Asian Statecraft in Comparative Context conference. Princeton University, 2005.

Su Bai 宿白. *Tang Song shiqi de diaoban yinshua* 唐宋时期的雕版印刷. Beijing: Wenwu chubanshe, 1999.

Sudō Yoshiyuki 周藤吉之. *Sōdai kanryōsei to daitochi shoyū* 宋代官僚制と大土地所有. Shakai kōseishi taikei. Tokyo: Nihon hyōronsha, 1950.

Sudō Yoshiyuki 周藤吉之 and Nakajima Satoshi 中島敏. *Chūgoku no rekishi*, vol. 5, *Godai–Sō* 中国の歴史: 五代–宋. Tokyo: Kōdansha, 1974.

Sun Qutian 孫蕖田. *Chen Wenjie gong Fuliang nianpu* 陳文節公傅良年譜. Xinbian Zhongguo mingren nianpu jicheng 新編中國名人年譜集成. Taibei: Taiwan Shangwu yinshuguan, 1982 (1929).

Sun Qutian 孫蕖田 and Wu Hongze 吳洪澤. *Chen Wenjie gong nianpu* 陳文節公年譜. Song ren nianpu congkan. Chengdu: Sichuan daxue chubanshe, 2003.

Suzuki Torao 鈴木虎雄. *Fushi taiyō* 賦史大要. Tokyo: Fuzanbō, 1936.

———. "Hakkobun no enkaku oyobi keishiki" 八股文の沿革及び形式. In *Shina bungaku kenkyū* 支那文學研究, 695–716. Kyoto: Kōbundō shobō, 1925.

Takahashi Yoshirō 高橋芳郎. "Sōdai no shijin mibun ni tsuite" 宋代の士人身分について. *Shirin* 史林 69, no. 3 (1986): 39–70.

Takatsu Takashi 高津孝. "The Selection of the 'Eight Great Prose Masters of the T'ang and Sung' and Chinese Society in the Sung and Later." *Acta Asiatica* 84 (2003): 1–19.

———. "Sō Gen hyōten kō" 宋元評點考. *Kagoshima daigaku hōbungakubu kiyō—jinbungakka ronshū* 鹿兒島大學法文學部紀要—人文學科論集 31 (1990): 127–56.

———. "Song chu xingjuan kao" 宋初行卷考. In *Keju yu shiyi: Songdai wenxue yu shiren shehui* 科舉與詩藝: 宋代文學與士人社會, ed. Takatsu Takashi 高津孝, 1–24. Shanghai: Shanghai guji chubanshe, 2005 (1992).

Takatsu Takashi 高津孝, ed. *Keju yu shiyi: Songdai wenxue yu shiren shehui* 科舉與詩藝: 宋代文學與士人社會. Riben Song xue yanjiu liuren ji. Shanghai: Shanghai guji chubanshe, 2005.

Tanaka Kenji 田中謙二. "Shu Ki to kakyo" 朱喜と科舉. *Chūgoku bunmei sen* 中国文明選 15 (1976): 6–8.

———. "Shumon deshi shiji nenkō" 朱門弟子師事年攷. *Tōhō gakuhō* 東方学報 44 (1973): 147–218.

———. "Shumon deshi shiji nenkō zoku" 朱門弟子師事年攷續. *Tōhō gakuhō* 東方学報 48 (1975): 261–357.

———. "*Shushi gorui* 'Gainin' hen shōchū (1–6)" 朱子語類外任篇詳注. 3 pts. *Tōyōshi kenkyū* 東洋史研究 28 (1969): 80–101; 29 (1970): 94–108; 30 (1972): 22–39.

Tao Xiang 陶湘. *Sheyuan suo jian Song ban shuying* 涉園所見宋版書影. N.p.: Wujin Tao shi sheyuan, 1937.

Teng Ssu-yu 鄧嗣禹. *Zhongguo kaoshi zhidu shi* 中國考試制度史. Taibei: Xuesheng shuju, 1957.

Terada Gō 寺田剛. *Sōdai kyōikushi gaisetsu* 宋代教育史概説. Tokyo: Hakubunsha, 1965.

Teraji Jun 寺地遵. *Nan Sō shoki seijishi kenkyū* 南宋初期政治史研究. Hiroshima: Keisuisha, 1988.

Tillman, Hoyt Cleveland. *Ch'en Liang on Public Interest and the Law.* Monograph no. 12, Society for Asian and Comparative Philosophy. Honolulu: University of Hawaii Press, 1994.

———. "Ch'en Liang on Statecraft: Reflections from Examination Essays Preserved in a Sung Rare Book." *HJAS* 44, no. 2 (1988): 403–31.

———. *Confucian Discourse and Chu Hsi's Ascendancy.* Honolulu: University of Hawaii Press, 1992.

———. "Encyclopedias, Polymaths, and Tao-Hsueh Confucians: Preliminary Reflections with Special Reference to Chang Ju-Yu." *JSYS* 22 (1990–92): 89–108.

———. "A New Direction in Confucian Scholarship: Approaches to Examining the Differences Between Neo-Confucianism and Tao-Hsueh." *Philosophy East and West* 42, no. 3 (1992): 455–74.

———. "Proto-Nationalism in Twelfth Century China? The Case of Chen Liang." *HJAS* 39, no. 2 (1979): 403–28.

———. "Reflections on Classifying 'Confucian' Lineages: Reinventions of Tradition in Song China." In *Rethinking Confucianism: Past and Present in China, Japan, Korea, and Vietnam,* ed. Benjamin A. Elman, John B. Duncan, and Herman Ooms, 33–64. Los Angeles: UCLA Asian Pacific Monograph Series, 2002.

———. *Utilitarian Confucianism: Ch'en Liang's Challenge to Chu Hsi.* Cambridge: Council on East Asian Studies, Harvard University, 1982.

Toda Kenjirō 土田健次郎. "Isen Ekiden no shisō" 伊川易伝の思想. In *Sōdai no shakai to bunka* 宋代の社會と文化. Tokyo: Kyūko shoin, 1983.

Tōhoku daigaku *Shushi gorui* kenkyūkai 東北大學朱子語類研究会. "*Shushi gorui* 'Honchō jinbutsu hen' yakuchū" 朱子語類本朝人物篇譯注. 37 pts. *Shūkan Tōyōgaku* 集刊東洋學 42 (1979): 100–105; 43 (1980): 110–16; 44 (1980) 100–115; 45 (1981): 77–83; 46 (1981): 75–81; 47 (1982): 98–103; 48 (1982): 93–100; 49 (1983): 99–108; 50 (1983): 99–109; 51 (1984): 134–42; 52 (1984): 119–27; 53 (1985): 124–33; 54 (1985): 137–45; 55 (1986): 99–113; 56 (1986): 120–30; 57 (1987): 142–46; 58 (1987): 96–114; 59 (1988): 131–42; 60 (1988): 152–60; 61 (1989): 127–31; 62 (1989): 158–65; 63 (1990): 137–49; 64 (1990): 165–72; 65 (1991): 83–90; 66 (1991): 158–68; 67 (1992): 124–29; 68 (1992): 136–45; 69 (1993): 122–31; 70 (1993): 125–33; 71 (1994): 113–19; 72 (1994): 123–32; 73 (1995):

81–87; 74 (1995): 159–64; 75 (1996): 17–22; 76 (1996): 122–29; 77 (1997): 130–142; 78 (1997): 104–11.

Tsien, Tsuen-hsuin. *Paper and Printing. Science and Civilization in China*, vol. 5, pt. I, ed. Joseph Needham. Cambridge: Cambridge University Press, 1985.

Tsuchida Kenjirō 土田健次郎. *Dōgaku no keisei* 道学の形成. Tōyōgaku sōsho. Tokyo: Sōbunsha, 2002.

———. "Dōtōron sai kō" 道統論再考. In *Chūgoku no bukkyō to bunka: Kamata Shigeo hakushi kanreki kinen ronshū* 中国の仏教と文化: 鎌田茂雄博士還暦記念論集, ed. Kamata Shigeo hakushi kanreki kinen ronshū kankōkai 鎌田茂雄博士還暦記念論集刊行会, 613–29. Tokyo: Daizō shuppan, 1988.

———. "Shū Tei juju sai kō" 周程授受再考. *Tōyō no shisō to shūkyō* 東洋の思想と宗教 13, no. 3 (1996): 26–39.

———. "Sōdai shishōshi jō ni okeru Shū Ton'i no ichi" 宋代思想史上における周敦頤の位置. In *Tōhō gakkai sōritsu gojisshūnen kinen Tōhōgaku ronshū* 東方学会創立五十周年記念東方学論集, ed. Tōhō gakkai 東方学会, 875–87. Tokyo: Tōhō gakkai, 1997.

Tu, Ching-i. "The Chinese Examination Essay: Some Literary Considerations." *Monumenta Serica* 31 (1974–75): 393–406.

Tucker, John Allen. "An Onto-hermeneutic and Historico-hermeneutic Analysis of Chu Hsi's Political Philosophy as Found in the *Chin Ssu-Lu*." *Chinese Culture* 24, no. 4 (1983): 1–25.

Twitchett, Denis. *Printing and Publishing in Medieval China*. New York: Frederic C. Beil, 1983.

Twitchett, Denis, ed. *The Cambridge History of China*, vol. 3, *Sui and T'ang China, 589–906*, pt. I. Cambridge: Cambridge University Press, 1979.

Twitchett, Denis, and Michael Loewe, eds. *The Cambridge History of China*, vol. I, *The Ch'in and Han Empires, 221 B.C.–A.D. 220*. Cambridge: Cambridge University Press, 1986.

Umehara Kaoru 梅原郁. *Sōdai kanryō seido kenkyū* 宋代官僚制度研究. Kyoto: Dōhōsha, 1985.

van Ess, Hans. "The Compilation of the Works of the Ch'eng Brothers and Its Significance for the Learning of the Right Way of the Southern Sung Period." *T'oung Pao* 90, no. 4–5 (2004): 264–98.

———. *Von Ch'eng I zu Chu Hsi: die Lehre vom rechten Weg in der Überlieferung der Familie Hu*. Asien- und Afrika-Studien der Humboldt-Universität zu Berlin 13. Wiesbaden: Harrassowitz, 2003.

Van Zoeren, Steven. *Poetry and Personality: Reading, Exegesis, and Hermeneutics in Traditional China*. Stanford: Stanford University Press, 1991.

von Glahn, Richard. "Chu Hsi's Community Granary in Theory and Practice." In *Ordering the World: Approaches to State and Society in Sung Dynasty China*, ed. Robert P. Hymes and Conrad Schirokauer, 221–54. Berkeley: University of California Press, 1993.

————. *The Sinister Way: The Divine and the Demonic in Chinese Religious Culture*. Berkeley: University of California Press, 2004.

Waley, Arthur. *Po Chü-I*. London: G. Allen & Unwin, 1949.

Waltner, Anne. "Building on the Ladder of Success: The Ladder of Success in Imperial China and Recent Work on Social Mobility." *Ming Studies* 17 (1983): 30–36.

Walton, Linda A. *Academies and Society in Southern Sung China*. Honolulu: University of Hawaii Press, 1999.

————. "Education, Social Change, and Neo-Confucianism in Sung-Yuan China: Academies and the Local Elite in Ming Prefecture (Ningpo)." Ph.D. diss., University of Pennsylvania, 1978.

————. "The Institutional Context of Neo-Confucianism: Scholars, Schools, and Shu-Yuan in Song-Yuan China." In *Neo-Confucian Education: The Formative Stage*, ed. Wm. Theodore de Bary and John W. Chaffee, 457–92. Berkeley: University of California Press, 1989.

Wang Deyi 王德毅. *Songdai xianliang fangzheng ke ji cike kao* 宋代賢良方正科及詞科考. Hong Kong: Chongwen shudian, 1971.

————. "Songdai xianliang fangzheng ke kao" 宋代賢良方正科考. *Taiwan daxue wenshizhe xuebao* 臺灣大學文史哲學報 14 (1965): 301–55.

Wang Jianqiu 王建秋. *Songdai Taixue yu Taixuesheng* 宋代太學與太學生. Zhongguo xueshu zhuzuo jiangzhu weiyuanhui congshu. Taibei: Taiwan Shangwu yinshuguan; Jinghua yinshuguan, 1965.

Wang Lan 王岚. *Song ren wenji bianke liuchuan congkao* 宋人文集编刻流传丛考. Zhongguo dianji yu wenhua yanjiu congshu. Nanjing: Jiangsu guji chubanshe, 2003.

Wang Ruisheng 王瑞生. "Jin cun Songdai zongji kao" 今存宋代總集考. *Tainan shizhuan xuebao* 臺南師專學報 9 (1976): 49–88.

Wang Shuizhao 王水照. *Songdai wenxue tonglun* 宋代文学通论. Songdai yanjiu congshu. Kaifeng: Henan daxue chubanshe, 1997.

Wang Wenjin 王文進. *Wenlu tang fangshu ji* 文祿堂訪書記. Beijing: Wenlu tang, 1942.

Wang Yu 王宇. "Nan Song kechang yu Yongjia xuepai de jueqi—Yi Chen Fuliang yu *Chunqiu* shiwen wei gean" 南宋科场与永嘉学派的崛起—以陈傅良与春秋时文为个案. *Zhejiang shehui kexue* 浙江社会科学, 2004, no. 2: 151–56.

Wang Yunwu 王雲五. *Song Yuan jiaoxue sixiang* 宋元教學思想. Taibei: Taiwan Shangwu yinshuguan, 1971.

Wechsler, Howard J. "The Confucian Teacher Wang T'ung (584?–617): One Thousand Years of Controversy." *T'oung Pao* 63 (1977): 225–72.

Weng Tongwen 翁同文. "Yinshua duiyu shuji chengben de yingxiang" 印刷對於書籍成本的影響. *Qinghua xuebao* 清華學報 6, no. 1–2 (1967): 35–41.

Wilhelm, Richard. *The I Ching*. Trans. Cary F. Baynes. Bollingen Series 19. Princeton: Princeton University Press, 1950.

Wilson, Thomas A. *Genealogy of the Way: The Construction and Uses of the Confucian Tradition in Late Imperial China*. Stanford: Stanford University Press, 1995.

————. "The Indelible Mark of an Overlooked Scholar: Toward a *Restructuring* of Sinological Hermeneutics." Paper presented at Subjects, Dialogues, Histories: An International Conference in Memory of Professor Edward T. Ch'ien, Taibei, 1997.

Wong, R. Bin. "Social Order and State Activism in Sung China: Implications for Later Centuries." *JSYS* 26 (1996): 229–50.

Woodside, Alexander. "The Divorce Between the Political Center and Educational Creativity in Late Imperial China." In *Education and Society in Late Imperial China, 1600–1900*, ed. Benjamin A. Elman and Alexander Woodside, 458–92. Berkeley: University of California Press, 1994.

————. "Territorial Order and Collective-Identity Tensions in Confucian Asia: China, Vietnam, Korea." *Daedalus* 127, no. 3 (1998): 191–220.

Wu Chunshan 吳春山. "Lü Zuqian yanjiu" 呂祖謙研究. Ph.D. diss., Guoli Taiwan daxue, 1978.

Wu Hongze 吳洪澤. *Song ren nianpu jimu—Song bian Song ren nianpu xuankan* 宋人年譜集目—宋編宋人年譜選刊. Quan Song wen yanjiu ziliao congkan. Chengdu: Ba Shu shushe, 1995.

Wu Kwang-tsing. "Chinese Printing Under Four Alien Dynasties (916–1369 A.D.)." *HJAS* 13, no. 3–4 (1950): 447–523.

Wu Wanju 吳萬居. *Songdai shuyuan yu Songdai xueshu de guanxi* 宋代書院與宋代學術之關係. Taibei: Wenshizhe chubanshe, 1991.

Xia Jianwen 夏健文. "Nan Song Yongjia Yongkang xuepai zhi jingshi zhiyong lun" 南宋永嘉永康學派之經世致用論. Master's thesis, Guoli Zhengzhi daxue, 1991.

Xie Shuishun 谢水顺 and Li Ting 李斑. *Fujian gudai keshu* 福建古代刻书. Fuzhou: Fujian renmin chubanshe, 1997.

Xiong Chengdi 熊承滌, ed. *Zhongguo gudai jiaoyu shiliao xinian* 中国古代教育史料系年. Beijing: Renmin jiaoyu chubanshe, 1985 (1991).

Xu Gui 徐規. "Chen Fuliang zhi kuan minli shuo" 陳傅良之寬民力説. *Zhejiang xuebao* 浙江學報 1, no. 1 (1947): 41–48.

Xu Gui 徐規 and Zhou Mengjiang 周夢江. "Chen Fuliang de zhuzuo ji qi shigong sixiang shulue" 陈傅良的著作及其事功思想述略. In *Chen Fu-*

*liang danchen babai liushi zhounian jinian ji* 陈傅良诞辰八百六十周年纪念集, ed. Jinian Chen Fuliang danchen babai liushi zhounian chouweihui 纪念陈傅良诞辰八百六十周年筹委会, 8–27. N.p., 1997?

Xu Naichang 徐乃昌. *Song Yuan keju san lu* 宋元科舉三錄. N.p.: Nanling Xu shi, 1923.

Xu Ruzong 徐儒宗. *Wuxue zhi zong—Lü Zuqian pingzhuan* 婺学之宗—吕祖谦传. Hangzhou: Zhejiang renmin chubanshe, 2005.

Xu Youshou 徐有守. *Zhongwai kaoshi zhidu zhi bijiao* 中外考試制度之比較. Taibei: Zhongyang wenwu gongyingshe, 1984.

Yamanoi Yū 山井湧. "Jingshu he zaopo" 經書和糟粕. In *Riben xuezhe lun Zhongguo zhexueshi* 日本學者論中國哲學史, 405–26. Beijing: Zhonghua shuju, 1986.

Yang Chengjian 杨成鉴 and Jin Taosheng 金涛声. *Zhongguo kaoshixue* 中国考试学. Beijing: Shumu wenxian chubanshe, 1995.

Yang Shiwen 楊世文. *Xue Jixuan nianpu* 薛季宣年譜. Songren nianpu congkan. Chengdu: Sichuan daxue chubanshe, 1993.

Yang Xuewei 杨学为, Liu Peng 刘芃, and Jiaoyubu kaoshi zhongxin 教育部考试中心. *Zhongguo kaoshi shi wenxian jicheng* 中国考试史文献集成. Beijing: Gaodeng jiaoyu chubanshe, 2003.

Yang Yongan 楊永安. *Wang Tong yanjiu* 王通研究. Hong Kong: University of Hong Kong, 1992.

Yang Yuxun 楊宇勛. "Nan Song Lizong zhong- wanchao de zhengzheng (A.D. 1233–1264)—cong Shi Miyuan zu hou zhi xiangwei gengti lai guancha" 南宋理宗中–晚朝的政爭 (A.D. 1233–1264)—從史彌遠卒後之相位更替來觀察. Master's thesis, Chenggong daxue, 1991.

Ye Dehui 葉德輝. *Shulin qinghua* 書林清話. Taibei: Wenshizhe, 1988 (1917).

———. *Xiyuan dushu zhi* 郋園讀書志. Shanghai: Danyuan, 1928.

Ye Guoliang 葉國良. *Song ren yijing gaijing kao* 宋人疑經改經考. Guoli Taiwan daxue wenshi congkan. Taibei: Tianyi chubanshe, 1980.

Yi Pu 易蒲 and Li Jinling 李金苓. *Hanyu xiucixue shigang* 汉语修辞学史纲. Changchun: Jilin jiaoyu chubanshe, 1989.

Yin Dexin 尹德新, ed. *Lidai jiaoyu biji ziliao—Song, Liao, Jin, Yuan bufen* 历代教育笔记资料—宋辽金元部分. Beijing: Zhongguo laodong chubanshe, 1991.

Yin Gonghong 尹恭弘. *Pianwen* 骈文. Zhongguo gudai wenti congshu. Beijing: Renmin wenxue chubanshe, 1994.

Yu Jingxiang 余景祥. *Tang Song pianwen shi* 唐宋骈文史. Shenyang: Liaoning renmin chubanshe, 1991.

Yu, Pauline. "Canon Formation in Late Imperial China." In *Culture and State in Chinese History*, ed. Theodore Huters, R. Bin Wong, and Pauline Yu, 83–104. Stanford: Stanford University Press, 1997.

Yü Ying-shih 余英時. "Intellectual Breakthroughs in the T'ang Sung Transition." In *The Power of Culture: Studies in Chinese Cultural History*, ed. Willard J. Peterson, Andrew H. Plaks, and Yü Ying-shih, 158–71. Hong Kong: Chinese University Press, 1994.

———. *Zhu Xi de lishi shijie: Songdai shidafu zhengzhi wenhua de yanjiu* 朱熹的歷史世界: 宋代士大夫政治文化的研究. 2 vols. Taibei: Yunchen, 2003.

Yu Zhaopeng 俞兆鹏. "Xie Dieshan xiansheng xinian yaolu" 谢叠山先生系年要录. *Jiangxi daxue xuebao: shehui kexue ban* 江西大学学报社会科学版, 1987, no. 1: 62–67.

Yüan Ts'ai. *Family and Property in Sung China: Yüan Ts'ai's "Precepts for Social Life."* Trans. Patricia Buckley Ebrey. Princeton Library of Asian Translations. Princeton: Princeton University Press, 1984.

Yuan Zheng 袁征. "The Grade System of Schools in Eleventh-to-Thirteenth Century China." *Chinese Studies in History* 25, no. 2 (1991): 17–52.

———. *Songdai jiaoyu* 宋代教育. Guangzhou: Guangdong gaodeng jiaoyu chubanshe, 1991.

Zach, Erwin von, trans. *Hsiung Yang's "Fa-Yen" (Worte strenger Ermahnung)*. Sinologische Beitrage. Batavia: Drukkerij Lux, 1939.

Ze, David Wei. "Printing as an Agent of Social Stability: The Social Organization of Book Production in China During the Sung Dynasty." Ph.D. diss., Simon Fraser University, 1995.

Zeng Xiangqin 曾祥芹, ed. *Wenzhangxue yu yuwen jiaoyu* 文章学与语文教育. Shanghai: Shanghai jiaoyu chubanshe, 1995.

Zeng Zaozhuang 曾枣庄. "Lun Songdai de siliu wen" 论宋代的四六文. *Wenxue yichan* 文学遗产 3 (1995).

Zhang Bowei 张伯伟. "Pingdian silun" 评点四论. *Zhongguo xueshu* 中国学术 6, no. 2 (2001): 1–40.

Zhang Dihua 張滌華. *Leishu liubie* 類書流別. Beijing: Shangwu yinshuguan, 1985.

Zhang Haiou 张海鸥 and Sun Yaobin 孙耀斌. "*Lunxue shengchi* yu Nan Song luntiwen ji Nan Song lunxue" 《论学绳尺》与南宋论体文及南宋论学. *Wenxue yichan*, 2006, no. 1: 90–101.

Zhang, Jiacai 张加才. "*Beixi ziyi* banben yuanliu lice" 北溪字义版本源流蠡测. *Beifang gongye daxue xuebao* 北方工业大学学报 11, no. 2 (1999): 80–88.

———. *Quanshi yu jian'gou: Chen Chun yu Zhuzixue* 诠释与建构: 陈淳与朱子学. Beijing: Renmin chubanshe, 2004.

Zhang Xiqing 张希清. "Bei Song de keju qushi yu xuexiao xuanshi" 北宋的科举取士与学校选士. In *Song shi yanjiu lunwen ji: Guoji Song shi yantaohui ji Zhongguo Song shi yanjiuhui di jiu jie nianhui biankan* 宋史研究论文集: 国际宋史研讨会暨中国宋史研究会第九届年会编刊, ed. Qi Xia 漆侠, 183–203. Baoding: Hebei daxue, 2002.

———. "Bei Song gongju dengke renshu kao" 北宋贡举登科人数考. *Guoxue yanjiu* 国学研究 1994, no. 2: 393–425.

———. "Lun Songdai keju zhong de tezouming" 论宋代科举中的特奏名. In *Song shi yanjiu lunwen ji* 宋史研究论文集, ed. Deng Guangming 邓广铭 et al., 77–93. Shijiazhuang: Hebei jiaoyu chubanshe, 1989.

———. "Lun Wang Anshi de gongju gaige" 论王安石的贡举改革. *Beijing daxue xuebao* 北京大学学报 4 (1986): 66–77.

———. "Nan Song gongju dengke renshu kao" 南宋贡举登科人数考. *Guji zhengli yu yanjiu* 古籍整理与研究 5 (1991): 129–46.

———. "Qin Gui yu keju" 秦檜与科举. *Yue Fei yanjiu* 岳飞研究 3 (1992): 246–57.

———. "Songdai dianshi zhidu shulun" 宋代殿试制度述论. *Beijing daxue xuebao* 北京大学学报 1992, no. 2: 22–34.

———. "Songdai gongju kemu shulun" 宋代贡举科目述论. In *Guoji Song shi yantaohui lunwen xuanji* 国际宋史研讨会论文选集, ed. Deng Guangming 邓广铭 and Qi Xia 漆侠, 320–41. Baoding: Hebei daxue chubanshe, 1993.

———. *Zhongguo keju kaoshi zhidu* 中国科举考试制度. Beijing: Xinhua chubanshe, 1993.

Zhang Xiumin 张秀民. *Zhongguo yinshua shi* 中国印刷史. Shanghai: Renmin chubanshe, 1989.

Zhang Yide 张义德. *Ye Shi pingzhuan* 叶适评传. Zhongguo sixiangjia pingzhuan congshu. [Nanjing]: Nanjing daxue chubanshe, 1994.

Zhang Yuan 張元. "Songdai lixuejia de lishiguan: Yi *Zizhi tongjian gangmu* wei li" 宋代理學家的歷史觀: 以資治通鑑綱目爲例." Ph.D. diss., National Taiwan University, 1975.

Zhang Yuanji 張元濟. *She yuan xuba jilu* 涉園序跋集錄. Shanghai: Gudian wenxue chubanshe, 1957.

Zhang Zhigong 张志公. *Chuantong yuwen jiaoyu chutan* 传统语文教育初探. Shanghai: Shanghai jiaoyu chubanshe, 1962.

Zheng Qinren 鄭欽仁. *Liguo de honggui* 立國的宏規. Zhongguo wenhua xinlun. Ed. Liu Dai 劉岱. Taibei: Lianjing, 1982.

Zhou Mengjiang 周夢江. "*Song shi* 'Chen Fuliang zhuan' buzheng" 宋史"陈傅良传"补正. In *Chen Fuliang danchen babai liushi zhounian jinian ji* 陈傅良诞辰八百六十周年纪念集, ed. Jinian Chen Fuliang danchen babai liushi zhounian chouweihui 纪念陈傅良诞辰八百六十周年筹委会, 28–38. N.p., 1997? (1988).

———. *Ye Shi yu Yongjia xuepai* 叶适与永嘉学派. Hangzhou: Zhejiang guji chubanshe, 1992.

Zhou Xuewu 周學武. "Liang Song Yongjia xueshu zhi bianqian" 兩宋永嘉學術之變遷. *Shumu jikan* 書目季刊 10, no. 2 (1976): 27–47.

———. *Ye Shuixin xiansheng nianpu* 葉水心先生年譜. Song ren nianpu congkan 11. Chengdu: Sichuan daxue chubanshe, 2003.

Zhou Yafei 周亚非. *Zhongguo lidai zhuangyuan lu* 中国历代状元录. Shanghai: Shanghai wenhua chubanshe, 1995.

Zhou Yanwen 周彦文. "Lun lidai shumu zhong de zhiju lei shuji" 論歷代書目中的制舉類書籍. *Zhongguo shumu jikan* 中國書目季刊 31, no. 1 (1997): 1–13.

Zhou Yuwen 周愚文. *Songdai de zhou-xianxue* 宋代的州縣學. Taibei: Guoli bianyiguan, 1996.

———. *Songdai ertong de shenghuo yu jiaoyu* 宋代兒童的生活與教育. Taibei: Shida shuyuan, 1996.

Zhou Zhenfu 周振甫. *Wenzhang lihua* 文章例話. Beijing: Zhongguo qingnian chubanshe, 1983.

Zhu Chuanyu 朱傳譽. *Songdai xinwen shi* 宋代新聞史. Taibei: Zhongguo xueshu zhuzuo jiangzhu weiyuanhui, 1967.

Zhu Hong 朱鴻. "Zhen Dexiu ji qi dui shizheng de renshi" 眞德秀及其對時政的認識. *Shihuo yuekan fukan* 食貨月刊復刊 9, no. 5–6 (1979): 217–24.

Zhu Minche 祝敏彻. *"Zhuzi yulei" jufa yanjiu* 朱子语类句法研究. Wuhan: Changjiang wenyi chubanshe, 1991.

Zhu Ruixi 朱瑞熙. "Baguwen de xingcheng yu moluo" 八股文的形成与没落. *Lishi yuekan* 歷史月刊 1995, no. 3: 108–14.

———. "Lun Zhu Xi de gongsi guan" 论朱熹的公私观. *Shanghai shifan daxue bao* 上海师范大学学报 1995, no. 4: 94–97.

———. "Songdai Lixuejia Tang Zhongyou" 宋代理學家唐仲友. In *Collected Studies on Song History Dedicated to Professor James T. C. Liu in Celebration of His Seventieth Birthday*, ed. Tsuyoshi Kinugawa, 43–53. Tokyo: Dōhōsha, 1989.

———. "Song Yuan de shiwen—Zhongguo baguwen de chuxing" 宋元的时文—中国八股文的雏形. *Lishi yanjiu* 歷史研究 1990, no. 3: 29–43.

———. *Zhongguo zhengzhi zhidu tongshi: Songdai* 中國政治制度通史: 宋代. Ed. Bai Gang 白鋼. Beijing: Renmin chubanshe, 1996.

———. "Zhu Xi dui shiwen-baguwen chuxing de pipan" 朱熹对时文–八股文雏形的批判. *Zhuzi xue kan* 朱子学刊 4, no. 2 (1991): 63–74.

Zhu Shangshu 祝尚書. *Songdai keju yu wenxue kaolun* 宋代科舉与文学考论. Zhengzhou: Da xiang chubanshe, 2006.

———. *Song ren bieji xu lu* 宋人別集叙錄. 2 vols. Beijing: Zhonghua shuju, 1999.

———. *Song ren zongji xu lu* 宋人總集敘錄. Beijing: Zhonghua shuju, 2004.

Zhu Zhongsheng 朱重聖. "Songdai Taixue zhi qushi ji qi zuzhi" 宋代太學之取士及其組織. In *Songshi yanjiu ji* 宋史研究集, no. 18, 211–60. Taibei: Guoli bianyiguan, 1988.

Žižek, Slavoj. *Mapping Ideology.* London and New York: Verso, 1994.

# Index

Note: Emperors are listed under the dynasty name, e.g., Song Huizong. Books by known authors are usually listed under the author's name by the Chinese title.

*Harvard East Asian Monographs*
(*out-of-print)

# Harvard East Asian Monographs

178. John Solt, *Shredding the Tapestry of Meaning: The Poetry and Poetics of Kitasono Katue (1902–1978)*

179. Edward Pratt, *Japan's Protoindustrial Elite: The Economic Foundations of the Gōnō*

180. Atsuko Sakaki, *Recontextualizing Texts: Narrative Performance in Modern Japanese Fiction*

181. Soon-Won Park, *Colonial Industrialization and Labor in Korea: The Onoda Cement Factory*

182. JaHyun Kim Haboush and Martina Deuchler, *Culture and the State in Late Chosŏn Korea*

183. John W. Chaffee, *Branches of Heaven: A History of the Imperial Clan of Sung China*

184. Gi-Wook Shin and Michael Robinson, eds., *Colonial Modernity in Korea*

185. Nam-lin Hur, *Prayer and Play in Late Tokugawa Japan: Asakusa Sensōji and Edo Society*

186. Kristin Stapleton, *Civilizing Chengdu: Chinese Urban Reform, 1895–1937*

187. Hyung Il Pai, *Constructing "Korean" Origins: A Critical Review of Archaeology, Historiography, and Racial Myth in Korean State-Formation Theories*

188. Brian D. Ruppert, *Jewel in the Ashes: Buddha Relics and Power in Early Medieval Japan*

189. Susan Daruvala, *Zhou Zuoren and an Alternative Chinese Response to Modernity*

*190. James Z. Lee, *The Political Economy of a Frontier: Southwest China, 1250–1850*

191. Kerry Smith, *A Time of Crisis: Japan, the Great Depression, and Rural Revitalization*

192. Michael Lewis, *Becoming Apart: National Power and Local Politics in Toyama, 1868–1945*

193. William C. Kirby, Man-houng Lin, James Chin Shih, and David A. Pietz, eds., *State and Economy in Republican China: A Handbook for Scholars*

194. Timothy S. George, *Minamata: Pollution and the Struggle for Democracy in Postwar Japan*

195. Billy K. L. So, *Prosperity, Region, and Institutions in Maritime China: The South Fukien Pattern, 946–1368*

196. Yoshihisa Tak Matsusaka, *The Making of Japanese Manchuria, 1904–1932*

197. Maram Epstein, *Competing Discourses: Orthodoxy, Authenticity, and Engendered Meanings in Late Imperial Chinese Fiction*

198. Curtis J. Milhaupt, J. Mark Ramseyer, and Michael K. Young, eds. and comps., *Japanese Law in Context: Readings in Society, the Economy, and Politics*

199. Haruo Iguchi, *Unfinished Business: Ayukawa Yoshisuke and U.S.-Japan Relations, 1937–1952*

200. Scott Pearce, Audrey Spiro, and Patricia Ebrey, *Culture and Power in the Reconstitution of the Chinese Realm, 200–600*

201. Terry Kawashima, *Writing Margins: The Textual Construction of Gender in Heian and Kamakura Japan*

202. Martin W. Huang, *Desire and Fictional Narrative in Late Imperial China*